Benchmark Papers in Ecology

Series Editor: Frank B. Golley
University of Georgia

Related Titles in BENCHMARK PAPERS IN BEHAVIOR Series

**Benchmark Papers
in Ecology / 10**

A BENCHMARK® Books Series

TROPICAL ECOLOGY

Edited by
CARL F. JORDAN
University of Georgia

Hutchinson Ross Publishing Company

Stroudsburg, Pennsylvania

LIBRARY OF CONGRESS CATALOGING IN PUBLICATION DATA
Main entry under title:
Tropical ecology.
  (Benchmark papers in ecology; v. 10)
  Includes indexes.
  1. Ecology—Tropics—Addresses, essays, lectures.
I. Jordan, Carl F.   II. Series.
QH84.5.T84   574.5'2623   81-4260
ISBN 0-87933-398-7   AACR2

Distributed world wide by Academic Press,
a subsidiary of Harcourt Brace Jovanovich,
Publishers.

CONTENTS

Contents

PART II: FUNCTIONING OF TROPICAL ECOSYSTEMS

SERIES EDITOR'S FOREWORD

Ecology—the study of interactions and relationships between living systems and environment—is an extremely active and dynamic field of science. The great variety of possible interactions in even the most simple ecological system makes the study of ecology compelling but difficult to discuss in simple terms. Further, living systems include individual organisms, populations, communities, and ultimately the entire biosphere; there are thus numerous subspecialties in ecology. Some ecologists are interested in wildlife and natural history, others are intrigued by the complexity and apparently intractable problems of ecological systems, and still others apply ecological principles to the problems of man and the environment. This means that a Benchmark Series in Ecology would be subdivided into innumerable subvolumes that represented these diverse interests. However, rather than take this approach, I have tried to focus on general patterns or concepts that are applicable to two particularly important levels of ecological understanding: the population and the community. I have taken the dichotomy between these two as my major organizing concept in the series.

In a field that is rapidly changing and evolving, it is often difficult to chart the transition of single ideas into cohesive theories and principles. In addition, it is not easy to make judgments as to the benchmarks of the subject when the theoretical features of a field are relatively young. These twin problems—the relationship between interweaving ideas and the elucidation of theory, and the youth of the subject itself—make development of a Benchmark series in the field of ecology difficult. Each of the volume editors has recognized this inherent problem, and each has acted to solve it in his or her unique way. Their collective efforts will, we anticipate, provide a survey of the most important concepts in the field.

The Benchmark series is especially designed for libraries of colleges, universities and research organizations that cannot purchase the older literature of ecology because of costs, lack of staff to select from the hundreds of thousands of journals and volumes, or from the unavailability of the reference materials. For example, in developing countries where a science library must be developed *de novo* I have seen where the Benchmark series can provide the only background literature available to the students and staff. Thus, the intent of the series is to provide an authoritative selection of literature, which can be read in the original form, but that is cast in a matrix of thought provided by the editor. The volumes are designed to explore the historical development of a concept in ecology and point the way toward new developments, without being a historical study. We hope that even though the Benchmark Series in Ecology is a library oriented series and bears an appropriate cost it will also be of sufficient utility so that many professionals will place it in their personal library. In a few cases the volumes have even been used as text books for advanced courses. Thus we expect that the Benchmark Series in Ecology will be useful not only to the student who seeks an authoritative selection of original literature but also to the professional who wants to quickly and efficiently expand his or her background in an area of ecology outside his special competence.

Carl F. Jordan has begun a new direction for the Benchmark Papers in Ecology series with his volume *Tropical Ecology*. Our movement away from an exclusive concern with principles toward a geographic organization of ecology is warranted by widespread interest in biogeographical tropics. The tropics are a region of immense current concern and are the subject of the first Benchmark volume on a geographical region. Jordan is very well qualified to provide us with an introduction to the ecology of this complex subject. He has worked in tropical ecology throughout his professional career and has been personally involved in several major tropical research programs, including the El Verde Project in Puerto Rico and the San Carlos Project in Venezuela. Carl Jordan is presently located at The University of Georgia Institute of Ecology where he holds the rank of Ecologist and Adjunct Professor of Botany. He is a member of the Board of the Organization of Tropical Studies and is active in many other tropical-oriented organizations and activities. His research programs have stressed productivity and nutrient cycling of tropical forests.

FRANK B. GOLLEY

PREFACE

The first question facing me upon beginning this Benchmark volume on tropical ecology was, "How does tropical ecology differ from other kinds of ecology?" There are, after all, many ecological studies in tropical regions that are fundamentally no different from studies carried out in other regions. For example, a study of the effect of drought on animal behavior carried out in Ecuador should not be considered "tropical ecology," since a similar study carried out in England would not be considered "temperate ecology."

There are certain subfields of ecology that have regional names, such as "Mediterranean ecology" and "alpine ecology," but these subfields have well-recognized paradigms, for example, the physiological adaptations of organisms to the types of environments typified by the Mediterranean region and by mountain-top regions. The tropics, however, include ecosystem types ranging from deserts and savannahs through deciduous forests and montane forests to lowland tropical rain forests. Therefore, there is no paradigm for tropical ecology based on habitat.

Despite the heterogeneity of tropical ecosystems, there are two ways in which each type of tropical ecosystem (forest, lake, and so forth) differs from counterpart ecosystems at other latitudes: (1) tropical ecosystems have more species; (2) process rates in tropical ecosystems are higher.

In this book I trace the development of ideas that resulted from studies of species richness and process rates in the tropics. In Part I, I present papers that indicate that the species richness of the tropics led to insights that helped give rise to the field of ecology. I also show that this species richness stimulated the development of evolutionary theory and that the ideas of evolutionary biologists working in the species-rich tropics are now contributing to a field of applied ecology concerned with size and management of ecological reserves.

In Part II, I first present papers that provide evidence for high process rates in the tropics and that show the influence this evidence has had on ideas for tropical development. Subsequent papers explain why the high process rates in natural tropical forests do not mean high economic crop productivity when natural forests are converted to plantation, pasture, or cropland, show how an understanding of the functioning of tropical ecosystems can lead to sustained yield management systems, and explore ways of looking at tropical forests so that values other than productivity are considered in the management of tropical lands.

Editing a Benchmark volume is a difficult task. The first problem is selecting the "benchmark" papers. No matter which criteria are used for selection, some important papers will not be included. The second difficult task is developing a theme that unites a group of important but disparate papers. Developing a book by first selecting the papers and then developing a uniting theme is like building a house by first selecting the bricks and then deciding what kind of house can be built with those bricks. Having a master plan first is a better approach. Therefore, instead of selecting papers first, I developed the themes first, and then selected papers relevant to those themes. The selection was based not so much on whether a paper is a classic, but instead on whether a paper contributed toward organizing tropical ecology into the themes.

In some cases, only one portion of the work selected contributes directly to the development of the theme. In cases where the original works are extremely lengthy, I have reproduced only those portions directly relevant. Where the original works are shorter, the whole work is included, even if it contains major portions not directly relevant.

CARL F. JORDAN

CONTENTS BY AUTHOR

Pero luego empezò a sentir que la grandeza estaba en la in-
finidad, en la repetición obsesionante de un motivo único
al parecer. Arboles, Arboles, Arboles. . . ! Una sola bóveda
verde sobre miríadas de columnas afelpadas de musgos, tiñosas
de líquenes, cubiertas de parásitas y trepardoras, trenzadas
y estranguladas por bejucos tan gruesos como troncos de
árboles. Barreras de árboles, murallas de árboles, macizos
de árboles! Siglos perennes desde la raíz hasta los copos,
fuerzas descommunales en la absoluta inmovilidad aparente,
torrente de savia corriendo en silencio. Verdes abismos cal-
lados. . . Bejucos, marañas. . . Arboles! Arboles!

Canaima, Rómulo Gallegos

TROPICAL ECOLOGY

Part I

SPECIES RICHNESS OF THE TROPICS

The variety of lines and forms in tropical forests surely exceeds what all surrealists together have been able to dream of, and many of these lines and forms are endowed with dynamism and with biological meaningfulness that are lacking, as far as I am able to perceive, in the creations exhibited in museums of modern art.

"Evolution in the Tropics,"
T. Dobzhansky

Regarding tropical forests:

Unknown are the autumn tints, the bright browns and yellows of English woods; much less the crimsons, purples, and yellows of Canada, where the dying foliage rivals, nay excels, the expiring dolphin in splendour. Unknown is the cold sleep of winter; unknown the lovely awakening of vegetation at the first gentle touch of spring. A ceaseless round of ever-active life weaves the fairest scenery of the tropics into one monotonous whole, of which the component parts exhibit in detail untold variety and beauty.

"The Naturalist in Nicaragua,"
Belt

Editor's Comments
on Papers 1 and 2

1 JORDAN
The Birth of Ecology: An Account of Alexander von Humboldt's Voyage to the Equatorial Regions of the New Continent

2 DARWIN
Excerpts from *Galapagos Archipelago*

THE BIRTH OF ECOLOGY

A fascinating aspect of the tropics is it species richness. Compared to an ecosystem such as a forest at higher latitudes, a counterpart ecosystem in the tropics almost always contains many more species (MacArthur, 1972). The abundance of species often makes a deep impression on biologists from other latitudes visiting the tropics, sometimes stimulating them to view nature in a new perspective. The first major thesis of this book is that the high species diversity in the tropics was a major impetus in the development of ecology. To illustrate the point, I have chosen selections illustrating the work of Alexander von Humboldt and Charles Darwin.

Alexander von Humboldt, during his travels in equatorial America between 1799 and 1804, wrote about what he called "harmony in nature," a reference to the interacting system of plants and animals with the soil, air, and water that surrounds them. We don't know, of course, why von Humboldt did not recognize the interacting systems in his northern European homeland. It could have been familiarity with the European landscape, or perhaps human-caused disturbances obscured relationships between organisms and habitat. Paper 1 suggests that it was the sharp differentiation among the series of life zones from tropical rain forest to alpine zone on the side of an Andean volcano that prompted von Humboldt to recognize the relationship between climate, soils, and biota.

My father, Dr. Emil Jordan, has had a life-long interest in von Humboldt. Using von Humboldt's original *Voyage to the Equator-*

ial Regions of the New Continent, a 30-volume, 12,000-page account of travels, explorations, and discoveries in the Canary Islands, South and Central America, and Cuba, Emil Jordan has written a popular account with pertinent quotations illustrating von Humboldt's influence in the formation of ecology as a science.

Tropical travel also influenced the thinking of other pioneers in ecology, among them Charles Darwin. Paper 2 is part of the chapter "Galapagos Archipelago" from Darwin's book *Journal of Researches into the Natural History and Geology of the Countries Visited During the Voyage of H.M.S. Beagle Round the World.* This chapter particularly illustrates the impressions that the high diversity of finch species made on Darwin.

Note particularly the passage "I never dreamed that islands about 50 or 60 miles apart, and most of them in sight of each other, formed of precisely the same rocks, placed under a quite similar climate, rising to a nearly equal height, would have been differently tenanted; but we shall soon see that this is the case" (p. 389). Also, "Reviewing the facts here given, one is astonished at the amount of creative force, if such an expression may be used, displayed on these small, barren, and rocky islands; and still more so, at its diverse yet analogous action on points so near to each other" (p. 393).

Naturalists before Darwin had realized that there were struggles between species and that these struggles resulted in natural selection. The new insight that came to Darwin after his observations of the Galapagos finches was that natural selection occurred within a species, and that a single species could give rise to several different species (Mayr, 1977).

REFERENCES

MacArthur, R. H., 1972, *Geographical Ecology: Patterns in the Distribution of Species*, Harper & Row, New York, 269 p.

Mayr, E., 1977, Darwin and Natural Selection, *Am. Sci.* **65**:321–327.

1

This article was prepared especially for this Benchmark volume

THE BIRTH OF ECOLOGY

An Account of Alexander von Humboldt's Voyage to the Equatorial Regions of the New Continent

E. L. Jordan
Rutgers University

On a bright spring day of 1799, a slender young man stood on the quarterdeck of the frigate Pizarro in the harbor of la Coruña in northern Spain. The wind blew through his light-brown hair, his gray eyes scanned the sea beyond the port, and great expectations filled his mind.

He stood at the threshold of a new life; bright new horizons beckoned. "Is all this real?" he wondered. "How did I reach this point, this strange port, this Spanish ship?"

His family's old castle at Tegel, in the heart of Prussia, far removed from seaports and shipping lanes, seemed a most unlikely starting point for a world traveler. Yet here he was, and as his thoughts wandered back to his youth, he remembered the happenings that had stirred his young mind: the lure of faraway continents as yet unexplored and the forces of nature as yet unknown.

As a small boy he had roamed the half-wild park of the family's mansion, had caught frogs and squirrels, and had collected eggs and birds' nests, fossils and rocks, plants and insects to the dismay of his straight-laced mother and to the delight of his warm-hearted father. One father-and-son excursion, which he never forgot, was a trip to the Royal Botanical Garden in Berlin; there the boy beheld the gigantic dragon tree from the Canary Islands. A tower had been built around it to shield it from the northern winter winds, and he imagined how wonderful it would be to see the huge plant in its natural habitat.

"Well," the traveler on the Pizarro mused, "with luck I shall stand in front of the biggest and most famous of all dragontrees, the one in Orotava, a fortnight from now. This ship is scheduled to call there. I shall remember my father."

*

Do childhood impressions fashion the course of our lives? If so, then Robinson Crusoe ought to be added to the dragon-tree, as the young boy's symbols of travel, adventure, and nature. For it happened that his tutor Campe translated, for a local publisher, Defoe's Robinson Crusoe into German. Alexander not only read the book, eager and excited, but also built a secret tree house à la Crusoe in the park of Tegel; it was so skillfully hidden that the gardeners did not discover it until months later.

He and his brother Wilhelm grew up in the care of successive private tutors who provided the basic instruction; they also learned the skills and manners of their class. To be both aristocratic and rich was a rare status in poor and spartan Prussia--a privilege which created self-assurance and independence. At no time did the brothers confine themselves to their social group, and freely they cultivated many and varied friendships. Frequently, for instance, they paid visits to the home of Dr. Herz, a prominent Jewish physician and scientist in Berlin, who lectured on physics and performed interesting experiments. Both boys enjoyed the stimulating intellectual circle that gathered in the doctor's house and became fond of his beautiful, young wife Henrietta, a hostess of great charm and wit who did not mind a bright teenager's ardent admiration. Alexander taught her the latest dance, the Minuet à la Reine, and in turn she introduced him to the basics of the Hebrew language; he turned out to be a most eager pupil who was soon able to write her flowery letters in Hebrew script.

At that time Alexander spoke French well, and in the course of the following years picked up, without great effort, Italian, Spanish, and English. Latin and Greek formed the basis of his classical education. In South America he studied various Indian idioms, particularly Quechua, the language of the Incas, and when in his old age he suspected that his letters were opened (some of his enemies in high positions spied on him because of his liberal views) he defied and mocked his censors by writing to his scholarly friends in Sanskrit.

But his linguistic flair represented only a small by-product of his genius. While young Wilhelm leaned toward the humanities and a diplomatic-scholarly career, Alexander developed his early interest in the natural sciences. Of the various educational institutions he attended, the University of Göttingen offered him a broad and solid foundation. The international reputation of the university was reflected in the student body; Humboldt's fellow students included, within an enrollment of only 812, the Duke of Cumberland, the future King of Hanover; the Duke of Essex who, later on, as president of the Royal Society corresponded with Humboldt on problems of terrestrial magnetism; the French Duke de Broglie; and the Count, later Prince, Metternich.

However, these super-noble classmates did not impress him; he found far more interesting a young scientist who lived in Göttingen, Georg Forster, the son-in-law of his favorite professor Heyne. Forster had been lucky; he had accompanied Captain Cook on his second voyage around the world and could talk fascinatingly of Tahiti and the New Hebrides, of atolls and coral reefs, of flying fish, and of animals that carried their young in a breast pouch. What a jewel to discover in a small college town! An immediate rapport sprang up between the two; they saw each other often, traveled together to France, Holland, and England, and became life-long friends.

After the years in Göttingen, Alexander received his professional education at the mining academy in Freiberg where his principal teacher, Werner, was the leading geologist of his time. Also the academy had an international reputation, and he had met fellow students from every corner of the world; some enduring friendships resulted.

It happened in Göttingen and Freiberg that his mind began to expand and unfold. He read widely and studied and experimented in almost all natural sciences known at the time. His brain fairly buzzed with ideas, theories, plans, and projects. He graduated as a mining engineer at 23, and the Prussian government promptly appointed him to the position of managing director of the mines of Ansbach-Bayreuth, a Prussian-owned duchy in southern Germany. As a recognized expert in mining and mineralogy, this youngest of all mining officials had an opportunity to travel widely, to Silesia, Poland, Austria, Switzerland, and Upper Italy, in order to exchange views with other mineralogists, to compare methods of mining, and to observe nature on the way.

He gladly remembered that at that time he had proved to be also a practical success. Under his management the output of the mines rose spectacularly, and a school for miners, which he established, increased production, safety, and morale.

Still the restrictions and jealousies of an entrenched bureaucracy, to which the young official was subjected, annoyed and discouraged him; for his colleagues he was too young, too rich, too brilliant, too independent. So when his mother died in 1796 and left him an impressive fortune (his father had passed away previously) he chose freedom; he resigned from his official position and began to prepare for a scientific career of his own.

At that time Paris was the world's scientific center, and there he settled with a congenial group of friends. One of them, Aimée Bonpland, was strongly drawn to him, became his assistant, and agreed to accompany him on his travels. Amiable, eager, and highly intelligent, Bonpland had earned the degree of Doctor of Medicine but was bored with the daily rounds of a general practitioner. He had become passionately involved in plants and plant growth, and as a natural-born botanist, explorer, and adventurer he would be an ideal collaborator in the wilderness.

At that time the scientific community of Paris was excited about a project of the French government. Two research ships were to be equipped with the latest precision instruments and staffed with outstanding scientists for a voyage of exploration around the world under the command of Captain Baudin. Humboldt and Bonpland applied for membership, but before a final decision could be reached, the project was canceled. The wars in Central Europe proved so costly that no money could be made available for non-military purposes.

By now the idea of a voyage of scientific exploration stirred his blood; he wanted action. As he surveyed the European situation, it became apparent that Spain with its vast colonial empire could be the key to their future; he and Bonpland proceeded to Madrid and soon moved in diplomatic circles with connections to the court. When introduced to the king, Humboldt was graciously given permission to travel anywhere in the Spanish colonies. The fact that the young scientist did not ask for a subsidy--he would pay the expenses of the expedition out of his own pocket--and the prospect that he might discover new gold and silver mines, helped to expedite all preparations. Eagerly the court provided him with two passports, one from the government in Madrid and one from the Council of the Indies in Cadiz. The passports listed his scientific instruments piece by piece as an official acknowledgment. This was thought to be advisable because stupid colonial officials and overly zealous padres might confiscate them as tools of the devil.

As the two young friends set sail for America, Humboldt felt both uneasiness and happiness at the hour of fulfillment. He left Europe for the first time and wondered what new state of existence awaited him. But happiness prevailed. He climbed down to his cabin and to his friend Freiesleben who had been his teacher at the Mining Academy at Freiberg, he wrote:

> My head is dizzy with joy. I am sailing on the Spanish frigate Pizarro. We shall land on the Canary Islands and on the coast of Caracas in South America.... More from there. Within a few hours we shall sail around Cape Finisterre.

> I shall collect plants and fossils, and with the best of instruments make astronomical observations.

Yet this is not the main purpose of my voyage. I shall endeavor to find out how nature's forces act upon each other, and in which way the geographic environment exerts its influence on animals and plants. In short, I must find out about Harmony in Nature.

*

In the course of his voyage this "Harmony in Nature" became the basis of a new science.

* * * * * * *

Editorial note   After explorations in Venezuela and Colombia, von Humboldt traveled south through the Andes to Quito, Ecuador. The following pages are about his stay in Quito.

* * * * * * *

Although the Quiteños lived on top of a volcano, Humboldt found them to be gay, lively, and unconcerned; fiestas, bullfights, and fireworks abounded, and among the Spanish nobility an atmosphere of luxury and ease prevailed. They lived dangerously and merrily.

Here Humboldt stayed for eight months, from January to August of 1802, as guest of a Spanish aristocrat with the formidable name of Don Juan Pio Aguirre y Montufar, Marques de Selva Alegre, a social and political leader. The visitor fitted easily into this circle; his pleasant manners and stimulating conversation made him a natural member of the household, but the family of the Marques thought him a bit queer. Doña Rosa, the daughter of the house who acted as hostess, remembered in her old age how greatly the handsome young European nobleman had attracted the young ladies of the city's elite; but he turned out to be as elusive as he was charming. Just as they hoped to enjoy his company, he jumped on his horse and disappeared into the mountains; he did not return until all hours of the night with plant specimens, pieces of rocks, and mud on his boots. He seemed always to be glad to be outdoors; at night, long after all had retired, he observed the stars. The young ladies did not understand his way of life and were a bit annoyed by his double image of charisma and aloofness.

It was unfortunate that Doña Rosa met Humboldt at a time of crisis and disappointment, when he was forced to change his plans and had to make far-reaching decisions. To fathom his character was not easy at any time, and doubly so during his stay in Quito.

Why was Humboldt so restlessly active during these months, so intent on redoubling his effort? Why did his drive for achievement seem to be stronger than ever?

What filled his mind with creative excitement were a letter from Paris and a challenging idea. The letter reported that the expedition of Captain Baudin had changed its course. Instead of circumnavigating the globe in a westerly direction, it did so on an easterly course, sailing around Africa and the Cape of Good Hope to the Orient. It would not touch Peru and therefore could not accept Humboldt and Bonpland as members. Humboldt's first reaction was naturally one of disappointment, but he consoled himself quickly. The Cordilleras had been practically unexplored by scientists, and when he surveyed his plant collections, his drawings, his notes on barometric and astronomical observations, his geographic measurements, and his atmospheric experiments, he felt that he had accomplished a great deal. It also occurred to him that so far he had only touched the surface; these mountains offered still innumerable opportunities for exploration and research. He saw a challenge which did not depend on outside help.

The new idea that occupied his mind arose from the environment in which he worked. He remembered the letter he had written to his friend and former teacher Freiesleben from la Coruña, when he was about to sail on the Pizarro: "I shall endeavor to find out how nature's forces act upon each other and in which way the geographic environment exerts its influence on animals and plants. In short, I must find out about Harmony in Nature."

During his excursions from Quito he began to look at the surrounding nature with an inquiring eye, and it dawned upon him that the Cordilleras offered a perfect laboratory for such an examination. At first his mountaineering excursions--he scaled in successive trips of two to three weeks the volcanos Pichincha, Cotopaxi, Antisana, and Illinica--were largely geographical and geological outings. But then he discovered something the jungle wilderness of the Orinoco had not disclosed. He saw the Andes as a vertical exhibit of the earth's climates, plants, and animals, from hot tropical lowlands to the eternal snow and ice of the towering summits which at that time were considered the highest roof of the world. The slopes disclosed a distinct number of floors, each life zone consisting of a definite grouping of soils, land forms, rainfall, temperature, plant species, and animal species. These environments--today we would call them ecosystems--had not assembled themselves accidentally or arbitrarily but had grown as communities with specific characteristics determined by interaction. All component forces continuously acted on each other and created a balanced unit.

*

He inspected the "exhibit" himself. He tramped through the deep valleys which produced palms and banana plants, sugar cane and cacao trees; he saw pumas and small animals inhabiting the jungle, but, on the whole, animal life remained scarce. In the foothills he sketched cotton bushes and coffee trees which flourished there, with a few deer and spectacled bears roaming the woods. In the rain forest he listed nut, rubber, and cinchona trees, also huge tree ferns. On the Altiplano he visited the Indian villages; the Indians cultivated barley and potatoes, wheat and corn; they kept cattle and horses which now and then were plagued by vampire bats. Animal life increased on the highest plains beyond the tree line, in a cold and windy climate; besides flocks of sheep, several typical South American species lived on the Alpine grasses and herbs: llamas and guanacos, vicuñas, alpaccas, and chinchillas.

The term "ecology" had not yet been coined, but Humboldt's field research in "interaction" and "harmony in nature," backed by his essays on plant geography and vegetation profiles, clearly formed the foundation for the concept of modern ecology.

*

Humboldt was probably the first conservationist in the New World; on numerous occasions he emphasized the need for preserving the continent's natural resources. To the fishermen of Araya he pointed out that not the sound of the oars, as they imagined, had driven away the oysters, but that their ruthless overfishing of the oyster beds had destroyed the once lucrative pearl fisheries.

When the worried landholders around Lake Victoria in Venezuela wondered why the lake was shrinking year after year, he told them that the phenomenon was of their own doing. By denuding the nearby mountain sides and cutting the trees that used to surround the lake, they had caused the erosion which in turn was diminishing the water supply of the springs that fed the lake.

He told the padres of Caripe that their "mine of fat" which involved the annual killing of thousands of guacharo birds, would lead to the near-extinction of the species. He warned the Franciscan missionaries who supervised the turtle egg harvest on the islands of the Orinoco about exhausting a source of food without a thought of the future. If they did not leave part of the island beaches untouched, so that a sufficient number of turtles were hatched every year, there would be no egg harvest before long.

He admired the beautiful groves of cinchona trees, the producers of quinine, on the Cordilleras of Peru; but the fact that thousands of trees were felled for their bark without any attempt at reforestation dismayed him. It meant that the Spanish colony would lose an important resource. As could be expected, his pessimistic forecasts came true.

His own era disregarded the warnings; it took another 150 years to make the world conservation-conscious.

*

Among the studies of Humboldt's successors, Darwin's Theory of Evolution was primarily ecological. Darwin acknowledged his debt to Humboldt; he told the botanist Hooker (and asked him to let Humboldt know about it) that his researches and findings sprang directly from reading and rereading Humboldt's Personal Narrative. "I used to admire Humboldt," he wrote on board the H.M.S. Beagle, "now I almost worship him."

*

In the great tasks of ecology, those additional branches of science which deeply interested Humboldt in the New World, from geology to meteorology, have become essential cooperative ecological tools. He could not foresee that in this context his studies would play a part, 150 years later, as the Environmental Sciences, but his approach to nature led to this result.

Before the knowledge explosion of the 19th and 20th centuries, and before scientific specialization became the rule, a few exceptionally gifted men were able to master the whole field as it was known in their day. For Humboldt this broad horizon proved a challenge and a boon. Far from being a handicap (as it would appear in this age of scientific fragmentation) it enabled him to study and encompass the environment as a whole, both in terns of physical geography and in terms of ecology.

During his American odyssey his daily chores were legion, but he never lost sight of his original great goal as he had described it in his letter to his friend and teacher Freiesleben. He did not just care about detecting and registering new species of plants and animals and other isolated phenomena, he meant to find out about Harmony in Nature. He reached his goal, not by philosophical speculation, but by scientific investigation. In its deepest and broadest implications, only the world of today is able to recognize his achievement.

He could not have gained his insights in his native Europe. There every arable plot had been farmed for 1000 years obliterating the virgin character of the land and the soil. The forests had largely been cut, and little remained of the original wildlife. For basic ecological investigations central Europe was hardly the proper ambiance.

In contrast, the Americas still offered an environment little disturbed by human hands, a wilderness in a virgin state with towering mountains and lush valleys--an ideal workshop for a young naturalist. He himself realized his good luck and eagerly reported his chance to his brother Wilhelm.

Today, when Humboldt is relatively unknown, his physical-geographic and ecological research and his broad, searching view of the environment turn out to be of a far greater impact than it appeared a century ago. At that time he became a world-famous celebrity, not exactly for the wrong reasons, for his achievements certainly loomed large. But his contemporaries could not speculate on the far-reaching consequences of his work; they could not know that he had initiated a breakthrough in the approach to nature, that he had discovered a new key to the environment and prepared the tools for a new era that is turning from exploitation to preservation.

*   *   *   *   *   *   *

BIBLIOGRAPHY

<u>Humboldt's</u> <u>Original</u> <u>Travel</u> <u>Notebooks</u>. Considered lost for many decades, the diaries have been retrieved from various sources and are now kept in the manuscript collection of the Deutsche Staatsbibliothek in Berlin. At present they are edited and prepared for publication by the Humboldt Commission of the German Academy of Sciences. The Roman numerals refer to the various volumes as marked by Humboldt.

I.   Voyage D'Espagne aux Canaries et à Cumana. Obs. astron. de Juin à Oct. 1799.

II. Et VI. Voy. à Caripe 1799. Obs. astron. Apure-Orenoque - Batabano (Cuba) à Sinu, Carthagène et Turbaco, 1801 - Quito 1802, Meteorologie - de Paris à Toulon, Oct. 1798 - Voy. d'Italie avec Gay-Lussac - Obs. Magn. 1805.

III. Voyage de Cumana à Caracas, Calabozo et San Fernando de Apure. De Nov. 1799 à Mars, 1799 (?).

IV.   Journal de la Navigation sur l'Apure, l'Orenoque, le Cassiquiare et le Rio Negro. Voy. par les Llanos de Caracas à San Fernando de Apure - Statistique de Cumana - pta. Araya.

V.   Reise von Cumana nach der Havana. Altes von der Reise Dresden, Wien, Salzburg.

VII a et VII b. Rio de la Magdalena - Bogota - Quindiu - Papayan - Quito (Antisana, Pichincha) - Pasto Volcan - Tolima.

Works Based on Humboldt's American Experiences.

Voyage to the Equatorial Regions of the New Continent. 30 volumes, 12,000 pages, 1425 illustrations (many hand-painted) and maps, published between 1807 and 1833 in French, German, and English editions. This work represented the zenith of his reputation.

Personal Narrative of the Voyage to the Equatorial Regions of the New Continent. German, English, and French editions published in the 1810s and 20s.

Essay Sur la Geographie des Plants. Paris 1807.

Ansichten der Natur. Tubingen 1807, translated into English as Views of Nature, a collection of nature studies based partly on his European and partly on his American experiences.

Essay Politique sur le Royaume de la Nouvelle Espagne avec un Atlas Physique et Geographique. 4 volumes; Paris 1810; French, German, and English editions.

Researches Concerning the Institutions and Monuments of the Ancient Inhabitants of America. 2 volumes, London 1814. Also French and German editions.

Des Lignes Isothermes et de la Distribution de La Chaleur sur le Globe. Paris 1817. Also a German edition.

Essay Geognostique des Roches dans les Deux Continents. Paris 1823. Also English and German editions.

Examen Critique de L'Histoire de la Geographie du Nouveau Continent et de Progres de L'Astronomie Nautique en 15e et 16e Siecles. 3 volumes, Paris 1836-39. In this historic-scientific work Humboldt traces the origin of the name America.

Kosmos, Entwurf einer physischen Weltbeschreibung. (Outline of a physical description of the universe); 5 volumes plus atlas. Tubingen 1845-62. Translated into eight other languages.

A Selection of Books on Alexander von Humboldt

Kellner, L. Alexander von Humboldt, London 1963. A short British biography.

Beck, Hanno. Alexander von Humboldt. Vol. I, Von der Bildungsreis zur Forschungsreise (from educational travel to scientific expedition). Vol. II, From Reisewerk zum Kosmos (from his travel report to Kosmos); Wiesbaden, 1959-61. This is the most comprehensive contemporary biography.

Dolan, Edward Jr. Green Universe, The Story of Alexander von Humboldt. New York 1959.

De Terra, Helmut. Humboldt, The Life and Times of Alexander von Humboldt. New York 1955.

Krammer, Mario. Alexander von Humboldt. Berlin 1951.

Möbius, Willy. Alexander von Humboldt. Wilmersdorf 1948.

Robles, Alessio Vito. Alejandro de Humboldt, su Vida y su Obra. Mexico City 1945.

Hagen, Wolfgang von. South America Called Them. New York 1945.

Rojas, Aristides. Humboldtiana. Caracas 1942.

Bonpland, Aimee. Archives Inedites de A. Bonpland. Trabajos de la Universidad Nacional, Vol. XXXI, Buenos Aires 1914-24.

May, Walter. Alexander von Humboldt und Darwin. Preussisches Jahrbuch, Vol. CV, Heft 2, 1901.

Dove, H. W. Die Forsters und die Humboldts. Leipzig 1881.

Bruhns, Karl. 3 volumes, Leipzig 1872. This was the most extensive Humboldt biography of the 19th century and is considered a failure. Bruhns had parceled out the various sectors of Humboldt's researches and explorations to ten specialists in the corresponding branches of science.

Each of these scholars proceeded on his own without regard to the others. The resulting fragmentation did not do justice either to Humboldt's personality or to his genius and accomplishments. The general assessment of Humboldt was written by Alfred Dove, 25 years old at that time; it showed an abysmal lack of understanding. An English edition appeared in London in 1873.

Lambert, A. B. Baron de Humboldt's Account of the Cinchona Forests of South America. London 1821.

2

Reprinted from pp. 367–376 and 389–393 of *Journal of Researches into the Natural History and Geology of the Countries Visited During the Voyage of H.M.S. Beagle Round the World*, Appleton, New York, 1855, 524 p.

GALAPAGOS ARCHIPELAGO

C. Darwin

The whole Group Volcanic—Numbers of Craters—Leafless Bushes—Colony at Charles Island—James Island—Salt-lake in Crater—Natural History of the Group—Ornithology, curious Finches—Reptiles—Great Tortoises, habits of—Marine Lizard, feeds on Sea-weed—Terrestrial Lizard, burrowing habits, herbivorous—Importance of Reptiles in the Archipelago —Fish, Shells, Insects—Botany—American Type of Organization—Differences in the Species or Races on different Islands—Tameness of the Birds—Fear of Man, an acquired Instinct.

September 15th.—This archipelago consists of ten principal islands, of which five exceed the others in size. They are situated under the Equator, and between five and six hundred miles westward of the coast of America. They are all formed of volcanic rocks; a few fragments of granite curiously glazed and altered by the heat, can hardly be considered as an exception. Some of the craters, surmounting the larger islands, are of immense size, and they rise to a height of between three and four thousand feet. Their flanks are studded by innumerable smaller orifices. I scarcely hesitate to affirm, that there must be in the whole archipelago at least two thousand craters. These consist either of lava and scoriæ, or of finely-stratified, sandstone-like tuff. Most of the latter are beautifully symmetrical; they owe their origin to eruptions of volcanic mud without any lava: it is a remarkable circumstance that every one of the twenty-eight tuff-craters which were examined, had their southern sides either much lower than the other sides, or quite broken down and removed. As all these craters apparently have been formed when standing in the sea, and as the waves from the trade wind and the swell from the open Pacific here unite their forces on the southern coasts of all the islands, this singular uniformity in the broken state of the craters, composed of the soft and yielding tuff, is easily explained.

Considering that these islands are placed directly under the equator, the climate is far from being excessively hot; this seems chiefly caused by the singularly low temperature of the surrounding water, brought here by the great southern Polar current. Excepting during one short season, very little rain falls, and even then it is irregular; but the clouds generally hang low. Hence, whilst the lower parts of the islands are very sterile, the upper parts, at a height of a thousand feet and upwards, possess a damp climate and a tolerably luxuriant vegetation. This is especially the case on the windward sides of the islands, which first receive and condense the moisture from the atmosphere.

In the morning (17th) we landed on Chatham Island, which, like the others, rises with a tame and rounded outline, broken here and there by scattered hillocks, the remains of former craters. Nothing could be less inviting than the first appearance. A broken field of black basaltic lava, thrown into the most rugged waves, and crossed by great fissures, is everywhere covered by stunted, sun-burnt brushwood, which shows little signs of life. The dry and parched surface, being heated by the noon-day sun, gave to the air a close and sultry feeling, like that from a stove: we fancied even that the bushes smelt unpleasantly. Although I diligently tried to collect as many plants as possible, I succeeded in getting very few; and such wretched-looking little weeds would have better become an arctic than an equatorial Flora. The brushwood appears, from a short distance, as leafless as our trees during winter; and it was some time before I discovered that not only almost every plant was now in full leaf, but that the greater number were in flower. The commonest bush is one of the Euphorbiaceæ: an acacia and a great odd-looking cactus are the only trees which afford any shade. After the season of heavy rains, the islands are said to appear for a short time partially green. The volcanic island of Fernando Noronha, placed in many respects under nearly similar conditions, is the only other country where I have seen a vegetation at all like this of the Galapagos Islands.

The Beagle sailed round Chatham Island, and anchored in several bays. One night I slept on shore on a part of the island, where black truncated cones were extraordinarily numerous: from one small eminence I counted sixty of them, all surmounted by craters more or less perfect. The greater number consisted merely of a ring of red scoriæ or slags, cemented together: and their height above the plain of lava was not more than from fifty to a hundred feet; none had been very lately active. The entire surface of this part of the island seems to have been permeated, like a sieve, by the subterranean vapours: here and there the lava, whilst soft, has been blown into great bubbles; and in other parts, the tops of caverns similarly formed have fallen in, leaving circular pits with steep sides. From the regular form of the many craters, they gave to the country an artificial appearance, which vividly reminded me of those parts of Staffordshire, where the great iron-foundries are most numerous. The day was glowing hot, and the scrambling over the rough surface and through the intricate thickets, was very fatiguing; but I was well repaid by the strange Cyclopean scene. As I was walking along I met two large tortoises, each of which must have weighed at least two hundred pounds: one was eating a piece of cactus, and as I approached, it stared at me and slowly walked away; the other gave a deep hiss, and drew in its head. These huge reptiles, surrounded by the black lava, the leafless shrubs, and large cacti, seemed to my fancy like some antediluvian animals. The few dull-coloured birds cared no more for me than they did for the great tortoises.

23rd.—The Beagle proceeded to Charles Island. This archipelago has long been frequented, first by the bucaniers, and latterly by whalers, but it is only within the last six years, that a small colony has been established here. The inhabitants are between two and three hundred in number; they are nearly all people of colour, who have been banished for political crimes from the Republic of the Equator, of which Quito is the capital. The settlement is placed about four and a half miles inland, and at a height probably of a thousand feet. In the first part of the road we passed through leafless thickets, as in Chatham Island. Higher up, the woods gradually became greener; and as soon as we crossed the ridge of the island, we were cooled by a fine southerly breeze, and our sight refreshed by a green and thriving vegetation. In this upper region coarse grasses and ferns abound; but there are no tree-ferns: I saw nowhere any member of the Palm family, which is the more singular, as 360 miles northward, Cocos Island takes its name from the number of cocoa-nuts. The houses are irregularly scattered over a flat space of ground, which is cultivated with sweet potatoes and bananas. It will not easily be imagined how pleasant the sight of black mud was to us, after having been so long accustomed to the parched soil of Peru and northern Chile. The inhabitants, although complaining of poverty, obtain, without much trouble, the means of subsistence. In the woods there are many wild pigs and goats; but the staple article of animal food is supplied by the tortoises. Their numbers have of course been greatly reduced in this island, but the people yet count on two days' hunting giving them food for the rest of the week. It is said that formerly single vessels have taken away as many as seven hundred, and that the ship's company of a frigate some years since brought down in one day two hundred tortoises to the beach.

September 29th.—We doubled the south-west extremity of Albemarle Island, and the next day were nearly becalmed between it and Narborough Island. Both are covered with immense deluges of black naked lava, which have flowed either over the rims of the great caldrons, like pitch over the rim of a pot in which it has been boiled, or have burst forth from smaller orifices on the flanks; in their descent they have spread over miles of the sea-coast. On both of these islands, eruptions are known to have taken place; and in Albemarle, we saw a small jet of smoke curling from the summit of one of the great craters. In the evening we anchored in Bank's Cove, in Albemarle Island. The next morning I went out walking. To the south of the broken tuff-crater, in which the Beagle was anchored, there was another beautifully symmetrical one of an elliptic form; its longer axis was a little less than a mile, and its depth about 500 feet. At its bottom there was a shallow lake, in the middle of which a tiny crater formed an islet. The day was overpoweringly hot, and the lake looked clear and blue: I hurried down the cindery slope, and, choked with dust, eagerly tasted the water—but, to my sorrow, I found it salt as brine.

The rocks on the coast abounded with great black lizards, between three and four feet long; and on the hills, an ugly yellowish-brown species was equally common. We saw many of this latter kind, some clumsily running out of the way, and others shuffling into their burrows. I shall presently describe in more detail the habits of both these reptiles. The whole of this northern part of Albemarle Island is miserably sterile.

October 8th.—We arrived at James Island: this island, as well as Charles Island, were long since thus named after our kings of the Stuart line. Mr. Bynoe, myself, and our servants were left here for a week, with provisions and a tent, whilst the Beagle went for water. We found here a party of Spaniards, who had been sent from Charles Island to dry fish, and to salt tortoise-meat. About six miles inland, and at the height of nearly 2000 feet, a hovel had been built in which two men lived, who were employed in catching tortoises, whilst the others were fishing on the coast. I paid this party two visits, and slept there one night. As in the other islands, the lower region was covered by nearly leafless bushes, but the trees were here of a larger growth than elsewhere, several being two feet and some even two feet nine inches in diameter. The upper region being kept damp by the clouds, supports a green and flourishing vegetation. So damp was the ground, that there were large beds of a coarse cyperus, in which great numbers of a very small water-rail lived and bred. While staying in this upper region, we lived entirely upon tortoise-meat: the breast-plate roasted (as the Guachos do *carne con cuero*), with the flesh on it, is very good; and the young tortoises make excellent soup; but otherwise the meat to my taste is indifferent.

One day we accompanied a party of the Spaniards in their whale-boat to a salina, or lake from which salt is procured. After landing, we had a very rough walk over a rugged field of recent lava, which has almost surrounded a tuff-crater, at the bottom of which the salt-lake lies. The water is only three or four inches deep, and rests on a layer of beautifully crystallized, white salt. The lake is quite circular, and is fringed with a border of bright green succulent plants; the almost precipitous walls of the crater are clothed with wood, so that the scene was altogether both picturesque and curious. A few years since, the sailors belonging to a sealing-vessel murdered their captain in this quiet spot; and we saw his skull lying among the bushes.

During the greater part of our stay of a week, the sky was cloudless, and if the trade-wind failed for an hour, the heat became very oppressive. On two days, the thermometer within the tent stood for some hours at 93°; but in the open air, in the wind and sun, at only 85°. The sand was extremely hot; the thermometer placed in some of a brown colour immediately rose to 137°, and how much above that it would have risen, I do not know, for it was not graduated any higher. The black sand felt much hotter, so that even in thick boots it was quite disagreeable to walk over it.

The natural history of these islands is eminently curious, and well deserves attention. Most of the organic productions are aboriginal creations, found nowhere else; there is even a difference between the inhabitants of the different islands; yet all show a marked relationship with those of America, though separated from that continent by an open space of ocean, between 500 and 600 miles in width. The archipelago is a little world within itself, or rather a satellite attached to America, whence it has derived a few stray colonists, and has received the general character of its indigenous productions. Considering the small size of these islands, we feel the more astonished at the number of their aboriginal beings, and at their confined range. Seeing every height crowned with its crater, and the boundaries of most of the lava-streams still distinct, we are led to believe that within a period, geologically recent, the unbroken ocean was here spread out. Hence, both in space and time, we seem to be brought somewhat near to that great fact—that mystery of mysteries—the first appearance of new beings on this earth.

Of terrestrial mammals, there is only one which must be considered as indigenous, namely, a mouse (Mus Galapagoensis), and this is confined, as far as I could ascertain, to Chatham Island, the most easterly island of the group. It belongs, as I am informed by Mr. Waterhouse, to a division of the family of mice characteristic of America. At James Island, there is a rat sufficiently distinct from the common kind to have been named and described by Mr. Waterhouse; but as it belongs to the old-world division of the family, and as this island has been frequented by ships for the last hundred and fifty years, I can hardly doubt that this rat is merely a variety, produced by the new and peculiar climate, food, and soil, to which it has been subjected. Although no one has a right to speculate without distinct facts, yet even with respect to the Chatham Island mouse, it should be borne in mind, that it may possibly be an American species imported here; for I have seen, in a most unfrequented part of the Pampas, a native mouse living in the roof of a newly built hovel, and therefore its transportation in a vessel is not improbable: analogous facts have been observed by Dr. Richardson in North America.

Of land-birds I obtained twenty-six kinds, all peculiar to the group and found nowhere else, with the exception of one lark-like finch from North America (Dolichonyx oryzivorus), which ranges on that continent as far north as 54°, and generally frequents marshes. The other twenty-five birds consist, firstly, of a hawk, curiously intermediate in structure between a buzzard and the American group of carrion-feeding Polybori; and with these latter birds it agrees most closely in every habit and even tone of voice. Secondly, there are two owls, representing the short-eared and white barn-owls of Europe. Thirdly, a wren, three tyrant-flycatchers (two of them species of Pyrocephalus, one or both of which would be ranked by some ornithologists as only varieties), and a dove—all analogous to, but distinct from, American species. Fourthly, a swallow, which though differing from the Progne purpurea of both Americas, only in being rather duller coloured, smaller, and slenderer, is considered by Mr. Gould as specifically distinct. Fifthly, there are three species of mocking thrush—a form highly characteristic of America. The remaining land-birds form a most singular group of finches, related to each other in the structure of their beaks, short tails, form of body and plumage: there are thirteen species, which Mr. Gould has divided into four sub-groups. All these species are peculiar to this archipelago; and so is the whole group, with the exception of one species of the sub-group Cactornis, lately brought from Bow Island, in the low Archipelago. Of Cactornis, the two species may be often seen climbing about the flowers of the great cactus-trees; but all the other species of this group of finches, mingled together in flocks, feed on the dry and sterile ground of the lower districts. The males of all, or certainly of the greater number, are jet black; and the females (with perhaps one or two exceptions) are brown. The most curious fact is the perfect gradation in the size of the beaks in the different species of Geospiza, from one as large as that of a hawfinch

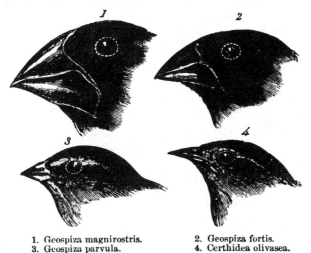

1. Geospiza magnirostris.
3. Geospiza parvula.

2. Geospiza fortis.
4. Certhidea olivasea.

to that of a chaffinch, and (if Mr. Gould is right in including his sub-group, Certhidea, in the main group) even to that of a warbler. The largest beak in the genus Geospiza is shown in Fig. 1, and the smallest in Fig. 3; but instead of there being only one intermediate species, with a beak of the size shown in Fig. 2, there are no less than six species with insensibly graduated beaks. The beak of the sub-group Certhidea, is shown in Fig. 4. The beak of Cactornis is somewhat like that of a starling; and that of the fourth sub-group, Camarhynchus, is slightly parrot-shaped. Seeing this gradation and diversity of structure in one small, intimately related group of birds, one might really fancy that from an original paucity of birds in this archipelago, one species had been taken and modified for different ends. In a like manner it might be fancied that a bird originally a buzzard, had been induced here to undertake the office of the carrion-feeding Polybori of the American continent.

Of waders and water-birds I was able to get only eleven kinds, and of these only three (including a rail confined to the damp summits of the islands) are new species. Considering the wandering habits of the gulls, I was surprised to find that the species inhabiting these islands is peculiar, but allied to one from the southern parts of South America. The far greater peculiarity of the land-birds, namely, twenty-five out of twenty-six being new species or at least new races, compared with the waders and web-footed birds, is in accordance with the greater range which these latter orders have in all parts of the world. We shall hereafter see this law of aquatic forms, whether marine or fresh-water, being less peculiar at any given point of the earth's surface than the terrestrial forms of the same classes strikingly illustrated in the shells, and in a lesser degree in the insects of this archipelago.

Two of the waders are rather smaller than the same species brought from other places: the swallow is also smaller, though it is doubtful whether or not it is distinct from its analogue. The two owls, the two tyrant-catchers (Pyrocephalus) and the dove, are also smaller than the analogous but distinct species, to which they are most nearly related; on the other hand, the gull is rather larger. The two owls, the swallow, all three species of mocking-thrush, the dove in its separate colours though not in its whole plumage, the

Totanus, and the gull, are likewise duskier coloured than their analogous species; and in the case of the mocking-thrush and Totanus, than any other species of the two genera. With the exception of a wren with a fine yellow breast, and of a tyrant-flycatcher with a scarlet tuft and breast, none of the birds are brilliantly coloured, as might have been expected in an equatorial district. Hence it would appear probable, that the same causes which here make the immigrants of some peculiar species smaller, make most of the peculiar Galapageian species also smaller, as well as very generally more dusky coloured. All the plants have a wretched, weedy appearance, and I did not see one beautiful flower. The insects, again, are small-sized and dull-coloured, and, as Mr. Waterhouse informs me, there is nothing in their general appearance which would have led him to imagine that they had come from under the equator.* The birds, plants, and insects have a desert character, and are not more brilliantly coloured than those from southern Patagonia; we may, therefore, conclude that the usual gaudy colouring of the intertropical productions, is not related either to the heat or light of those zones, but to some other cause, perhaps to the conditions of existence being generally favourable to life.

[*Editor's Note*:   Material has been omitted at this point.]

---

* The progress of research has shown that some of these birds, which were then thought to be confined to the islands, occur on the American continent. The eminent ornithologist, Mr. Sclater, informs me that this is the case with the Strix punctatissima and Pyrocephalus nanus; and probably with the Otus Galapagoensis and Zenaida Galapagoensis: so that the number of endemic birds is reduced to twenty-three, or probably to twenty-one. Mr. Sclater thinks that one or two of these endemic forms should be ranked rather as varieties than species, which always seemed to me probable.

† This is stated by Dr. Günther (Zoolog. Soc., Jan. 24th, 1859) to be a peculiar species, not known to inhabit any other country.

I have not as yet noticed by far the most remarkable feature in the natural history of this archipelago; it is, that the different islands to a considerable extent are inhabited by a different set of beings. My attention was first called to this fact by the Vice-Governor, Mr. Lawson, declaring that the tortoises differed from the different islands, and that he could with certainty tell from which island any one was brought. I did not for some time pay sufficient attention to this statement, and I had already partially mingled together the collections from two of the islands. I never dreamed that islands, about 50 or 60 miles apart, and most of them in sight of each other, formed of precisely the same rocks, placed under a quite similar climate, rising to a nearly equal height, would have been differently tenanted; but we shall soon see that this is the case. It is the fate of most voyagers, no sooner to discover what is most interesting in any locality, than they are hurried from it; but I ought, perhaps, to be thankful that I obtained sufficient materials to establish this most remarkable fact in the distribution of organic beings.

The inhabitants, as I have said, state that they can distinguish the tortoises from the different islands; and that they differ not only in size, but in other characters. Captain Porter has described * those from Charles and from the nearest island to it, namely, Hood Island, as having their shells in front thick and turned up like a Spanish saddle, whilst the tortoises from James Island are rounder, blacker, and have a better taste when cooked. M. Bibron, moreover, informs me that he has seen what he considers two distinct species of tortoise from the Galapagos, but he does not know from which islands. The specimens that I brought from three islands were young ones: and probably owing to this cause, neither Mr. Gray nor myself could find in them any specific differences. I have remarked that the marine Amblyrhynchus was larger at Albemarle Island than elsewhere; and M. Bribon informs me that he has seen two distinct aquatic species of this genus; so that the different islands probably have their representative species or races of the

* Voyage in the U. S. ship Essex, vol. i. p. 215.

Amblyrhynchus, as well as of the tortoise. My attention was first thoroughly aroused, by comparing together the numerous specimens, shot by myself and several other parties on board, of the mocking-thrushes, when, to my astonishment, I discovered that all those from Charles Island belonged to one species (Mimus trifasciatus); all from Albemarle Island to M. parvulus; and all from James and Chatham Islands (between which two other islands are situated, as connecting links) belonged to M. melanotis. These two latter species are closely allied, and would by some ornithologists be considered as only well-marked races or varieties; but the Mimus trifasciatus is very distinct. Unfortunately most of the specimens of the finch tribe were mingled together; but I have strong reasons to suspect that some of the species of the sub-group Geospiza are confined to separate islands. If the different islands have their representatives of Geospiza, it may help to explain the singularly large number of the species of this sub-group in this one small archipelago, and as a probable consequence of their numbers, the perfectly graduated series in the size of their beaks. Two species of the sub-group Cactornis, and two of the Camarhynchus, were procured in the archipelago; and of the numerous specimens of these two sub-groups shot by four collectors at James Island, all were found to belong to one species of each; whereas the numerous specimens shot either on Chatham or Charles Island (for the two sets were mingled together) all belonged to the two other species: hence we may feel almost sure that these islands possess their respective species of these two sub-groups. In land-shells this law of distribution does not appear to hold good. In my very small collection of insects, Mr. Waterhouse remarks, that of those which were ticketed with their locality, not one was common to any two of the islands.

If we now turn to the Flora, we shall find the aboriginal plants of the different islands wonderfully different. I give all the following results on the high authority of my friend Dr. J. Hooker. I may premise that I indiscriminately collected everything in flower on the different islands, and fortunately kept my collections separate. Too much confidence, however, must not be placed in the proportional results, as the small collections brought home by some other naturalists, though in some respects confirming the results, plainly show that much remains to be done in the botany of this group: the Leguminosæ, moreover, has as yet been only approximately worked out:—

Name of Island.	Total No. of Species.	No. of Species found in other parts of the world.	No. of Species confined to the Galapagos Archipelago.	No. confined to the one Island.	No. of Species confined to the Galapagos Archipelago, but found on more than the one Island.
James Island	71	33	38	30	8
Albemarle Island	46	18	26	22	4
Chatham Island	32	16	16	12	4
Charles Island	68	39 (or 29, if the probably imported plants be subtracted)	29	21	8

Hence we have the truly wonderful fact, that in James Island, of the thirty-eight Galapageian plants, or those found in no other part of the world, thirty are exclusively confined to this one island; and in Albemarle Island, of the twenty-six aboriginal Galapageian plants, twenty-two are confined to this one island, that is, only four are at present known to grow in the other islands of the archipelago; and so on, as shown in the above table, with the plants from Chatham and Charles Islands. This fact will, perhaps, be rendered even more striking, by giving a few illustrations:—thus, Scalesia, a remarkable arborescent genus of the Compositæ, is confined to the archipelago: it has six species: one from Chatham, one from Albemarle, one from Charles Island, two from James Island, and the sixth from one of the three latter islands, but it is not known from which: not one of these six species grows on any two islands. Again, Euphorbia, a mundane or widely distributed genus, has here eight species, of which seven are confined to the archipelago, and not one found on any two islands: Acalypha and Borreria, both mundane genera, have respectively six and seven species, none of which have the same species on two islands, with the exception of one Borreria, which does occur on two islands. The species of the Compositæ are particularly local; and Dr. Hooker has furnished me with several other most striking illustrations of the difference of the species on the different

28

islands. He remarks that this law of distribution holds good both with those genera confined to the archipelago, and those distributed in other quarters of the world: in like manner we have seen that the different islands have their proper species of the mundane genus of tortoise, and of the widely distributed American genus of the mocking thrush, as well as of two of the Galapageian sub-groups of finches, and almost certainly of the Galapageian genus Amblyrhynchus.

The distribution of the tenants of this archipelago would not be nearly so wonderful, if, for instance, one island had a mocking-thrush, and a second island some other quite distinct genus;—if one island had its genus of lizard, and a second island another distinct genus, or none whatever;—or if the different islands were inhabited, not by representative species of the same genera of plants, but by totally different genera, as does to a certain extent hold good: for, to give one instance, a large berry-bearing tree at James Island has no representative species in Charles Island. But it is the circumstance, that several of the islands possess their own species of the tortoise, mocking-thrush, finches, and numerous plants, these species having the same general habits, occupying analogous situations, and obviously filling the same place in the natural economy of this archipelago, that strikes me with wonder. It may be suspected that some of these representative species, at least in the case of the tortoise and of some of the birds, may hereafter prove to be only well-marked races; but this would be of equally great interest to the philosophical naturalist. I have said that most of the islands are in sight of each other: I may specify that Charles Island is fifty miles from the nearest part of Chatham Island, and thirty-three miles from the nearest part of Albemarle Island. Chatham Island is sixty miles from the nearest part of James Island, but there are two intermediate islands between them which were not visited by me. James Island is only ten miles from the nearest part of Albemarle Island, but the two points where the collections were made are thirty-two miles apart. I must repeat, that neither the nature of the soil, nor height of the land, nor the climate, nor the general character of the associatel beings, and therefore their action one on another, can differ much in the different islands. If there be any sensible difference in their climates, it must be between the Windward group (namely Charles and Chatham Islands), and that to leeward; but there seems to be no corresponding difference in the productions of these two halves of the archipelago.

The only light which I can throw on this remarkable difference in the inhabitants of the different islands, is, that very strong currents of the sea running in a westerly and W.N.W. direction must separate, as far as transportal by the sea is concerned, the southern islands from the northern ones; and between these northern islands a strong N.W current was observed, which must effectually separate James and Albemarle Islands. As the archipelago is free to a most remarkable degree from gales of wind, neither the birds, insects, nor lighter seeds, would be blown from island to island. And lastly, the profound depth of the ocean between the islands, and their apparently recent (in a geological sense) volcanic origin, render it highly unlikely that they were ever united; and this, probably, is a far more important consideration than any other, with respect to the geographical distribution of their inhabitants. Reviewing the facts here given, one is astonished at the amount of creative force, if such an expression may be used, displayed on these small, barren, and rocky islands; and still more so, at its diverse yet analogous action on points so near each other. I have said that the Galapagos Archipelago might be called a satellite attached to America, but it should rather be called a group of satellites, physically similar, organically distinct, yet intimately related to each other, and all related in a marked, though much lesser degree, to the great American continent.

[*Editor's Note:* In the original, material follows this excerpt.]

Editor's Comments
on Papers 3 Through 7

WHY ARE THERE SO MANY SPECIES IN THE TROPICS?

The question "Why are there so many species in the tropics?" has stimulated two symposia (Lowe-McConnell, 1969; Brookhaven, 1969) and at least four comprehensive reviews including those of Fisher (1960), Pianka (Paper 3), Baker (1970), and Rohde (1978). Despite the abundance of theories to explain the species richness of the tropics, none is completely satisfactory. Some of these theories rest on circular reasoning. For example, there are a large number of species in the tropics because tropical species have narrower niches. Narrow niches result from high competition. The reason for high competition is that there are so many species.

Another theory reviewed by Pianka in Paper 3 is that climatic stability in the tropics is the reason for high species diversity. The idea is that stable climate, that is, lack of strong seasonality, results in species occupying narrower niches in the tropics because they do not have to adapt to changing seasons and thus they can become more highly specialized. With narrower niches, more species can be accommodated in the tropical environment. It can be argued, however, that just because a climate has annual cycles does not mean that it is unstable. An unstable climate would be one that changes but not in periodic cycles to which organisms could adapt. In this sense, there is no evidence that temperature climates are any more unstable than tropical climates.

Recently, there have been arguments that it is instability rather than stability that gives rise to the species richness of the tropics. Vuilleumier (1971) has suggested that the tropics have probably been unstable for the last million years from the point of view of the biotas, and this instability resulted primarily from the Pleistocene climate. Connell (1978) and Huston (1979) have argued that nonequilibrium or unstable conditions in the environment on the time scale less than the life span of the organisms contributes to the high species diversity in tropical ecosystems.

However, these hypotheses about instability in the tropics still cannot explain differences in species diversity between tropical and temperature ecosystems. For instability in the tropics to explain differences, it would have to be shown that temperate zones were more stable, either on long or short time scales, than tropical zones.

Jordan and Murphy (1978) have suggested that instead of asking "Why there are so many species in the tropics?" we should be asking "Why are there so few species at higher latitudes?" They point out that Chaney (1947) has described Eocene fossil floras resembling tropical and subtropical species assemblages at latitudes as high as 49° in North America. Chaney's conclusion is that environmental conditions deteriorated late in the Tertiary period causing a migration of most plant populations toward the equator. Jordan and Murphy suggest that only the best competitors were able to survive the lower temperatures and shorter growing seasons of the higher latitudes, and, consequently, these latitudes were left with diminished flora and fauna.

This argument is not another version of the tropical stability/instability hypotheses. These hypotheses deal with a time period between one million years ago and the present, and the problem with them is that there is a lack of evidence that the tropics were either more stable or less stable than temperate zones during this period. In contrast, the idea that changing Tertiary climate reduced species diversity at high latitudes deals with a time span of up to sixty million years. The evidence is quite strong that the high latitudes experienced a much greater change in climate than tropical zones during this period (Chaney, 1947).

Despite the fact that the question of high diversity in tropical ecosystems, or preferably, low species diversity in temperate ecosystems, will probably go unresolved, the question has stimulated many lines of research. One of the lines, for example, is whether autogamy and genetic drift are important in the process of speciation in the tropics. The following papers illustrate how this line of research sprang from observations of tropical species diversity, and how it has led to an understanding of population dynamics of plants in the tropics.

Wallace, in his book *Tropical Nature and Other Essays* (Paper 12), compares temperate and tropical vegetation and states that for plants at high latitudes, "Their struggle for existence is against climate." In the tropics, however, he says there is a "never-ceasing struggle for existence between the various species in the same area...." Theodosius Dobzhansky (Paper 4) elaborates on these differences between tropical and temperate regions. For example, he states, "The process of adaptation for life in temperate and especially in cold zones consists, for man as well as for other organisms, primarily in coping with the physical environment and in securing food" (p. 220). The situation for the tropics, Dobzhansky says, is different, because in the tropics, competition is more important. He states, "The tremendous intensity of the competition for space among plants in tropical forests can be felt even by a casual observer. The apparent scarcity, concealment, and shyness of most tropical animals attest to the same fact of extremely keen competition among the inhabitants" (p. 220). Dobzhansky suggests that it is this competition that results in the high species diversity and consequent species rarity in tropical forests.

Fedorov (Paper 5) discusses the data from tropical forests published by Dobzhansky showing high species diversity and consequent rarity of many tropical species and asks, "What might be the result of the extremely low population density and the small size and isolation of populations in tropical forests?". He concludes that, "Apparently these conditions are favourable for the development of automatic genetic processes such as genetic drift" (p. 6).

Ashton (Paper 6) disputes Fedorov's arguments that genetic drift is more important in the tropics than at higher latitudes. After an analysis of data from a southeast Asian Dipterocarp forest, Ashton states, "Thus the evidence from the comparison of areas and vegetation types showing great floristic diversity allows an interpretation of genecological processes which does not differ fundamentally from those of other ecosystems." In other words, speciation in the tropics is not unique, and genetic drift cannot account for the species richness of tropical ecosystems.

The position that genetic drift is no more important in the tropics than in other regions is supported by Bawa's (1974) finding that many tropical tree species in a sample of lowland semideciduous forests are obligately outcrossed. In addition, evidence provided by Hubbell (Paper 7) that self-compatible species are no more frequent among rare species than among common species in a tropical forest argues against the importance of genetic drift in the tropics. Hubbell's paper is interesting in that his analysis of a forest community is a novel approach for testing many different ecological hypotheses, including the question of self-compatibility of tropical species.

REFERENCES

Baker, H. G., 1970, Evolution in the Tropics, *Biotropica* **2**:101–111.

Bawa, K. S., 1974, Breeding Systems of Tree Species of a Lowland Tropical Community, *Evolution* **28**:85–92.

Brookhaven Symposia in Biology No. 22, 1969, *Diversity and Stability in Ecological Systems*, Biology Department Brookhaven National Laboratory, Upton, New York, BIVL 50175 (C–56), 264 p.

Chaney, R. W., 1947, Tertiary Centers and Migration Routes, *Ecol. Monogr.* **17**:141–148.

Connell, J. H., 1978, Diversity in Tropical Rain Forests and Coral Reefs, *Science* **199**:1302–1310.

Fischer, A. G., 1960, Latitudinal Variation in Organic Diversity, *Evolution* **14**:64–81.

Huston, M., 1979, A General Hypothesis of Species Diversity, *Am. Nat.* **113**:81–101.

Jordan, C. F., and P. G. Murphy, 1978, A Latitudinal Gradient of Wood and Litter Production and Its Implication Regarding Competition and Species diversity in Trees, *Am. Midl. Nat.* **99**:415–434.

Lowe-McConnell, R. H., 1969, Speciation in Tropical Environments: Proceedings of a Symposium, Academic Press, New York, 246 p.

Rohde, K., 1978, Latitudinal Gradients in Species Diversity and Their Causes, I: A Review of the Hypotheses Explaining the Gradients, *Biol. Zentralbl.* **97**:393–403.

Vuilleumier, B. S., 1971, Pleistocene Changes in the Fauna and Flora of South America, *Science* **173**:771–780.

3

LATITUDINAL GRADIENTS IN SPECIES DIVERSITY: A REVIEW OF CONCEPTS

Eric R. Pianka

Department of Zoology, University of Washington, Seattle, Washington*

INTRODUCTION: DIVERSITY INDICES

The simplest index of diversity is the total number of species, usually of a specific taxon under investigation, inhabiting a particular area. Since this index does not take into account differing abundances of species, divergent communities may show similar "diversities." Because of this, more sophisticated measures have been proposed which weight the contributions of species according to their relative abundances. As early as 1922 Gleason described and discussed the now well known "species-area" curve (Gleason, 1922, 1925). Later, Fisher, Corbet, and Williams (1943) proposed an index, alpha, discussed in detail by C. B. Williams (1964), which can be shown to approximate Gleason's "exponential ratio" (H. S. Horn, personal communication). Margalef (1958) has also used a modification of this index "d," in phytoplankton diversity studies, as well as several other indices (Margalef, 1957). The most recent, and currently widely used diversity index, is the information theory measure, H, derived by Shannon (1948). This index, $-\Sigma p_i \log p_i$, in which p_i represents the proportion of the total in the i-th category, has been used to quantify the "dispersion" of the distribution of entities with no ordered sequence, such as species in a community, alphabetic letters on a page, etc. Unfortunately, there has as yet been little discussion of the application of statistical procedures to this quantity. However, even without statistical embellishments, H has been a useful and productive tool (Crowell, 1961, 1962; MacArthur, 1955, 1964; MacArthur and MacArthur, 1961; Margalef, 1957, 1958; Paine, 1963; and Patten, 1962).

The choice of the index used in any particular investigation depends on several factors, especially the difficulty of appraisal of species abundances, but also on the degree to which relative abundances shift during the period of study, and for many purposes the simplest index, the number of species present, may be the most useful measure of local or regional diversity. This index weights rare and common species equally, and is the logical measure of diversity in situations with many rare, but regular, species (such as desert lizard faunas, Pianka, in preparation).

THE PROBLEM: SPECIES DIVERSITY GRADIENTS

Latitudinal gradients in species diversity have been recognized for nearly a century, but only recently have some of these polar-equatorial

*Present address: Department of Biology, Princeton University, P.O. Box 704, Princeton, New Jersey.

trends been discussed in any detail (Darlington, 1959; Fischer, 1960; Simpson, 1964; Terent'ev, 1963). A few groups, such as the marine infauna (Thorson, 1957), and some fresh water invertebrates and phytoplankton appear not to follow this pattern, but many plant and animal taxa display latitudinal gradients. A phenomenon as widespread as this may have a general explanation, knowledge of which would be of considerable utility in making predictions about the operation of natural selection upon community organization. Because of the global scope of the problem, however, it has usually been impossible for a single worker to study a complete species diversity gradient.

Approaches to the study of diversity gradients have so far been mainly of two types, the method of gross geographic lumping with comparison of total species lists for a group (Simpson, 1964; Terent'ev, 1963), and the approach by synecological studies on a smaller scale, comparing the diversity of a taxon through many different habitats (MacArthur and MacArthur, 1961; MacArthur, 1964, and in press). Terent'ev and Simpson used the number of species as indices of diversity, and MacArthur and MacArthur used Shannon's information theory formula to calculate indices of faunal and environmental diversity. Simpson (1964) points out that diversity gradients indicated by the method of gross geographic lumping have two components, one due to the number of habitats sampled by a given quadrate (and thus to the topographic relief) and another component due to ecological changes of some kind. Low latitude regions have more kinds of habitats, (i.e., Costa Rica has a whole range of habitats from low altitude tropical to middle altitude temperate to high altitude boreal habitats; whereas regions of higher latitude progressively lose some of these habitats) and therefore the presence of more species there is neither surprising nor theoretically very interesting. The question of basic ecological interest is that of the second component of diversity—namely, what are the factors that allow ecological co-existence of more species at low latitudes? Ecological data relating to species diversity gradients are scant, and no one has yet attempted the logical step of merging the synecological with an autecological approach.

Despite the handicap of insufficient ecological data, or perhaps because of it, theorization and speculation as to the possible causes of diversity gradients has been frequent and varied (Connell and Orias, 1964; Darlington, 1957, 1959; Dobzhansky, 1950; Dunbar, 1960; Fischer, 1960; Hutchinson, 1959; Klopfer, 1959, 1962; Klopfer and MacArthur, 1960, 1961; MacArthur, 1964, and in press; Paine, in press; and C. B. Williams, 1964). These efforts have produced six more or less distinct hypotheses: (a) the time theory, (b) the theory of spatial heterogeneity, (c) the competition hypothesis, (d) the predation hypothesis, (e) the theory of climatic stability, and (f) the productivity hypothesis. It is instructive to consider each of these hypotheses separately, attempting to suggest possible tests and observations for each, even though only one pair represent mutually exclusive alternatives, and thus several of the proposed mechanisms of control of diversity could be operating simultaneously in a given situation.

In the following discussion, ecological and evolutionary saturation are defined as the ecological and evolutionary upper limits to the number of

species supported by a given habitat. The assumption of ecological saturation is implicit in ecological studies of species diversity gradients, an assumption without which the study of such gradients must be made in terms
of the history of the area. There is reasonable evidence that the majority
of habitats are ecologically saturated (Elton, 1958; MacArthur, in press).

The time theory

Proposed chiefly by zoogeographers and paleontologists, the theory of
the "history of geological disturbances" assumes that all communities tend
to diversify in time, and that older communities therefore have more species
than younger ones. (The evidence behind this assumption is scanty, and
the assumption may or may not be valid.) Temperate regions are considered
to be impoverished due to recent glaciations and other disturbances (Fischer,
1960). It is useful to distinguish between ecological and evolutionary
processes as subcategories of the theory. Ecological processes would be
applicable to those circumstances where a species exists which can fill a
particular position in the environment; but this species has not yet had time
enough to ,disperse into the relatively newly opened habitat space. Evolutionary processes apply to longer time spans, to those cases where a newly
opened habitat is not yet utilized, but will be occupied given time enough
for speciation and the evolution of an appropriate organism.

Tests of the ecological and evolutionary time theories are by necessity
indirect; but several authors have suggested possibilities for assessing the
importance of evolutionary time as a control of species diversity. Simpson
(1964) has argued that the warm temperate regions have had a long undisturbed history (from the Eocene to the present), long enough to become both
ecologically and evolutionarily saturated, and that since there are fewer
species in this zone than there are in the tropics, other factors must be in-
voked to explain the difference between tropical and temperate diversities.
Beyond this, he reasons that if the time theory were correct, the steepest
gradient in species diversities should occur in the recently glaciated temperate zone. Since, for North American mammals, at least, this zone shows
a fairly flat diversity profile, there is some evidence against the evolutionary time theory. Simpson (1964) emphasizes that temperate zones have
probably been in existence as long as have tropical ones. Newell (1962)
stresses that temperate areas at intermediate latitudes were probably not
eliminated during the glacial periods, but were simply shifted laterally
along with their floras and faunas, and that, if this is the case, they have
had as long a time to adapt as have the non-glaciated areas. R. H. Mac-
Arthur (personal communication) has suggested that possibilities exist for a
test of the effects of glaciation, by comparisons of areas in the glaciated
north temperate with their non-glaciated southern temperate counterparts.
However, the evolutionary time theory is not readily amenable to conclusive
tests, and will probably remain more or less unevaluated for some time.

The evidence relating to the ecological time theory, summarized in detail
by Elton (1958), and discussed by Deevey (1949), indicates that most con-

tinental habitats are ecologically saturated. Only in those cases where barriers to dispersal are pronounced can the ecological time theory be of importance in determining species diversity. Islands have sometimes been considered cases of historical accident in which maximal utilization of the biotope is often achieved by pronounced behavioral modifications of those species which have managed to inhabit them (Crowell, 1961, 1962; Lack, 1947). More recently, several theories for equilibrium in insular zoogeography have been proposed, and data given which shows strong dependence of species composition on island size, distance from "source" areas, and time available for colonization (Hamilton, Barth, and Rubinoff, 1964; MacArthur and Wilson, 1963; Preston, 1962). These papers indicate that predictable patterns of species diversity occur even on islands, and further lessen the probable importance of the ecological time theory.

The theory of spatial heterogeneity

Proponents of this hypothesis claim that there might be a general increase in environmental complexity as one proceeds towards the tropics. The more heterogeneous and complex the physical environment becomes, the more complex and diverse the plant and animal communities supported by that environment. Again it is useful to distinguish two subcategories of this theory, one on a macro-, the other on a micro- scale. The first is the factor Simpson (1964) calls topographic relief, discussed in more detail by Miller (1958). The factor of topographic relief is especially interesting in the study of speciation, and has been much discussed in books and symposia on that subject (Blair, 1961; Mayr, 1942, 1957, 1963). The component of total diversity due to topographic relief has been mentioned earlier in this paper.

In contrast to topographic relief, micro-spatial heterogeneity is on a local scale, with the size of the environmental elements corresponding roughly to the size of the organisms populating the region. Elements of the environmental complex in this class might be soil particle size, rocks and boulders, karst topography, or if one is considering the animals in a habitat, the pattern and complexity of the vegetation. Environmental heterogeneity of the micro-spatial type has been little studied by zoologists, and is more interesting to the ecologist than to the student of speciation (considerations of sympatric speciation processes will, however, involve micro-spatial attributes of the environment). It should be noted here that in the only study which relates species diversity of a taxon to environmental diversity in a quantitative way, the environmental diversity is of the micro-spatial type (MacArthur and MacArthur, 1961; MacArthur, 1964). These authors demonstrate that foliage height diversity is a good predictor of bird species diversity, and that knowledge of plant species diversity does not improve the estimate. Further tests of the theory of environmental complexity will probably follow similar lines, although it would be useful to consider alternative ways of examining this hypothesis.

Spatial heterogeneity has several shortcomings when applied to the explanation of global diversity patterns. The component of total diversity due to topographic relief and number of habitats (macro-spatial heterogeneity) certainly increases towards the tropics, but does not offer an explanation for diversity gradients within a given habitat-type. Vegetative spatial heterogeneity is clearly dependent on other factors and explanation of animal species diversity in terms of vegetative complexity at best puts the question of the control of diversity back to the control of vegetative diversity. Since there is no reason to suppose that micro-spatial heterogeneity of the physical environment changes with latitude, the theory of micro-spatial heterogeneity seems to explain only local diversity. Ultimately, resolution into independent variables will require consideration of non-biotic factors such as climate which change more or less continuously from pole to pole (see section on the climatic stability hypothesis).

The competition hypothesis

Advocated by Dobzhansky (1950) and C. B. Williams (1964), this idea is that natural selection in the temperate zones is controlled mainly by the exigencies of the physical environment, whereas biological competition becomes a more important component of evolution in the tropics. Because of this there is' greater restriction to food types and habitat requirements in the tropics, and more species can co-exist in the unit habitat space. Competition for resources is *keener* and niches 'smaller' in more diverse communities. Dobzhansky emphasizes that natural selection takes a different course in the tropics, because catastrophic indiscriminant mortality factors (density-independent), such as drought and cold, seldom occur there. He notes that catastrophic mortality usually causes selection for increased fecundity and/or accelerated development and reproduction, rather than selection for competitive ability and interactions with other species. Dobzhansky predicts that tropical species will be more highly evolved and possess finer adaptations than will temperate species, due to their more directed mortality and the increased importance of competitive interactions. No statement has been given as to exactly why competition might be more important in the tropics, but the hypothesis is testable in its present form.

Because the predation hypothesis predicts very nearly the opposite mechanisms of control of diversity than does the competition hypothesis, I will briefly outline the predation hypothesis before proceeding with discussion. These two hypotheses are almost mutually exclusive alternatives, and the same tests, by and large, apply to both.

The predation hypothesis

It has been claimed that there are more predators (and/or parasites) in the tropics, and that these hold down individual prey populations enough to lower the level of competition between and among them (Paine, in press). This lowered level of competition then allows the addition and co-existence of new intermediate prey types, which in turn support new predators in the system, etc. The mechanism can apply to both evolutionary and dispersal additions of new species into the community. Paine (1966) argues that the upper limits on the process are set by productivity factors, which will here be considered separately.

According to this hypothesis, competition among prey organisms is *less* intense in the tropics than in temperate areas. Thus, a test between these two hypotheses is possible, provided that the intensity of competition can be measured. Several approaches to the quantification of competition might find application here (Connell, 1961a, 1961b; Elton, 1946; Kohn, 1959; MacArthur, 1958; Moreau, 1948). Also, if the predation hypothesis holds, community structure should shift along a diversity gradient, with an increase in the proportion of predatory species as the communities become more diverse. Evidence for such a shift in trophic structure along a diversity gradient is given by Grice and Hart (1962). These authors present data showing that the proportion of predatory species in the marine zooplankton increases along a latitudinal diversity gradient. A similar shift in community structure accompanies a terrestrial species diversity gradient in the deserts of western North America (Pianka, in preparation). Fryer (1959, 1965) has argued that predation enhances migration and speciation, thereby resulting in increased species diversity, in some African lake fishes. As will be pointed out in a later section, demonstration that species have either finer or more overlapping habitat requirements in the tropics could be used to support three of the six hypotheses, and is therefore not a powerful distinguishing tool.

The theory of climatic stability

According to this hypothesis, restated by Klopfer (1959), regions with stable climates allow the evolution of finer specializations and adaptations than do areas with more erratic climatic regimes, because of the relative constancy of resources. This also results in "smaller niches" and more species occupying the unit habitat space. Another way of stating this principle in terms of the organism, rather than the environment, is that, in order to persist and successfully exploit an environment, a species must have behavioral flexibility which is roughly inversely proportional to the predictability of the environment (J. Verner, personal communication). In recent years the theory of climatic stability has become a favorite for explaining the generality of latitudinal gradients in species diversity, but has as yet remained untested. Rainfall and temperature can be shown to vary less in the tropics than in temperate zones, but rigorous correlation with faunal diversity, let alone demonstration of causal connection, has not yet been possible. It should be realized that climatic factors could well determine directly floral and/or vegetative complexity, while being only indirectly related to the faunal diversity of the area.

Evidence that tropical species have more restricted habitat requirements than temperate species would support the competition hypothesis, the predation hypothesis, and the theory of climatic stability. Klopfer and Mac-

Arthur (1960) have attempted to test the hypothesis that "niches" are "smaller" in the tropics by comparing the proportion of passerine birds to non-passerines along a latitudinal gradient. Their thesis is that the non-passerines, possessing a more stereotyped behavior, are better adapted to exploit the more constant tropical environment than are the passerines, whose more plastic behavior allows them to inhabit less predictable habitats. Klopfer (in press, in preparation) has compared the degree of behavioral stereotypy in temperate and tropical birds, and tentatively concludes that "while tropical species are in fact 'stereotyped,' this is more likely an effect rather than a cause of their greater diversity."

An interesting variation on this theme is that of increased "niche overlap" in more diverse communities (Klopfer and MacArthur, 1961). Klopfer and MacArthur attempted to test this idea by comparing the ratios of bill lengths in congeners among several sympatric bird species in Panama and Costa Rica. Simpson (1964) notes that this ratio may come as close to 1.00 in temperate birds as it did in Klopfer and MacArthur's tropical species, but fails to realize that morphological character displacement is expected only in species occupying the same space (Brown and Wilson, 1956; Hutchinson, 1959; Klopfer and MacArthur, 1961; MacArthur, in press). Ratios of culmen lengths may often approach unity in species such as *Dendroica* which clearly divide up the biotope space (MacArthur, 1958), and thus demonstrate behavioral, rather than morphological character displacement. However, it is apparent that even if the comparison were valid, the test would not distinguish between "smaller niches" and increased "niche overlap" (C. C. Smith, personal communication). It is difficult to devise tests which will distinguish between these alternatives, and perhaps none can be suggested until "niche" has been operationally defined. Use of some of the dimensions of Hutchinson's (1957) multidimensional niche may allow partial testing between these alternatives, as in the work of Kohn (1959) on *Conus* in Hawaii.

Increased overlap in selected dimensions of the niche can imply either increased, decreased, or constant competition; the first if the overlapping resources are in short supply, the second if the overlapping resources are so abundant that sharing of them is only slightly detrimental to each species, and the third if independent environmental factors (such as more predictable production) allow increased sharing of the same amount of resource. Hence, because data supporting the niche overlap idea has ambiguous competitive interpretations, it does not distinguish between three (or four) hypotheses either.

Tests distinguishing between the theory of climatic stability and the competition hypothesis are especially difficult to devise, as there is considerable overlap between the two, and indeed, they are usually mixed when either is suggested. This similarity makes it all the more important to evaluate the importance of each, and in keeping with the rest of this paper they will be considered separately and at least two possible distinguishing tests suggested.

41

According to the theory of climatic stability, a unit of habitat will support the same number of individuals in the tropics and temperate regions, but since each of the species may be rarer (without becoming extinct) in the tropics, there can be more of them. The competition hypothesis implies that more individuals occupy the same habitat space, or else competition would not be increased. Considering a fixed areal dimension as a unit of habitat, abundance data from the tropics generally suggest that the number of individuals is relatively similar from temperate to tropics and therefore support the theory of climatic stability (Klopfer and MacArthur, 1960; Skutch, 1954). Another way in which these two theories might be separated is by examining the intensity of competition occurring along an increasing diversity gradient; if the level of competition remained constant, or decreased along the gradient, the prediction of an increased proportion of predatory species could be used to separate the predation hypothesis from the theory of climatic stability.

The productivity hypothesis

The most recent and most complete statement of this hypothesis is that of Connell and Orias (1964). They blend this hypothesis with the theory of climatic stability, distinguishing between the energetic cost of maintenance and the energy left for growth and reproduction. Their synthesis also includes aspects of the theory of spatial heterogeneity, and reasonably explains latitudinal trends in diversity, but the productivity hypothesis will be considered here in its "pure" form.

The productivity hypothesis states that greater production results in greater diversity, everything else being equal. Since it is patently impossible to hold everything else equal, the hypothesis can only be tested in crude or indirect ways. Experimental manipulation of nutrient levels in freshwater lakes, for instance, might provide a possible test. Such enrichments have often been made, both intentionally and accidentally (sewage), and quantitative data have been taken on the response of the biota. The data needed for calculating diversities probably exist, and it would be interesting to see such calculations performed. Qualitative indications are that enrichment usually causes an impoverished fauna (Patrick, 1949; L. G. Williams, 1964).

If productivity were of overwhelming importance in the regulation of species diversity, one would expect a correlation despite uncontrolled extraneous variables. Only one such correlation is known to me (Patten, Mulford, and Warinner, 1963), and in fact there may often be an inverse relation between species diversity and abundance or standing crop (which should usually be positively correlated with production) (Hohn, 1961; Hulburt, 1963; Yount, 1956; L. G. Williams, 1964; and my own observations on desert lizards). Those who would claim that the above studies are on nonequilibrium populations and thus not applicable to the problem at hand, would do well to search for data from "equilibrium" conditions which are relevant to the productivity hypothesis.

A common modification of the productivity hypothesis which has been claimed to be of importance in regulating species diversity is the notion of increased temporal heterogeneity in the tropics. The main argument is that the longer season of tropical regions allows the component species to partition the environment temporally as well as spatially, thereby permitting the coexistence of more species (MacArthur, in press). This notion has been rephrased by Paine (in press) who argues that the "stability of primary production" is a major determinant of the species diversity of a community. Paine integrates the predation hypothesis with this idea to form a sort of synergistic system controlling diversity. This hypothesis is also a blend of the stability and productivity theories, but in this case, the mixing suggests new observations, and a new mechanism of control of diversity than does either hypothesis alone. The mechanism for the regulation of species diversity by stability of primary production may be similar to the mechanism suggested for climatic stability, except that in this case, plants may buffer climatic variability by utilizing their own homeostatic adaptations and storage capacities to increase the stability of primary production.

These notions can be tested by analyses, such as that of MacArthur (1964), of the length of breeding seasons, but there are other ways of examining them as well. Thus, comparisons of the division of the day (or night) and season into discrete activity periods by different animal species might elucidate latitudinal trends. Another possible angle of approach is by means of the "stability of primary production," which can be measured directly and examined for latitudinal trends. Unfortunately, there are all too few reliable measures of primary production, let alone the variability in this quantity, and at this point it is difficult to assess the stability of primary production along a latitudinal gradient. The necessary data are simple enough in theory, but in practice a single determination of primary productivity is tedious (especially in terrestrial habitats). An indirect possibility for testing the hypothesis exists, however, for arid regions, where primary productivity is strongly positively correlated with precipitation (Pearson, 1965; Walter, 1939, 1955, 1962). In this environment, the amount and variability of precipitation can be used to estimate the amount and variability of primary production. Preliminary analysis of weather and lizard data for the deserts of western North America shows no correlation with either the average amount or the variability of rainfall and the number of lizard species (Pianka, in preparation).

Since clutch size is closely related to these ideas of increased temporal heterogeneity and stability of primary production, it may be profitably considered here. The fact of reduced clutch size in tropical birds (Skutch, 1954) has been discussed as a possible factor allowing the coexistence of more species in the tropics (MacArthur, in press). MacArthur argues that by lowering its clutch size, a species reduces its total energy requirements and is therefore able to survive in less productive areas which were formerly marginal habitats. He reasons that such reductions in total energy requirements will also allow the existence of more species, when the total amount of energy available is held constant. Apart from the problems of population replacement raised by these theoretical arguments, there are other reasons for doubting the importance of reduced clutch size as a determinant of increased tropical diversity. For instance, it is highly possible that tropical habitats never achieve food densities as high as those usual further north, because of their greater species diversity and the fact that most of the breeding birds do not migrate. In contrast, great blooms of production characterize the temperate regions, and most of the breeding birds are migrants. Thus it may be energetically impossible for tropical birds to raise as many young as can be supported in the more productive northern areas (that is, more productive on a short term basis, during the short growing season) (Orians, personal communication). Support of this notion comes from the large territory sizes of many tropical bird species (Skutch, 1954), which suggests that food may be scarce. If this is indeed the case, the smaller clutches of tropical birds would be a result, rather than a cause of, the greater diversity in the tropics. These criticisms of the reduced clutch size hypothesis are, however, in themselves largely theoretical, and it will be worthwhile to examine clutch sizes of other taxa along various diversity gradients. Clutch sizes of desert lizards vary latitudinally, but whether or not the largest clutches are from the south depends on the species concerned (Pianka, in preparation).

CONCLUSIONS

Obviously, there is room for considerable overlap between these different hypotheses, and several may be acting in concert or in series in any particular situation. Because of the preliminary state of knowledge on the subject of species diversity, for the sake of clarity, and in order to suggest tests of the various hypotheses, it is useful first to consider and assess each of the components of control of diversity in isolation, before attempting various mixtures. Once the relative importance of each factor has been assessed for many different diversity gradients, an attempt may be made to merge them. In general, the compounding of hypotheses is to be avoided, unless such blending suggests new tests not applicable to the isolated theories. As more and more parameters are included, the more complex hypothesis tends to "answer" all cases and becomes less and less testable and useful.

A fact often overlooked is that most of the hypotheses can be either supported or rejected by appropriate observations on a limited scale; any species diversity gradient might be a suitable study system. If the broader geographical gradients are found to be qualitatively different from local diversity patterns, this in itself would be interesting, and understanding the difference would ultimately require thorough knowledge of the control of local species diversity.

Finally, since ecologists can seldom structure their experiments except by their choice of observations and measurements, the natural system usually sets the bounds within which they must work. The basic technique of descriptive science is correlation, and it is well to keep in mind that correlation does not necessarily mean causation. This is especially true in the study of latitudinal gradients in species diversity, where many different factors vary along the gradient in a fashion similar to the taxon studied, and spurious correlations may be frequent. For these reasons all significant correlations must be carefully examined and attempts made to understand the mechanisms and causal connections (if any) between variates. Unambiguous demonstration of causality can only be attained by experimental manipulation of the independent variables in the system.

SUMMARY

The six major hypotheses of the control of species diversity are restated, examined, and some possible tests suggested. Although several of these mechanisms could be operating simultaneously, it is instructive to consider them separately, as this can serve to clarify our thinking, as well as assist in the choice of the best test situations for future examination.

ACKNOWLEDGMENTS

Many fruitful discussions preceded this effort, particularly those with H. S. Horn, G. H. Orians, R. T. Paine, and C. C. Smith. Drs. A. J. Kohn, R. H. MacArthur, G. H. Orians, and R. T. Paine have read the manuscript and made many valuable suggestions. I wish to acknowledge J. F. Waters for translating Terent'ev's paper. The work was supported by the National Institutes of Health, predoctoral fellowship number 5-F1-GM-16,447-01 to -03.

LITERATURE CITED

Blair, W. F. [Editor]. 1961. Vertebrate speciation. Univ. Texas Press, Austin, Texas.

Brown, W. L., Jr., and E. O. Wilson. 1956. Character displacement. Syst. Zool. 5: 49-64.

Connell, J. H. 1961a. Effects of competition, predation by *Thais lapillus*, and other factors on natural populations of the barnacle *Balanus balanoides*. Ecol. Monogr. 31: 61-106.

———. 1961b. The influence of interspecific competition and other factors on the distribution of the barnacle *Chthamalus stellatus*. Ecology 42(4): 710-723.

Connell, J. H., and E. Orias. 1964. The ecological regulation of species diversity. Amer. Natur. 98: 399-414.

Crowell, K. 1961. The effects of reduced competition in birds. Proc. Nat. Acad. Sci. 47: 240-243.

———. 1962. Reduced interspecific competition among the birds of Bermuda. Ecology 43: 75-88.

Darlington, P. J., Jr. 1957. Zoogeography; the geographical distribution of animals. John Wiley & Sons, Inc., New York and London.

———. 1959. Area, climate, and evolution. Evolution 13: 488-510.

Deevey, E. S., Jr. 1949. Biogeography of the Pleistocene. Bull. Geol. Soc. Amer. 60: 1315-1416.

Dobzhansky, T. 1950. Evolution in the tropics. Amer. Sci. 38: 209-221.

Dunbar, M. J. 1960. The evolution of stability in marine environments. Natural selection at the level of the ecosystem. Amer. Natur. 94: 129–136.

Elton, C. S. 1946. Competition and the structure of ecological communities. J. Anim. Ecol. 15: 54–68.

———. 1958. The ecology of invasions by animals and plants. Meuthen, London.

Fischer, A. G. 1960. Latitudinal variation in organic diversity. Evolution 14: 64–81.

Fisher, R. A., A. S. Corbet, and C. B. Williams. 1943. The relation between the number of species and the number of individuals in a random sample of an animal population. J. Anim. Ecol. 12: 42–58.

Fryer, G. 1959. Some aspects of evolution in Lake Nyasa. Evolution 13: 440–451.

———. 1965. Predation and its effects on migration and speciation in African fishes: A comment. Proc. Zool. Soc. London 144: 301–322.

Gleason, H. A. 1922. On the relation between species and area. Ecology 3: 158–162.

———. 1925. Species and area. Ecology 6: 66–74.

Grice, G. D., and A. D. Hart. 1962. The abundance, seasonal occurrence and distribution of the epizooplankton between New York and Bermuda. Ecol. Monogr. 32: 287–309.

Hamilton, T. H., R. H. Barth, Jr., and I. Rubinoff. 1964. The environmental control of insular variation in bird species abundance. Proc. Nat. Acad. Sci. 52: 132–140.

Hohn, M. H. 1961. The relationship between species diversity and population density in diatom populations from Silver Springs, Florida. Trans. Amer. Microscop. Soc. 80: 140–165.

Hulburt, E. M. 1963. The diversity of phytoplanktonic populations in oceanic, coastal, and esturine regions. J. Marine Res. 21: 81–93.

Hutchinson, G. E. 1957. Concluding remarks. Cold Spring Harbor Symp. Quant. Biol. 22: 415–427.

———. 1959. Homage to Santa Rosalia, or why are there so many kinds of animals? Amer. Natur. 93: 145–159.

Klopfer, P. H. 1959. Environmental determinants of faunal diversity. Amer. Natur. 93: 337–342.

———. 1962. Behavioral aspects of ecology. Prentice-Hall, Englewood Cliffs, N. J.

Klopfer, P. H., and R. H. MacArthur. 1960. Niche size and faunal diversity. Amer. Natur. 94: 293–300.

———. 1961. On the causes of tropical species diversity: niche overlap. Amer. Natur. 95: 223–226.

Kohn, A. J. 1959. The ecology of *Conus* in Hawaii. Ecol. Monogr. 29: 47–90.

Lack, D. 1947. Darwin's finches. Cambridge Univ. Press, Cambridge, England. Reprinted 1961 by Harper and Brothers, New York.

MacArthur, R. H. 1955. Fluctuations of animal populations, and a measure of community stability. Ecology 36: 553–536.

———. 1958. Population ecology of some warblers of north-eastern coniferous forests. Ecology 39: 599–619.

————. 1964. Environmental factors affecting bird species diversity. Amer. Natur. 98: 387–398.

————. 1965. Patterns of species diversity. Biol. Rev. (In press).

MacArthur, R. H., and J. W. MacArthur. 1961. On bird species diversity. Ecology 42: 594–598.

MacArthur, R. H., and E. O. Wilson. 1963. An equilibrium theory of insular zoogeography. Evolution 17: 373–387.

Margalef, D. R. 1957. Information theory in ecology. Gen. Syst. 3: 37–71. Reprinted 1958.

————. 1958. Temporal succession and spatial heterogeneity in phytoplankton. *In* Perspectives in marine biology. A. Buzzati-Traverso [ed.], Univ. California Press, Berkeley.

Mayr, E. 1942. Systematics and the origin of species. Columbia Univ. Press, New York. Reprinted 1964 by Dover Publications, Inc., New York.

————. 1957. [*Editor*], The species problem. Amer. Ass. Advance. Sci., Publ. No. 50.

————. 1963. Animal species and evolution. The Belknap Press of Harvard Univ. Press, Cambridge, Mass.

Miller, A. H. 1958. Ecologic factors that accelerate formation of races and species in terrestrial vertebrates. Evolution 10: 262–277.

Moreau, R. E. 1948. Ecological isolation in a rich tropical avifauna. J. Anim. Ecol. 17: 113–126.

Newell, N. D. 1962. Paleontological gaps and geochronology. J. Paleontol. 36: 592–610.

Paine, R. T. 1963. Trophic relationships of eight sympatric predatory gastropods. Ecology 44: 63–73.

————. 1966. Food web complexity and species diversity. Amer. Natur. 100: 65–75.

Patrick, Ruth. 1949. A proposed biological measure of stream conditions, based on a survey of the Conestoga Basin, Lancaster County, Pennsylvania. Proc. Acad. of Natur. Sci. Philadelphia 101: 277–341.

Patten, B. C. 1962. Species diversity in net phytoplankton of Raritan Bay. J. Marine Res. 20: 57–75.

Patten, B. C., R. A. Mulford, and J. E. Warinner. 1963. An annual phytoplankton cycle in the lower Chesapeake Bay. Chesapeake Sci. 4: 1–20.

Pearson, L. C. 1965. Primary production in grazed and ungrazed desert communities of eastern Idaho. Ecology 46(3); 278–286.

Preston, F. W. 1962. The canonical distribution of commonness and rarity. Part I: Ecology 43: 185–215. Part II: Ecology 43: 410–431.

Schoener, T. W. 1965. The evolution of bill size differences among sympatric congeneric species of birds. Evolution 19: 189–213.

Shannon, C. E. 1948. The mathematical theory of communication. *In* C. E. Shannon and W. Weaver, The mathematical theory of communication. Univ. Illinois Press, Urbana.

Simpson, G. G. 1964. Species density of North American recent mammals. Syst. Zool. 13: 57–73.

Skutch, A. F. 1954. Life histories of Central American birds. Vols. I and II. Cooper Ornithological Society Pacific Coast Avifauna Numbers 31 and 34.

Terent'ev, P. V. 1963. Opyt primeneniya analiza variansy k kachestvennomu bogatstvu fauny nazemnykh pozvonochnyk. Vestnik Leningradsk Univ. Ser. Biol. 18(21: 4); 19–26. English abstract in Biol. Abstr. 80822 (45).

Thorson, G. 1957. Bottom Communities (sublittoral or shallow shelf). *In* H. S. Ladd [ed.], Treatise on marine ecology and paleoecology. Geol. Soc. Amer. Mem. 67: 461–534.

Walter, H. 1939. Grasland, Savanne und Busch der arideren Teile Afikas in ihrer ökologischen Bedingtheit. Jahrbucher für wissenschaftliche Botanik 87: 750–860.

———. 1955. Le facteur eau dans les regiones arides et sa signification pour l'organisation de la vegetation dans les contrees sub-tropicales, p. 27–39. *In* Colloques Internationaux du Centre National de la Recherche Scientifique, Vol. 59; Les Divisions Ecologiques du Monde. Centre National de la Recherche Scientifique, Paris. 236 p.

———. 1962. Die Vegetation der Erde in ökologischer Betrachtung. Veb Gustav Fischer Verlag Jena. Jena, Germany.

Williams, C. B. 1964. Patterns in the balance of nature. Academic Press, New York and London.

Williams, L. G. 1964. Possible relationships between plankton-diatom species numbers and water-quality estimates. Ecology 45: 809–823.

Yount, J. L. 1956. Factors that control species numbers in Silver Springs, Florida. Limnol. Oceanogr. 1: 286–295.

4

Reprinted from *Am. Sci.* **38**:209–221 (1950)

EVOLUTION IN THE TROPICS

By THEODOSIUS DOBZHANSKY

Columbia University

BECOMING acquainted with tropical nature is, before all else, a great esthetic experience. Plants and animals of temperate lands seem to us somehow easy to live with, and this is not only because many of them are long familiar. Their style is for the most part subdued, delicate, often almost inhibited. Many of them are subtly beautiful; others are plain; few are flamboyant. In contrast, tropical life seems to have flung all restraints to the winds. It is exuberant, luxurious, flashy, often even gaudy, full of daring and abandon, but first and foremost enormously tense and powerful. Watching the curved, arched, contorted, spirally wound, and triumphantly vertical stems and trunks of trees and lianas in forests of Rio Negro and the Amazon, it often occurred to me that modern art has missed a most bountiful source of inspiration. The variety of lines and forms in tropical forests surely exceeds what all surrealists together have been able to dream of, and many of these lines and forms are endowed with dynamism and with biological meaningfulness that are lacking, so far as I am able to perceive, in the creations exhibited in museums of modern art.

Tropical rainforest impresses even a casual observer by the enormity of the mass of protoplasm arising from its soil. The foliage of the trees makes a green canopy high above the ground. Lianas, epiphytes, relatively scarce undergrowth of low trees and shrubs, and, finally, many fungi and algae form several layers of vegetational cover. Of course, tropical lands are not all overgrown with impenetrable forests and not all teeming with strange-looking beasts. One of the most perfect deserts in the world lies between the equator and the Tropic of Capricorn, in Peru and northern Chile. Large areas of the Amazon and Orinoco watersheds, both south and north of the equator, are savannas, some of them curiously akin to southern Arizona and Sonora in type of landscape. But regardless of the mass of living matter per unit area, tropical life is impressive in its endless variety and exuberance.

Since the animals and plants which exist in the world are products of the evolutionary development of living matter, any differences between tropical and temperate organisms must be the outcome of differences in evolutionary patterns. What causes have brought about the greater richness and variety of the tropical faunas and floras, compared to faunas and floras of temperate and, especially, of cold lands? How does life in tropical environments influence the evolutionary potentialities of the inhabitants? Should the tropical zone be regarded as an evolutionary cradle of new types of organization which sends out migrants to colonize the extratropical world? Or do the tropics serve as sanctuary for evolutionary old age where organisms that were widespread in the geological past survive as relics? These and related problems have never been approached from the standpoint of modern conceptions of the mechanism of evolutionary process. Temperate faunas and floras, and species domesticated by or associated with man, have supplied, up to now, practically all the material for studies on population genetics and genetical ecology.

Classical theories of evolution fall into two broad groups. Some assume that evolutionary changes are autogenetic, i.e., directed somehow from within the organism. Others look for environmental agencies that bring forth evolutionary changes. Although two eminent French biologists, Cuénot and Vandel, have recently espoused autogenesis, autogenetic theories have so far proved sterile as guides in scientific inquiry. Ascribing arbitrary powers to imaginary forces with fancy names like "perfecting urge," "combining ability," "telefinalism," etc., does not go beyond circular reasoning.

Environmentalist theories stem from Lamarck and from Darwin. Lamarck and psycholamarckists saw in exertion to master the environment the principal source of change in animals. Induction of changes in the body and the germ cells by direct action of physical agencies is the basis of mechanolamarckism. The organism is molded by external factors. A blend of psycholamarckist and mechanolamarckist notions, the latter borrowed chiefly from Herbert Spencer, has been offered by Lysenko as "progressive," "Michurinist," and "Marxist" biology. Advances of genetics have made Lamarckist theories untenable. There is not only no experimental verification of the basic assumptions of Lamarckism, but the known facts about the mechanics of transmission of heredity make these assumptions, to say the least, far-fetched.

Important developments and many changes have taken place in Darwinism since the publication of *The Origin of Species* in 1859. The essentials of the modern view are that the mutation process furnishes the raw materials of evolution; that the sexual process, of which Mendelian segregation is a corollary, produces countless gene patterns; that the possessors of some gene patterns have greater fitness than the possessors of other patterns, in available environments; that natural selection increases the frequency of the superior, and fails to perpetuate the adaptively inferior, gene patterns; and that groups of gene combinations of proved adaptive worth become segregated into closed genetic systems called species.

The role of environment in evolution is more subtle than was realized in the past. The organism does not suffer passively changes produced by external agents. In the production of mutations, environment acts as a trigger mechanism, but it is, of course, decisive in natural selection. However, natural selection does not "change" the organism; it merely provides the opportunity for the organism to react to changes in the environment by adaptive transformations. The reactions may or may

Fig. 1. Interior of equatorial rainforest near Belem, Brazil. The "terra firme" association. (Courtesy of Mr. Otto Penner, of the Instituto Agronomico do Norte, Belem do Pará.)

not occur, depending upon the availability of genetic materials supplied by the mutation and recombination processes.

Diversity of Species in the Tropics

Gause pointed out in 1934 that two or more species with similar ways of life can not coexist indefinitely in the same habitat, because one of them will inevitably prove more efficient than the others and will crowd out and eliminate its competitors. This "Gause principle" is a fruitful working hypothesis in studies on evolutionary patterns in tropical and temperate climates. The diversity of organisms which live in a given territory is a function of the variety of available habitats. The richer and more diversified the environment becomes, the greater should be the multiformity of the inhabitants. And vice versa: diversity of the inhabitants signifies that the environment is rich in adaptive opportunities.

Now, the greater diversity of living beings found in the tropical compared to the temperate and cold zones is the outstanding difference which strikes the observer. This is most apparent when tropical and temperate forests are compared. In temperate and cold countries, the forest which grows on a given type of terrain usually consists of masses of individuals of a few, or even of a single, species of tree, with only an admixture of some less common tree species and a limited assortment of shrubs and grasses in the undergrowth. The vernacular as well as the scientific designation of the temperate forest associations usually refers to these dominant species ("pine forest," "oak woodland," etc.). The forests of northern plains may, in fact, be monotonously uniform. In the forested belt of western Siberia one may ride for hundreds of miles through birch forests interrupted only by some meadows and bogs. Mountain forests are usually more diversified than those of the plains. Yet in the splendid forest of the Transition Zone of Sierra Nevada of California one rarely finds more than half a dozen tree species growing together. By contrast tropical forests, even those growing on so perfect a plain as that stretching on either side of the Amazon, contain a multitude of species, often with no single species being clearly dominant. Dr. G. Black, of the Instituto Agronomico do Norte in Belem, Brazil, made, in cooperation with Dr. Pavan and the writer, counts of individuals and species of trees 10 or more centimeters in diameter at chest height on one-hectare plots (100 $\times$ 100 meters) near Belem. On such a plot in a periodically inundated (igapó) forest 60 species were found among 564 trees. The numbers of species represented by various numbers of individuals on this plot were as follows:

Individuals	1	2	3	4	5	6	7	8	9
Species	22	9	7	2	2	2	4	2	1

Individuals	10	14	15	16	21	29	33	41	241
Species	1	1	1	1	1	1	1	1	1

The commonest species was the assaí palm (*Euterpe oleracea*), of which 241 individuals were found in the hectare plot, but as many as 22 species were represented by single individuals. On a plot of similar size, only a mile away but on higher ground (terra firma), 87 species were found among 423 trees. The numbers of individuals per species were as follows:

Individuals	1	2	3	4	5	6	7	9	12	17	20	25	37	49
Species	33	15	15	3	4	2	3	4	1	2	1	1	2	1

Here the commonest species was represented by only 49 individuals and as many as 33 species were found as single individuals. This high frequency of species represented by single individuals means that if

Fig. 2. The wet ground ("igapo") forest near Belem, Brazil. (Courtesy of Mr. Otto Penner, of the Instituto Agronomico do Norte, Belem do Pará.)

we had studied other hectare plots contiguous to the one actually examined, many new species would have been found. It is probable that each of the plant associations which we have sampled in the vicinity of Belem contains many more than 100 tree species, and, incidentally, only a few species occur in both associations. Similar results were obtained by Davis and Richards in the forests of British Guiana and by Beard in Trinidad.

The numbers of breeding species of birds recorded in the literature for territories in various latitudes, from arctic North America to equatorial Brazil, are as follows (data kindly supplied by Dr. E. Mayr):

Territory	Number of Species	Authority
Greenland	56	F. Salomonsen
Peninsular Labrador	81	H. S. Peters

Newfoundland	118	H. S. Peters
New York	195	K. C. Parkes
Florida	143	S. A. Grimes
Guatemala	469	L. Griscom
Panama	1100	L. Griscom
Colombia	1395	R. M. de Schauensee
Venezuela	1148	W. H. Phelps, Jr.
Lower Amazonia	738	L. Griscom

The numbers of recorded species of snakes are as follows (data obtained through the courtesy of Dr. C. M. Bogert):

Territory	*Number of Species*	*Authority*
Canada	22	Mills
United States	126	Steineger and Barbour
Mexico	293	Taylor and Smith
Brazil	210	Amaral

The progressive increase in diversity of species from the Arctic toward the equator is apparent in general, even though there are irregularities in this increase, resulting from such factors as how the different territories compare in size, how uniform or varied they are ecologically and topographically, and how intensively their faunas have been studied. There can be no doubt that, for most groups of organisms, tropical environments support a greater diversity of species than do temperate- or cold-zone environments.

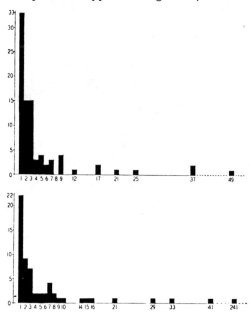

In order to survive and to leave progeny, every organism must be adapted to its physical and biotic environments. The former includes temperature, rainfall, soil, and other physical variables, while the latter is composed of all the organisms that live in the same neighborhood. A diversified biotic environment influences the evolutionary patterns of the inhabitants in several ways. The greater the diversity of inhabitants in a territory, the more adaptive opportunities exist in it. A tropical forest with its numer-

Fig. 3. Numbers of species of trees (ordinates) represented by different numbers of individuals (abscissae) on 1-hectare (100 × 100 meters) plots of equatorial rainforest near Belem, Brazil. "Terra firme" association (above) and "igapo" association (below). Only trees 10 or more cm. at chest height counted. (*Data of Black, Dobzhansky, and Pavan.*)

ous tree species supports many species of insects, each feeding on a single or on several species of plants. On the other hand, the greater the number of competing species in a territory, the fewer become the habitats open for occupancy by each of these species. In the absence of competition a species tends to fill all the habitats that it can make use of; abundant opportunity favors adaptive versatility. When competing species are present, each of them is forced to withdraw to those habitats for which it is best adapted and in which it has a net advantage in survival. The presence of many competitors, in biological evolution as well as in human affairs, can be met most successfully by specialization. The diversity of habitats and the diversity of inhabitants which are so characteristic of tropical environments are conflicting forces, the interaction of which will determine the evolutionary fates of tropical organisms.

Chromosomal Polymorphism in Drosophila

Adaptive versatility is most easily attained by a species' becoming adaptively polymorphic, i.e., consisting of two or more types, each possessing high fitness in a certain range of environments. One of the most highly polymorphic species in existence is *Homo sapiens,* and it is this diversity of human nature which has engendered cultural growth and has permitted man to draw his existence from all sorts of environments all over the world. Now, the adaptive polymorphism of the human species is conditioned, on the cultural level, chiefly by the ability to become trained and educated to perform different activities. In species other than man, adaptive polymorphism is attained chiefly through genetic diversification—formation of a group of genetically different types with different habitat preferences.

In *Drosophila* flies, organisms best suited for this type of study, adaptive polymorphism takes its chief form in diversification of chromosome structure. In some American and European species of these flies, natural populations are mixtures of several interbreeding chromosome types which differ in so-called inverted sections. These chromosomal types have been shown, both observationally and experimentally, to have different environmental optima. The situation in tropical species has been studied in various bioclimatic regions in Brazil by a group consisting of Drs. A. Brito da Cunha, A. Dreyfus, C. Pavan, and E. N. Pereira of the University of São Paulo, A. G. L. Cavalcanti and C. Malogolowkin of the University of Rio de Janeiro, A. R. Cordeiro of the University of Rio Grande do Sul, M. Wedel of the University of Buenos Aires, H. Burla of the University of Zürich, N. P. Dobzhansky, and the writer. During the school year 1948-1949, the work of this group was supported by grants from the University of São Paulo, the Rockefeller Foundation, and the Carnegie Institution of Washington.

The commonest species in Brazil is *Drosophila willistoni.* It is also the adaptively most versatile species, since it has been found in every one of the 35 localities in various bioclimatic regions of Brazil in which collection was made. Significantly enough, this species, taken as a whole, shows not only the greatest chromosomal polymorphism among Brazilian species but the greatest so far known anywhere. The species *Drosophila nebulosa* and *Drosophila paulistorum* are also very common, but somewhat more specialized than *Drosophila willistoni*. *Drosophila nebulosa* is at its best in the savanna environments where dry seasons alternate with rainy ones, and *Drosophila paulistorum,* conversely, in superhumid tropical climates. The species are rich in chromosomal polymorphism, but not so rich as *Drosophila willistoni* in this respect. We have also examined some less common and biotically more specialized species, all of which showed much less or no chromosomal polymorphism.

The comparison of chromosomal polymorphism in *Drosophila willistoni* in different bioclimatic zones of Brazil proved to be even more interesting. The greatest diversity is found in those parts of the valleys of the Amazon and its tributaries where *Drosophila willistoni* is the dominant species. The exuberant rainforests and savannas of the Amazon basin are remarkable in the rich diversity of their floras and faunas; furthermore, the Amazon basin appears to be the geographical center of the distribution of the species under consideration, where it has captured the greatest variety of habitats. Yet, wherever in this region *Drosophila willistoni* surrenders its dominance to competing species, the chromosomal polymorphism is sharply reduced. This has been found to happen in the forested zone of the territory of Rio Branco, near Belem, in the state of Pará, and in the savanna of Marajó Island. In the first and the second of the regions named, *Drosophila paulistorum,* and in the third region, *Drosophila tropicalis,* reduce *Drosophila willistoni* to the status of a relatively rare species. In the peculiar desert-like region of northeastern Brazil, called "caatingas," *Drosophila willistoni* reaches its limit of environmental tolerance; *Drosophila nebulosa* seems to be the only species which still flourishes in this highly rigorous environment. The chromosomal variability of *Drosophila willistoni* is much reduced on the caatingas. It is also reduced in southern Brazil where *Drosophila willistoni* approaches the southern limit of its distribution, and is, presumably, losing its grip on the habitats.

Any organism which lives in a temperate or a cold climate is exposed at different periods of its life cycle or in different generations to sharply different environments. The evolutionary implications of nature's annually recurrent drama of life, death, and resurrection have not been sufficiently appreciated. In order to survive and reproduce, any species must be at least tolerably well adapted to every one of the environments which it regularly meets. No matter how favored a strain may be in summer, it will be eliminated if it is unable to survive winters, and vice versa. Faced with the need of being adapted to diverse environments, the organism may be unable to attain maximum efficiency in any one of them. Changeable environments put the highest premium on versatility rather than on perfection in adaptation.

Adaptive Versatility in the Tropics

The widespread opinion that seasonal changes are absent in the tropics is a misapprehension. Seasonal variations in temperature and in duration and intensity of sunlight are, of course, smaller in the tropical than in the extratropical zones. However, the limiting factor for life in the tropics is often water rather than temperature. Some tropical climates, for example those of the caatingas of northeastern Brazil, have variations of such an intensity in the availability of water that plants and animals pass through yearly cycles of dry and wet environments which entail biotic changes probably no less serious than those brought about by the alternation of winter and summer in temperate lands. Absence of drastic seasonal changes in tropical environments is evidently a relative matter. In Belem, at the opening of the Amazon Valley to the Atlantic Ocean, the mean temperature of the warmest month, 26.2° C., is only 1.3° C. higher than that of the coolest month, and the highest temperature ever recorded, 35.1° C., is only 16.6° C. higher than the all-time low, which was 18.5° C. The wettest month has 458 mm. of precipitation, and the driest has 86 mm., which is still sufficient to prevent the vegetation from suffering from drought.

It might seem that the inhabitants of the relatively invariant tropical climates should be free from the necessity of being genetically adapted to a multitude of environments, and hence that evolution in the tropics would tend toward perfection and specialization, rather than to adaptive versatility. This is not the case, however. The climate of Espirito Santo Island, in the tropical Pacific, is seasonally one of the most constant in the world as far as temperature and humidity are concerned. Nevertheless, Baker and Harrison found that native plants have definite flowering and fruiting seasons, and animals have cycles of breeding activity in this climate. Our observations in Brazil show that populations of *Drosophila* flies undergo expansions or contractions from month to month, as well as changes in the relative abundance of different species. The magnitude and speed of these changes are quite comparable to those which occur in California, for example, or in the eastern part of the United States. Such pulsations have been observed even in the rainforests of Belem, about 1½° latitude south of the equator. They are caused mainly by seasonal variations in the availability of different kinds of fruits which are preferred by different species of *Drosophila*. Despite the apparent climatic uniformity, the biotic environment of tropical rainforests is by no means constant in time.

This writer observed several years ago that certain populations of the fly *Drosophila pseudoobscura* which live in the mountains of California undergo seasonal changes in the relative frequencies of chromosomal types. Dubinin and Sidorov found similar changes in the Russian *Drosophila funebris*. What happens is that some of the chromosomal types of these flies possess highest adaptive values in summer, and other types in winter or in spring environments. Natural selection augments the frequency of favorable types and reduces the frequency of unfavorable types. The populations thus react to changes in their environment by adaptive modifications. This is one of the rare occasions when evolutionary changes taking place in nature under the

Fig. 4. Giant trees and lianas in the "terra firme" rainforest near Belem. (Courtesy of Mr. Otto Penner, of the Instituto Agronomico do Norte, Belem do Pará.)

influence of natural selection can actually be observed in the process of happening. We may add that some of these changes have also been reproduced in laboratory experiments in which artificial populations of the species concerned were kept in special "population cages."

We have observed populations of *Drosophila willistoni* in three localities in southern Brazil for approximately one year. One of the localities, situated in the coastal rainforests south of São Paulo, has a rather uniform superhumid tropical climate. Periodic sampling of these populations has disclosed alterations in the incidence of chromosomal types similar in character to those observed in California and in Russian fly species. Adaptive alterations which keep living species attuned to their changing environments occur in tropical as well as in temperate-zone organisms. This constant evolutionary turmoil, so to speak, precludes evolutionary stagnation and rigidity of the adaptive structure of tropical and of temperate species equally.

It is nevertheless true that tropical environments are more constant than temperate ones, in a geological sense. Major portions of the present temperate and cold zones of the globe underwent drastic climatic and biotic changes owing to the Pleistocene glaciation. The present floras and faunas of the territories that were covered by Pleistocene ice are composed almost entirely of newcomers. The territories adjacent to the glaciated areas have passed through more or less radical climatic upheavals. Although the bioclimatic history of the tropical continents is still very little known, it is fair to say that their environments suffered less change.

The repeated expansions and contractions of the continental ice sheets, and the alternation of arid and pluvial climates in broad belts of land bordering on the ice, made large territories rather suddenly (in the geological sense) available for occupation by species that could evolve the necessary adaptations in the shortest possible time. This has not simply increased the rates of evolution, but often has favored types of changes which can be characterized collectively as evolutionary opportunisms. Such changes have the effect of conferring on the organism a temporary adaptive advantage at the price of loss or limitation of evolutionary plasticity for further change. Here belong the various forms of deterioration of sexuality observed in so many species of temperate- and cold-zone floras. An apomictic or an asexual species, with an unbalanced chromosome number and a heterosis preserved by loss of normal meiotic behavior of the chromosomes, may be highly successsful for a time but its evolutionary possibilities in the future are more limited than those of sexual and cross-fertilizing relatives. Polyploidy is also a form of evolutionary opportunism in so far as it produces at least a temporary loss of genetic variability. Although some plant species native in the tropics have also become stranded in these evolutionary blind alleys, the incidence of such species is higher in and near the regions which were glaciated.

Evolutionary Importance of Biotic Environment

The contradictory epithets of "El Dorado" and "Green Hell," so often used in descriptions of tropical lands, really epitomize the two aspects of tropical environments. The process of adaptation for life in temperate and especially in cold zones consists, for man as well as for other organisms, primarily in coping with the physical environment and in securing food. Not so in the tropics. Here little protection against winter cold and inclement weather is needed. In the rainforests, the amount of moisture is sufficient at any time to prevent the inhabitants from suffering from desiccation. Relatively little effort is necessary for man to secure food, and it seems that the amount of food is less often a limiting factor for the growth of populations of tropical animals than it is in the extratropical zones. But the biological environment in the tropics is likely to be harsh and exacting. Man must beware that his blood does not become infected with malarial plasmodia, his intestines with hookworms, and his skin with a variety of parasites always ready to pounce on him and rob him of his vitality if not of life itself. The tremendous intensity of the competition for space among plants in tropical forests can be felt even by a casual observer. The apparent scarcity, concealment, and shyness of most tropical animals attest to the same fact of extremely keen competition among the inhabitants.

Now, the processes of natural selection which arise from encounters between living things and physical forces in their environment are different from those which stem from competition within a complex community of organisms. The struggle for existence in habitats in which harsh physical conditions are the limiting factors is likely to have a rather passive character as far as the organism is concerned. Physical factors, such as excessive cold or drought, often destroy great masses of living beings, the destruction being largely fortuitous with respect to the individual traits of the victims and the survivors, except for traits directly involved in resistance to the particular factors. As pointed out by Schmalhausen, indiscriminate destruction is countered chiefly by development of increased fertility and acceleration of development and reproduction, and does not lead to important evolutionary advances. Physically harsh environments, such as arctic tundras or high alpine zones of mountain ranges, are inhabited by few species of organisms. The success of these species in colonizing such environments is due simply to the ability to withstand low temperatures or to develop and reproduce during the short growing season.

Where physical conditions are easy, interrelationships between competing and symbiotic species become the paramount adaptive problem. The fact that physically mild environments are as a rule inhabited by many species makes these interrelationships very complex. This is probably the case in most tropical communities. The effectiveness of natural selection is by no means proportional to the severity of the struggle for existence, as has so often been implied, especially by some early Darwinists. On the contrary, selection is most effective when, instead of more or less random destruction of masses of organisms, the survival and elimination acquire a differential character. Individuals that survive and reproduce are mostly those that possess combinations of traits which make them attuned to the manifold reciprocal dependences in the organic community. Natural selection becomes a creative process which may lead to emergence of new modes of life and of more advanced types of organization.

The role of environment in evolution may best be described by stating that the environment provides "challenges" to which the organisms "respond" by adaptive changes. The words "challenge" and "response" are borrowed from Arnold Toynbee's analysis of human cultural evolution, although not necessarily with the philosophical implications given to the terms by this author. Tropical environments provide more evolutionary challenges than do the environments of temperate and cold lands. Furthermore, the challenges of the latter arise largely from physical agencies, to which organisms respond by relatively simple physiological modifications and, often, by escaping into evolutionary blind alleys. The challenges of tropical environments stem chiefly from the intricate mutual relationships among the inhabitants. These challenges require creative responses, analogous to inventions on the human level. Such creative responses constitute progressive evolution.

REFERENCES

1. Anonymous. Normais Climatologicas. Serviço da Meteorologia, Ministerio da Agricultura. Rio de Janeiro, 1941.
2. Baker, J. R. The seasons in a tropical forest. Part 7. *Jour. Linn. Soc. London 41*, 248-258, 1947.
3. Beard, J. S. The natural vegetation of Trinidad. *Oxford Forestry Mem. 20*, 1-155, 1946.
4. Black, G. A., Dobzhansky, Th., and Pavan, C. Some attempts to estimate the species diversity and population density of trees in Brazilian forests. *Bot. Gaz.*, 1950. (In press.)
5. da Cunha, A. B., Burla, H., and Dobzhansky, Th. Adaptive chromosomal polymorphism in *Drosophila willistoni. Evolution*, 1950. (In press.)
6. Davis, T. A. V., and Richards, P. W. The vegetation of Moraballi Creek, British Guiana. *J. Ecol. 22*, 106-155, 1934.
7. Dobzhansky, Th. Observations and experiments on natural selection in *Drosophila. Proc. 8 Internat. Cong. Genetics, Hereditas Suppl.*, 210-224, 1949.
8. Dobzhansky, Th., and Pavan, C. Local and seasonal variations in relative frequencies of species of *Drosophila* in Brazil. *Jour. Animal. Ecol.*, 1950. (In press.)
9. Dubinin, N. P., and Tiniakov, G. G. Inversion gradients and natural selection in ecological races of *Drosophila funebris. Genetics 31*, 537-545, 1946.
10. Gause, G. F. The struggle for existence. Baltimore, 1934.
11. Lack, D. Darwin's finches. Cambridge, 1947.
12. Patterson, J. T. The Drosophilidae of the Southwest. *Univ. Texas Publ. 4313*, 7-216, 1943.
13. Schmalhausen, I. I. Factors of evolution. Philadelphia, 1949.
14. Stebbins, G. L. Variation and evolution in plants, New York, 1950. (In press.)
15. Toynbee, A. J. A study of history. New York and London, 1947.
16. Vandel, A. L'homme et l'évolution. Paris, 1949.
17. Vavilov, N. I. Studies on the origin of cultivated plants. Leningrad, 1926.

5

Reprinted from *J. Ecol.* **54**:1–11 (1966)

THE STRUCTURE OF THE TROPICAL RAIN FOREST AND SPECIATION IN THE HUMID TROPICS*

By AN. A. FEDOROV

In the course of investigations into the distribution of plants of the Temperate and Cold Zones of the Northern Hemisphere it was long ago established that taxonomically close, and therefore closely allied, species of plants, do not as a rule occupy the same area, but are most usually geographically isolated from one another, the areas of these species being either distant from each other, or adjacent, or arranged as a kind of mosaic consisting of intermingled contiguous areas. Vicarism is also a related phenomenon, studied and described particularly thoroughly by Vierhapper (1919).

The wide distribution of the phenomenon of geographical isolation in the Holarctic even served as a basis for a special morphologico-geographical method in plant taxonomy the authors of which were Kerner (1865, 1869) and von Wettstein (1896, 1898) in Austria, and almost at the same time Bunge (1872), Korshinsky (1892) and Komarov (1908) in Russia.

It was also established that the isolation of closely allied species may be not only geographical, but ecological as well, as when such species are found in different habitats within one area.

It is important to point out that both the geographical and the ecological types of isolation characteristic of holarctic plant species are good illustrations of the principle of incompatibility of closely allied species under the same environmental conditions, ensuing from Darwin's theory of natural selection. However, an entirely different situation with respect to the distribution of closely allied species was discovered in the course of investigations of the flora and vegetation of humid tropics, in particular, of the Tropical Rain forest.

So far as I know, Richards (1945) was the first to note that a great number of closely allied species occur side by side within the same community of a tropical rain forest. Van Steenis (1957) made the same point more explicitly. Sukachev (1958) was the next to observe this phenomenon, which suggested an absence of isolation between closely allied species in tropical forest. He concluded that 'it is of prime importance for the understanding of the process of speciation in nature'. Naturally, during my observations in the tropics of China (Fedorov 1957) and subsequently in Ceylon (Fedorov 1961) and Indonesia my attention was also arrested by the remarkable fact of the existence of uninterrupted series of closely allied species in tropical rain forest and in the humid tropics in general. Takhtajan (1957) generalized this phenomenon, pointing out the existence, in the humid tropics, of continuous series not only of allied species, but also of taxa of higher ranks, such as genera and families. However, no explanation of this phenomenon has so far been found, as was also noted by van Steenis (1957), one of the most eminent authorities on the flora of the tropics. It is possible that we are now on the way to explaining this phenomenon. In this connection let us draw attention to some basic relevant facts, using as our first example the family Dipterocarpaceae which is typical of the tropics of Asia and extends to Madagascar and Africa. From the work of

---

* Presented at the Xth International Botanical Congress, Edinburgh, August 1964.

many botanists, particularly of Foxworthy (1911, 1927, 1932, 1938), van Slooten (1926–32), and Symington (1941), we know that not only the large genera of this family, containing many species, such as *Dipterocarpus*, *Shorea* and *Hopea*, but in fact almost all the other genera, including those which are oligotypic, e.g. *Dryobalanops*, and monotypic (*Upuna*) comparatively seldom transgress in their distribution the limits of the tropical rain forest. The numbers of species belonging to the main genera of Dipterocarpaceae are very indicative: *Shorea* comprises about 167 species, *Hopea* about 100, *Dipterocarpus* about 80, *Vatica* about 87, *Balanocarpus* about 20, *Anisoptera* about 14, *Doona* (Ceylon) about 12.*

There are many other tropical families, in which the overwhelming majority of species are represented in the tropical rain forest community. These are: Annonaceae (particularly the genera *Mitrephora* with 25, and *Polyalthia* with 70), Burseraceae (particularly *Protium* with 60 in South America), Bombacaceae (*Bombax* 60, *Durio* 27), Caesalpiniaceae (*Bauhinia* 250, *Cynometra* 40, *Dialium* 20), Combretaceae (*Combretum* 350, *Terminalia* 120), Clusiaceae (*Calophyllum* 80, *Garcinia* 200, *Mammea* 26), Elaeocarpaceae (*Elaeocarpus* 90, *Sloanea* 45), Lauraceae (*Actinodaphne* 128, *Beilschmiedia* 236, *Cryptocarya* 318, *Endiandra* 95, *Litsea* 474, *Neolitsea* 93, *Ocotea* 697, *Persea* 240, *Phoebe* 174), Lecythidaceae (*Eschweilera* 80, *Lecythis* 45), Meliaceae (*Aglaia* 125, *Dysoxylum* 140), Mimosaceae (*Abarema* 44, *Albizzia* 50, *Piptadenia* 45, *Pithecolobium* 60, *Parkia* 20), Myristicaceae (*Knema* 40, *Myristica* 85), Moraceae (*Artocarpus* 60, *Ficus* over 900), Myrtaceae (*Eugenia* 750, *Syzygium* 140), Sapindaceae (*Allophyllus* 120, *Nephelium* 25), Sapotaceae (*Mimusops* 65, *Palaquium* 65, *Sideroxylon* 100).

Some genera of these families comprising especially large numbers of species are of particular interest. For instance, according to Corner (1958), the genus *Ficus* within the limits of Asia and Australasia contains 900 species most of which are encountered in the tropical rain forest, the strangling figs (that become independent trees after having passed the epiphytic stage) often reaching the upper canopy of the forest and sometimes attaining large numbers per unit area. Corner, a specialist in this genus, is justified in his opinion that its significance for the knowledge of the plant world in the Eastern tropics is quite as important as that of the family Dipterocarpaceae.

The above-mentioned numbers of species of the large genera belonging to the typical tropical families are certainly understated, except for the quite exact data for Lauraceae taken from the most recent work of Kostermans (1957). All the other data were taken mainly from Willis's *Dictionary* (1931), an excellent book, but rather out of date. There is no doubt that for many genera the actual numbers of species are about twice as high.

In mountain rain forests a great diversity of species having the character of a complete series of species is observed in the genera of Ericaceae (particularly in *Rhododendron*), Fagaceae, Magnoliaceae, Lauraceace, Symplocaceae, Theaceae and of some other families. The family Fagaceae includes several allied genera, each of which contains a large number of species. These genera are: *Castanopsis, Cyclobalanopsis, Lithocarpus, Pasania, Quercus* (s. str.). In the uper part of the forest belt of the tropical mountains there also occur all the other genera of Fagaceae, except *Castanea* (s. str.), viz. *Fagus*, a monotypic genus *Trigonobalanus* (northern Thailand to Celebes) and an oligotypic genus *Nothofagus* (New Guinea, more widespread farther south in the Antarctic Region).

---

* At present intensive studies of the family Dipterocarpaceae are in progress. Every year numerous descriptions of new species are published. A multitude of new species have been described for British Borneo alone by Ashton (1962) and the taxonomy of the family is being made more precise (Ashton 1963).

The fact that the species belonging to the above-mentioned families and genera form not merely random groups or mixtures, but natural series and other groups of affinity, can be seen from the analysis of the above-mentioned family Dipterocarpaceae.

Unfortunately, as yet there is no completely elaborated system for this family, as is the case in fact for many other extremely important tropical families of plants. Taxa of a rank lower than section, i.e. of such a rank as series, have almost never been described; however, should such a genus as *Shorea* be considered at the level of section, it would be found that in the tropical rain forest of the Malay Peninsula alone this genus is represented by almost all its sections, viz. Shorea, Anthoshorea, Richetioides, Brachyptera, Mutica (Symington 1941).* More or less the same holds true also for such genera as *Dipterocarpus, Hopea, Vatica* and others. Thus it is perfectly obvious that (as reflected in the system) the genera of the family Dipterocarpaceae and other tropical families are concentrated in the tropical rain forest as a multitude of clearly allied species, grouped into various series and other groups of allied taxa up to the rank of the family, which, as it has already been mentioned, is itself almost entirely confined in its distribution to the limits of the tropical rain forest.

It might appear possible to assume that all the multitude of allied species of the genus *Shorea* and of the other genera of Dipterocarpaceae is distributed in a tropical rain forest among different strata and synusiae, and that previously developed ecological isolation provides the possibility of the co-existence of several closely allied species within the same area without being isolated geographically. In fact, however, only a small proportion of species is isolated in this way. An overwhelming majority of species of, for instance, the above-mentioned genus *Shorea*, are found within the same stratum, in this particular instance in the upper storey (stratum A, according to the terminology of Richards, 1952) or even above this stratum, among 'outstanding' or emergent trees.

Therefore it must be assumed that a great number of closely allied species can co-exist side by side in the tropical rain forest because there had been established some special kind of isolation, each species being confined to a corresponding niche, and most species from their very origin forming populations which were very small, but stable with respect to the numbers of constituent individuals.

Symington (1941), who studied the Dipterocarpaceae of the Malay Peninsula in detail, obtained some evidence from which the following principles of distribution of species belonging to the genera *Dipterocarpus, Hopea, Shorea* and *Vatica* in the tropical rain forests of this part of south-eastern Asia can be inferred: (1) a certain proportion of species, particularly some of those endemic in the Malay Peninsula, occur only in a few localities, (2) an appreciable number of species, both endemic and more widespread, occur as small groups of individuals, as it were in isolated dots, (3) some species are characteristic of mountain ridges or, on the contrary, of patches of lowlands; many species are dispersed all over the forest area in the Peninsula. However, they form no perceptible accumulations and the density of their populations never exceeds one or two individuals per acre (0·4 ha). It might be concluded that both partly geographical and partly ecological isolation are observed here. However, the nature of this phenomenon is different. Even if both these types of isolation do exist, it still cannot be imagined that scores and hundreds of species belonging to the genera of Dipterocarpaceae, characteristic of the tropical rain forest of the Malay Peninsula, are distributed in such a way as to indicate that there was a special niche for each of them within the same plant community. For such a tremendous number of species of Dipterocarpaceae (156 species belong to ten

* The names of the sections of *Shorea* have been amended as in Ashton (1963).

genera) it is impossible to assume the existence of such a great diversity of physico-geographical and ecological conditions within the area of so small a patch of land as the Malay Peninsula. Although it is frequently alleged that a tropical rain forest is 'a conglomerate of habitats', nevertheless, so far as large, or even gigantic, trees are concerned, such as most species of Dipterocarpaceae are, the diversity of habitats for them is confined to one or two upper strata of the forest. On the contrary, it would be more appropriate to call to mind in this connection the smoothness, uniformity and ubiquitous 'paradisiacal' optimality of environment for the growth and development of trees in the tropical rain forest.

Several taxonomically close species of such genera as *Shorea* or *Dipterocarpus* can be found on practically any standard sample area in a tropical rain forest of any region of south-eastern Asia where Dipterocarp forests are widespread; however, the population density of each of these species usually does not exceed one or two individuals per acre (0·4 ha). In other words, the fact of geographical compatibility of distribution areas of closely allied species is beyond doubt. The compatibility of distribution areas of many species of Dipterocarpaceae becomes particularly obvious from the numbers of these species in but one forest type, the Lowland Dipterocarp Forest that is characteristic of the Malay Peninsula from sea level up to an altitude of only 1000 ft (300 m). In this forest there are 59 species of *Shorea*, 31 species of *Dipterocarpus*, 35 species of *Hopea*, 22 species of *Vatica* and 9 species of *Anisoptera*. The genus *Shorea* is represented here by four sections: (1) Shorea, (2) Anthoshorea, (3) Richetioides, and (4) Mutica and Brachyptera, comprising 13, 13, 10 and 23 species respectively.

An essential characteristic feature of the interesting phenomenon of coincidence of distribution areas and habitats of a large number of closely allied species in the tropical rain forest, without any geographical or ecological isolation between these species, has thus been established. This feature is the very low value of the population density, usually not exceeding one or two individuals per acre (0·4 ha).

Therefore, for the further elucidation of the problem studied it is necessary to take into consideration the available data on the numbers of species for the tropical rain forest.

The works of Dobzhansky and his colleagues (Black *et al.* 1950; Pires *et al.* 1953) are of particular interest in this respect. Working in the Amazonian rain forest, these investigators made counts of trees with trunks over 10 cm in diameter in three 1 ha (100 × 100 m) plots in the vicinity of Tefé, State of Amazonas, in a tropical rain forest of the type called 'terra firme', and later in the forest of the 'Igapo' type near Belém, State of Pará. In these three plots 50, 87 and 79 tree species were represented by 564, 423 and 230 individuals respectively.

It was shown by analysis of the data obtained that only about half of the total number of species can be recorded in this way, since most rare, or even merely uncommon, species are usually missing in the plots. For most species present in such plots in the Amazonian tropical rain forest the population density does not exceed one individual per hectare, there being observed an inverse relationship between the number of species and the numbers of individuals. Thus, 22–35 species present in a plot are represented by not more than one individual per hectare, and only one or two species have a population density attaining 40–50 and even up to 200 or still more individuals per hectare. The transition between these numbers in both directions is quite gradual, the series of the number of species being, as it were, balanced by the inversely progressing series of the numbers of individuals.

Dobzhansky and his colleagues arrived at the conclusion that 'the population density of a half or more of the tree species . . . is likely to be less than one individual per hectare. This is a low population density by the standards of temperate zone forests. It is probable that at least some of the rare species of trees are cross-pollinated. Provided that the distribution of pollen by insects and by wind is about equally efficient in tropical and temperate zone forest, the reproductive communities in tropical species of trees will contain on the average fewer individuals than reproductive communities in temperate zone species. If so, the genetically effective population sizes in many tropical species are likely to be limited. This would have a profound influence on their evolutionary patterns' (Black *et al.* 1950, p. 425).

Having resumed, after a 3 years' interval, their investigations in an Amazonian forest, Dobzhansky and his colleagues (Pires *et al.* 1953) chose an experimental plot in 'a luxuriant and virgin terra firme forest' also in the state of Pará, but in another locality, near the village of Tres de Outubro, between Castanhal and the River Guama. This plot had an area of 3·5 ha and was divided into smaller sections for the sake of accurateness of tree counts. In this study also, the general character of the results established earlier remained unchanged, but the actual numbers observed were somewhat different. It is remarkable that even with a larger plot size about seventy species known for the forest studies were missing from the plot. Thus the minimal area is apparently so large in a tropical rain forest that it cannot be covered even by the largest possible plot which would permit precise investigations and accurate counts.

Most of the recent data obtained by studies of the number of species and the abundance of individuals in tropical rain forests are summarized in the book by Cain & Castro (1959).

Only a small number of species in the tropical rain forest can attain a relatively large population density. Usually these are species occupying the lower strata, but occasionally the same phenomenon is observed also among the species of the uppermost storey (stratum A). Such instances were described by Richards (1952) for the forests of British Guiana (the forests with separate dominance of *Eperua falcata, Mora excelsa, M. gonggrijpii, Ocotea rodiaei*). Monodominance is known in tropical rain forests of Central Africa, Ceylon and of some other regions of the tropics. I was lucky to find a tropical rain forest with dominance of *Dipterocarpus retusus* on Sumbawa Island in Indonesia. Another forest I have visited, belonging to this type of monodominant community, is the famous Tjibodas forest on the slopes of the volcano Gedeh in Java. The dominant species here is *Altingia excelsa*: this forest, however, belongs to a somewhat different type of montane rain forest.

For all that, a pure, one-species canopy is almost never observed in the tropical rain forest and the dominant species never suppresses any other tree species of the forest. The general floristic composition always remains very diverse and rich, containing an immense multitude of species in any one part of the forest.

Dominant tree species most frequently also form relatively small patches or consociations. In other words, not only rare species, but dominant species too, usually form small populations.

Thus the main principles underlying the numbers of species and the abundance of individuals within species in a tropical rain forest are clear: the number of species per unit area is extremely large, but all the species are represented by sparse populations and the population density of most species is, as a rule, extremely low.

Another remarkable feature of the tropical rain forest, having a direct bearing on the problem considered in this study, is the absence of climatic seasons* and the consequent irregular rhythm of flowering and fructification of most tree species. Many species of trees flower very seldom, sometimes only once in 5 or even 10 years, each time possibly at a different season of the year, with almost no relation to any particular season. Coincidence in the time of flowering is rare not only in any two or more closely allied species, but also in different individuals of the same species. If the extremely small population density of most tree species be taken into consideration, it must be assumed that the possibilities for cross-pollination are so scanty in the tropical rain forest that self-pollination presumably prevails here.

What might be the result of such a combination of circumstances as the small size and isolation of populations, the extremely low population density and the difficulty of cross-pollination? Apparently these conditions are favourable for the development of automatic genetic processes such as genetic drift.

In botanical literature the great significance of this phenomenon for speciation and even for the rapid origin of differences distinguishing higher taxa was pointed out by Takhtajan (1961). It was by the favourable combined effect of the genetic drift and natural selection that he tried to explain the mystery of the origin of the angiosperms. An attempt will be made here to apply the concept of genetic drift to the problem considered in this paper, the problem of speciation in the tropical rain forest.

As is generally known, the probability of the meeting of genetically identical gametes is higher in small populations; this results in homozygous forms appearing with increasing frequence and accumulating. The elimination of deleterious genes and the accumulation of beneficial genes by natural selection increases. Alongside this the role of chance in the accumulation of separate genotypes also becomes more important. The frequency of mutant genes or the 'saturation with mutations' in small populations is very high and may increase with increasing numbers in these populations.

It is interesting that Dubinin (1940), one of the authors of the theory of genetic drift, a long time ago advanced a suggestion about an important significance of these 'genetico-automatic processes' for speciation in the tropics. He wrote: 'It is quite possible that the relatively small size of the populations of tropical organisms provides the conditions for the rapid initiation of new races and species by means of the accumulation both of adaptive characters and of diverse indifferent characters fixed by means of genetico-automatic processes.'

Thus considerable genetic differences may accumulate in separate small populations of any tree species of the tropical rain forest, differences that as a consequence of the fortuitous and rare exchanges of genes owing to the above-mentioned difficulty of cross-pollination, may become a source of still greater and more rapid shifts towards speciation.

In short, in the tropical rain forest there exist conditions favourable for the relatively rapid origin of species by means of genetic drift, which explains why many genera in this community have very large numbers of closely allied species, not exhibiting in their distribution any ecological or geographical isolation. In consequence of the low population density and partial, though considerable, biotic isolation due to the limited possibilities of cross-pollination, they all co-exist here *in loco natali*, side by side, without any mutual supersession.

* Certainly there are some partly seasonal rhythms of vegetation in certain tree species in the Equatorial Zone of the Tropics. This seasonal character of rhythms increases towards the northern and the southern limits of the Tropical Zone and becomes quite distinct by 20° N and S. See Aubréville (1949), Foggie (1947) and also the present author (Fedorov 1958). Nevertheless in the Equatorial Zone rhythms of leaf change, flowering and fructification not synchronous with seasons obviously prevail over seasonal phenomena.

Thus a tropical rain forest becomes gradually flooded with a multitude of small new populations, of complete series of closely allied species originated within it and possessing various characters that are, however, neither beneficial, nor deleterious under the given conditions and play no significant part in adaptive evolution. However, as the distribution areas of these species expand or in consequence of climatic and other physico-geographical changes, some of the morphological changes initiated in the species that had originated by means of genetic drift may subsequently acquire considerable adaptive significance.

It was shown by studies in the size of leaf-blade in tree species of the Amazonian rain forest, carried out by Cain *et al.* (1956), that the prevailing type of leaf is the mesophyll about 15 cm long and about 8 cm wide. Apparently the trees with such leaves are best adapted to the conditions of the tropical rain forest. But alongside these there are also some tree species with leaves of the microphyll type and some with leaves of the macro- and megaphyll type. Nevertheless these leaf characters have no distinct adaptive significance in the tropics, as is the case in fact with many other specific distinguishing features. Evidently there are many such species in the tropical rain forest, particularly among closely allied species of the same genus, that differ in characters that are perfectly indifferent, though clearly distinct. But this is exactly the manifestation of the genetic drift, in the course of which natural selection does not eliminate indifferent mutations and their distribution takes place purely at random.

Let us assume that mutations appear inducing increased branching. Under conditions in which genetic drift prevails over natural selection these would very rapidly lead to splitting up of large boughs into small branches, which in its turn would involve a whole series of new morphological changes, possibly even if there were no new genetical reconstructions. The splitting of branches would inevitably lead to a considerable reduction in the size of leaves, flowers, fruits and other organs borne on ramified shoots. Megaphyll and macrophyll leaves might be reduced in size to mesophyll and microphyll respectively. It is interesting to note that flowers and fruits may retain their large size while leaves are reduced in size if the progenitor species was already characterized by cauliflory. Cauliflorous flowers and fruits are not affected by the splitting up of branches. This, by the way, is one of the many remarkable conclusions drawn by Corner (1949, 1954) from his Durian theory.

The existence in the tropical rain forest of series of species developing in the direction of splitting up of branches, determined by the mutation process under the conditions of the genetic drift, becomes evident from an examination of the species composition of such genera as *Dipterocarpus* or *Shorea*, or even only of their species growing in the forests of the Malay Peninsula. In most of these species there is a conspicuous correlation between the leaf and fruit size. Species having large leaves have, as a rule, large fruits, and vice versa, small-leaved species are small-fruited. Thus there are several scores of species of *Dipterocarpus* having leaves and fruits varying in length from 20 to 30 cm and from 3 to 5 cm respectively. Especially large fruits and leaves are found in *Dipterocarpus retusus*, *D. cornutus*, while the smallest are those of *D. semivestitus* and *D. verrucosus*.

The hypothesis proposed is open to criticisms based on the existence of many interspecific differences (besides those involving the ratios of leaf, flower and fruit size) between certain closely allied species of some sections of *Dipterocarpus* or *Shorea* (or in fact any other tropical genus comprising a multitude of species). How can such differences be explained from the positions of this hypothesis? It appears most probable to me that ‘ the origin of at least some of these new characters might be also explained in this way.

Let us assume that in some macrophyllous species with sparse pubescence a mutation occurs that causes the splitting of branches. Let us further assume that another species exists similar to the original one, but with small leaves and dense pubescence (which actually is frequently the case). The appearance of such a species is quite natural. Although the particular mutation that causes a more intense branching and a smaller leaf size does not affect directly the shape of hairs or the character of pubescence, the density of the pubescence will increase considerably, simply in consequence of the morphogenesis peculiar to the leaves of the new species. Since the area of the leaf-blade surface diminishes, the hairs grow closer to one another and the pubescence becomes dense. Thus, wide qualitative interspecific differences involving various plant parts may arise in the progenies of mutant parents not as the direct results of the corresponding mutations, but rather as the sequel to the consequent changes in the morphogenesis. Various deviations in the morphogenesis alone appear to be responsible for the most diverse changes of shape (contours) of leaves (or any other organs), of the venation, etc., if the mutation involves the processes governing the growth rates of different plant parts. This appears to be the mechanism of the initiation and development of a large number of specific characters of no selective advantage for the plant. Similarly mutations affecting other characters besides branching obviously entail swarms of morphological changes giving rise to distinct specific features of most various kinds in the newly originating forms of plants.

It was as early as the thirties that Koltzoff (1936) and Snell (1931) pointed out that there is no need to assume that a large number of mutations are necessary for any drastic morphological change. A very considerable phenotypic effect may be attained by means of quite slight changes in the genotype. Thus, according to the view of Koltzoff (1936) and shared by Takhtajan (1964), neoteny was the way of initiation as separate forms, of certain taxa, even of some high-rank taxa, although no more than a single mutation might have served to initiate this process.

At this point we inevitably approach a very important new problem which is not only of purely morphological and ecologic-geographical significance, but also particularly important for the the evolution of forms of the taxonomical rank of species in the tropical rain forest.

It becomes obvious that this significance is by no means the same in the tropics as in the Cold or Temperate Zone. If we abstract ourselves from the taxonomic rank of a species, however firmly the latter be established morphologically for tropical species, then, as compared to the species of the non-tropical zones, the tropical species representing any series of closely allied forms may be regarded as a single disintegrating species, as it were *in statu nascendi*, there being no geographical distribution of the different newly initiating species among different areas. In other words, a very widespread concept of the accelerated course of speciation under extreme or adverse environmental conditions, e.g. in the mountains, is in disagreement with the observations in the tropics. It is, under the optimum conditions of the humid tropics, apparently inhabited by the most ancient representatives of the plant kingdom, that the actual centre of speciation is located, the most cogent evidence for this being the immense number of species and, at the same time, the obviously primordial small numbers of individuals of each.

The extremely low population density of most tree species of the tropical rain forest affects in the first place competition between the constituents of the forest canopy.

No competitive relations ever develop within the populations of most species, except in some rare cases when a species forms more or less dense associations, and there is no elimination of individuals within populations such as would ensue from the overpopulation of any area by individuals of the same species. But undoubtedly there is violent competition for light and space among the representatives of a great many different genera and families forming the strata of forest, particularly the stratum A. The integrating effect of natural selection is very pronounced here, the result being that among the trees of any particular stratum the species belonging to quite different genera and families are very difficult to distinguish from one another, even for the experienced botanist, so much are their characters levelled down by biological convergence.

SUMMARY AND CONCLUSIONS

On the basis of all that has been said above, the following conclusions may be drawn:

1. The floristic composition of the tropical rain forest is remarkable in containing complete, uninterrupted series of taxa, particularly of species connected by indubitable affinity. Further, many entire plant families (with all their genera and species) and many very large genera are frequently confined within the limits of such a community, e.g. Dipterocarpaceae in the tropics of Asia, the rich genera of Lecythidaceae in South America, etc.

2. Besides this, it has been established by many investigators that most plant species (especially trees) of the tropical rain forest are characterized by very small numbers of individuals, being represented by populations with very low densities (one, two or three individuals per hectare). The lowest population density is characteristic both of the stratum of the 'outstanding' or emergent trees and of the uppermost storey situated below it. The highest population density is observed in the lower strata; here also, however, it is seldom found that a certain species prevails over the others; as a rule the species composition here is also mixed.

3. Occasional instances of the dominance of a certain species in the upper storey (stratum A) do not break the general principle, viz. the existence of a very large number of tree species always represented by small populations in any part of the tropical rain forest.

4. Small size of populations of tree species, low population density and partial, though considerable, biotic isolation between populations and even between separate individuals, associated with the absence of seasonal rhythm and extreme irregularity of flowering and the consequent difficulty of cross-pollination—all these factors create favourable conditions for the process of speciation in which the role of genetic drift prevails over that of natural selection. Under such conditions mutant genes accumulate in populations and contribute to the relatively rapid origin of series of closely allied species, differing considerably from one another morphologically but possessing many 'indifferent' characters.

5. Theoretically it might be assumed that the origin of a series of closely allied species was the result of only a few mutations causing changes directly in a relatively small number of characters, but leading indirectly, through the consequent changes in morphogenesis involving different plant organs, to changes in many other characters. Thus, a single mutation responsible for the intensification of branching leads, by means of the changes in the morphogenesis, to a more or less pronounced decrease in the size of leaves, flowers and fruits growing on branching shoots. Accordingly the shape and

venation of leaves, the density of pubescence, etc., can be changed. The shape of the flower, the fruit and of the other organs can be changed in a similar way.

6. It may be concluded that in the tropical rain forest, in the absence of any considerable disturbances, there is no intraspecific competition and no species is ever superseded by other species closely allied to it. The only existing competition is between ecologically similar representatives of various families and genera which because of biological convergence, are externally hardly distinguishable.

REFERENCES

Ashton, P. S. (1962). Some new Dipterocarpaceae from Borneo. *Gdns' Bull., Singapore*, **19**, 253–319.

Ashton, P. S. (1963). Taxonomic notes on Bornean Dipterocarpaceae. *Gdns' Bull., Singapore*, **20**, 229–84.

Aubréville, A. (1949). *Climats, Forêts et Désertification de l'Afrique Tropicale*. Paris.

Black, G. A., Dobzhansky, Th. & Pavan, C. (1950). Some attempts to estimate species diversity and population density of trees in Amazonian forests. *Bot. Gaz.* **111**, 413–25.

Bunge, A. (1872). Die Gattung *Acantholimon*. *Zap. imp. Akad. Nauk*, Ser. 7, **18**,

Cain, S. A. & Castro, G. M. de O. (1959). *Manual of Vegetation Analysis*. New York.

Cain, S. A., Castro, G. M. de O., Pires, J. M. & de Silva, N. T. (1956). Application of some phytosociological techniques to Brazilian Rain forest. *Am. J. Bot.* **43**, 911–41.

Corner, E. J. H. (1949). The Durian Theory of the origin of the modern tree. *Ann. Bot.* N.S. **13**, 367–414.

Corner, E. J. H. (1954). The evolution of the tropical forest. In: *Evolution as a Process* (Ed. J. Huxley, A. C. Hardy and E. B. Ford). London.

Corner, E. J. H. (1958). An introduction to the distribution of *Ficus*. *Reinwardtia*, **4**, 15–45.

Dubinin, N. P. (1940). Darwinism and genetics of populations. (Russian). *Usp. sovrem. Biol.* **13**, 257.

Fedorov, An. A. (1957). The flora of south-western China and its significance to the knowledge of the plant-world of Eurasia. (Russian). *Komarov. Chten.* **10**, 20–50.

Fedorov, An. A. (1958). The tropical rain forest of China. (Russian with English summary). *Bot. Zh. SSSR*, **43**, 1385–408.

Fedorov, An. A. (1960). The Dipterocarp equatorial rain forest of Ceylon. (Russian with English summary). *Trudy mosk. Obshch. ispyt. Prir.* **3**, 306–32.

Fedorov, An. A. (1961). Preface to the Russian translation of P. W. Richards's *The Tropical Rain Forest*. Moscow.

Fedorov, An. A. (1964). Structure of a tropical rain forest and speciation in the humid tropics. (Abstract). *Abstracts Xth int. bot. Congr.*, p. 518.

Foggie, A. (1947). Some ecological observations on a tropical forest type in the Gold Coast. *J. Ecol.* **34**, 88–106.

Foxworthy, F. W. (1911). Philippine Dipterocarpaceae. *Philip. J. Sci.* **6**, 67.

Foxworthy, F. W. (1927). Commerical timber trees of the Malay Penninsula. *Malay. Forest Rec.* **3**.

Foxworthy, F. W. (1932). Dipterocarpaceae of the Malay Penninsula. *Malay. Forest Rec.* **10**.

Kerner, A. (1865). Gute und schlechte Arten. *Öst bot. Z.* **15**.

Kerner, A. (1869). *Die Abhängigkeit der Pflanzengestalt von Klima und Boden*. Innsbruck.

Koltzoff, N. K. (1936). *The Organisation of the Cell*. (Russian). Moscow and Leningrad.

Komarov, V. L. (1908). *Prolegomena ad floras Chinae necnon Mongoliae*. (Russian and Latin). *Trudy imp. S-peterb. bot. Sada*, **29**, 1–2.

Korshinsky, S. I. (1892). *Flora of the East of European Russia in its Taxonomical and Geographical Connections*. (Russian). Tomsk.

Kostermans, A. J. G. H. (1957). Lauraceae. *Reinwardtia*, **4**, 193–256.

Pires, J. M., Dobzhansky, Th. & Black, G. A. (1953). An estimate of the number of species of trees in an Amazonian forest community. *Bot. Gaz.* **114**, 467–77.

Richards, P. W. (1945). The floristic composition of primary tropical rain forest. *Biol. Rev.* **20**, 1–13.

Richards, P. W. (1952). *The Tropical Rain Forest*. Cambridge.

Slooten, D. F. van (1926–32). The Dipterocarpaceae of the Dutch East Indies. I–VI. *Bull. Jard. bot. Buitenz.* Ser. 3; **8**, 1–17, 263–352, 370–80; **9**, 67–137; **10**, 393–400; **12**, 1–45.

Slooten, D. F. van (1961). Sertulum Dipterocarpacearum Malayensium. VII. *Reinwardtia*, **5**, 457–79.

Snell, G. D. (1931). Inheritance in the house mouse, the linkage relation of short-ear, hairless and naked. *Genetics, Princeton*, **16**, 42–74.

Steenis, C. G. G. J. van (1957). Specific and infraspecific delimitation. In: *Flora Malesian*, Ser. 1, Vol. 5, pp. clxvii–ccxxxiv.

Sukachev, V. N. (1958). On the tropical forests of China. *Vest. Akad. Nauk SSSR*, **5**, 106–13.

Symington, C. F. (1938a). Notes on Malayan Dipterocarpaceae. IV. *Gdns' Bull. Straits Settl.* **9**, 321, 353.

Symington, C. F. (1938b). Notes on Malayan Dipterocarpaceae. VI. *J. Malayan Brch R. Asiat. Soc.* **19**, 139–68.

Symington, C. F. (1941). Foresters' manual of Dipterocarps. *Malay. Forest Rec.* **16**.

Takhtajan, A. L. (1957). On the origin of the temperate flora of Eurasia. (Russian with English summary). *Bot. Zh. SSSR*, **42**, 1635–52.

Takhtajan, A. L. (1961). *The Origin of Angiosperms.* (Russian). Moscow.

Takhtajan, A. L. (1964). *The Principles of the Evolutionary Morphology of Angiosperms.* (Russian). Moscow and Leningrad.

Vierhapper, F. (1919). Ueber echten und falschen Vikarismus. *Öst. bot. Z.* **68**.

Wettstein, R. von (1896). Die europaeischen Arten der Gattung *Gentiana*, sect. Endotricha. *Denkschr. Akad. Wiss., Wien, natp.-nat. Kl.* **64**.

Wettstein, R. von (1898). *Grundzüge der geographisch-morphologischen Methode der Pflanzensystematik.* Jena.

Willis, J. C. (1931). *A Dictionary of the Flowering Plants and Ferns*, 6th edn. Cambridge.

(*Received* 28 *January* 1965)

6

Reprinted from pp. 189–194 of *Biol. J. Linn. Soc.* **1**:155–196 (1969)

SPECIATION AMONG TROPICAL FOREST TREES: SOME DEDUCTIONS IN THE LIGHT OF RECENT EVIDENCE

P. S Ashton

[*Editor's Note:* In the original, material precedes this excerpt.]

CONCLUSIONS

Fedorov supposed that 'in tropical rain forest a group of allied forms can be considered as a single disintegrated species *in statu nascendi*, there being no geographical distribution of different newly initiating species among different areas. In other words the concept of accelerated evolution under extreme or adverse conditions is in disagreement with observations in the tropics'.

I have endeavoured to bring together evidence that tropical tree taxa, as exemplified by south-east Asian Dipterocarpaceae, possess the following characteristics:

(1) Highly specialized adaptation to their biotic and physical environment, notwithstanding that a single microhabitat may be filled by one of several alternative species in part by chance.

(2) Limited efficiency of fruit dispersal imposing contagious distribution of individuals within their habitats and inability to cross all but the narrowest dispersal barriers.

(3) An unspecialized pollination system in which autogamy is usual, but outcrossing between individuals of a clump, and to a lesser but significant extent, between clumps of a population, frequent enough to allow gene exchange throughout populations in a continuous habitat.

(4) Allopatric differentiation between populations in response to differential selective pressures.

(5) Rarity of hybrid populations.

(6) Great morphological constancy within taxa even when with a widespread but disjunct distribution; rarity of clinal variation.

(7) Long life-cycles combined with low number, but high numerical constancy, of chromosomes.

The complexity of the rain forest ecosystems has been explained in terms of:

(*a*) The seasonal and geological stability of the climate which has led to selection for mutual avoidance, and, through increased specialization, to increasingly narrow ecological amplitudes, leading to complex integrated ecosystems of high productive efficiency. As the complexity increases the numbers of biotic niches into which evolution can take place increases but they become increasingly narrow.

(*b*) Their great age.

The case against random drift has already received sufficient discussion, but it is necessary to provide an alternative interpretation of speciation which can not only account for the characteristics outlined but which causally relates them to one another.

The rain-forest species share with temperate closed-forest tree species protracted life span of individuals reaching maturity, high and selective seedling mortality and modest to poor dispersal capacity (Heslop-Harrison, 1964). They differ in their apparently narrow ecological range; this can be explained in terms of the optimal climatic environment and diversion of selection pressures to edaphic and biotic specialization. The type of morphological variation, however, requires further interpretation, for it is discontinuous yet allopatric and apparently mainly adaptive. The constancy of morphological characters among disjunct widespread populations could be explained if they were equivalent to biotypes. That this could have occurred through random drift has been rejected; they could have evolved, though, under conditions of intensive selection pressure among small isolated populations such that selective gene fixation would occur; such conditions are described by Carson (1959) as homoselection. Yet even such disjunct yet morphologically constant populations as those of *Shorea polyandra* (Fig. 1) consist of large numbers of contagiously distributed individuals throughout which there appears to be no long-term barrier to gene exchange. There therefore seems to be a single alternative explanation; that though outcrossing is frequent and homozygosity fairly low the total genetic variability of these populations is low. This could occur without increase in homozygosity if intense selection severely restricts the number of viable recombinations; each generation will produce a full range of variation through free recombination and high heterozygosity, but selection will ensure that the very few seedlings which reach maturity out of the large number that germinate are phenotypically uniform.

Such an interpretation would in fact fit neatly into the present evidence. The increasing physiological specialization imposed on populations by an increasingly complex yet narrow habitat would progressively reduce the range of viable genotypes, thus echoing Dobzhansky's (1950) conclusions based on chromosomal studies in Brazilian *Drosophila*. Nevertheless, when occasions do arise for migration into a vacant niche and subsequent isolation of a segregate population, then conditions exist for rapid spread of advantageous mutations. Restriction of variability in a species population, accompanied by narrowing of ranges of individual taxa and restriction of the habitats to which they are adapted, coupled with poor dispersal and seedling establishment, is used by Stebbins (1942) to explain the lack of diversification among temperate trees. But these characters are all common also to rain-forest trees, and the age, optimality and stability of the ever-wet tropical climate would appear by deduction to have led to the extreme differences in floristic diversity between the two types of forest.

The history of the Dipterocarpaceae in south-east Asia could be read in reverse thus:

(1) Present-day speciation is taking place between allopatric populations of widespread species. Differentiation appears to occur in characters which are either adaptive or linked to adaptive characters. Though it is impossible to evaluate the mode of differentiation between allopatric populations without experimentation, the most probable process by which it has been achieved can be suggested by deduction. The origin of diversification could have taken place either:

(*a*) Through the spatial isolation of an original cline in variation, leading through gene exchange throughout each population segregate to morphological uniformity within it; or

(*b*) founder principle phenomena operating in the peripheral zones of an initially uniform expanding population; or

(*c*) adaptative diversification through mutation within small populations isolated from the rest of a population-segregate and later reuniting before diversification had proceeded too far for successful hybridization and flow of selectively advantageous gene-complexes.

The first alternative is not supported by the rarity of clinal variation. The second seems unlikely when the longevity of seed-bearing is taken into account; it is difficult to envisage a barrier frequently arising which would allow the passage over hundreds of years of so few seeds that founder population conditions would prevail, with the possible exceptions of the major sea barriers mentioned above. The third possibility appears most likely, as the processes of local river-capture and erosion might be expected to have taken place over periods adequate for this degree of diversification. This process has not been confined to recent geological time but is probably the only one existing at present.

(2) Much of the present diversification is occurring among species which had evolved from very widespread predecessors, particularly into new habitats provided

by the tectonic instability of the margins of the Sunda Shelf; notably along the north-west Borneo neogeosyncline, where areas of young physiography, and great soil immaturity and partially in consequence diversity, have been of sufficient expanse to allow speciation to take place. This was the second wave of speciation in the area postulated by Croizat.

(3) The widespread predecessors, many of which are still extant, are those upland or sublittoral species already referred to when discussing the implications of fruit dispersal. These, the product of Croizat's first era of speciation, have remained through early edaphic specialization confined to the oldest habitats, to the old granitic mountains and mature physiography of Sundaland and on undulating surfaces on rocks providing similar soils elsewhere on the one hand, and on the other to the shifting yet ancient swamps and deltaic sands of the Proto-Mekong basin, that is the coasts of East Sumatra, Malaya, and south-west to north-west Borneo.

(4) The meagre palynological evidence available suggests that the Dipterocarpaceae, whose emergent crowns impose the characteristic structure by which western Malesian rain forests can be distinguished from all others, arrived in the tropical rain forests of south-east Asia and rapidly became abundant in the mid-Tertiary, arriving in northern Borneo, for instance, in about the Upper Oligocene. I have given reasons elsewhere (1963c) for believing them to have originated further west and to have reached an advanced state of evolution before their spread into the western Malesian rain forests. Most, if not all, of the genera are likely to have already been in existence. They spread into an ecosystem where they could at maturity occupy a 'vacant' habitat, for their great height and large leptocaul crowns even now find few competitors; this enabled them through copious seed production to secure numerical advantages in the under-storey also. Whether or not the early colonizers were genetically diverse, the rapid spread that might be expected would have built up a high genetic diversity that was subsequently divided up as differentiation and speciation through interspecific competition became increasingly intense. At this stage of rapid divergent evolution genetic drift among what must, even then owing to their means of dispersal, have been small semi-isolated founder populations, might well have contributed a role, in combination with selection, in evolution as Takhtajan (1966) and Stebbins (1966) have suggested; but this is not a phenomenon confined to rain-forest conditions.

Heslop-Harrison theorized 'that the history of the tropical rain forest reveals a long-maintained, favourable environment with no trend of secular change does not mean evolutionary stagnation. This is presumably because there must always be ways in which an organism can steal a march on others even, as it were in Utopia.' This the dipterocarps did when first they spread into ever-wet Malesia; but since then the pattern of evolution has been more for mutual avoidance so that species growing together are ecologically complementary. He is right when he further points out that, under such latter circumstances, an evolutionary change in one species produces a new environment for others, so that it may be supposed that the patterns of selective forces will be in continuous flux.

Stebbins (1950) states that a variable physical environment promotes long-term evolution; but in the rain forest the biotic environment alone has changed and become more complex through the processes of evolution, and this has, with the occurrence of temporary (in geological time) barriers to dispersal, itself accelerated diversification through increasing selection pressures, fractionizing vacant habitats and creating new ones. There is no theoretical reason why this phenomenon should not be general but one must go to the ancient stable environment of the tropics to observe it most clearly.

What has just been described conforms broadly to Simpson's (1953) concept of an evolutionary episode, of initial expansion under conditions of reduced competition, leading to adaptive diversification, gradual increase in competition and in selection pressures, and finally slowing down towards an evolutionary climax. It is questionable, however, whether an 'evolutionary climax', implying a halt to further diversification, will ever be attained even under these stable conditions. As Wilson & Taylor (1967) indicated, an equilibrium can be defined in 'ecological' or evolutionary time; the former is a quasi-equilibrium whereas the latter is never reached; therefore the total potential number of species which can occupy a habitat is never reached.

Corner (1949 and elsewhere) and others have argued that the essentials of angio-sperm evolution were complete by the end of the Cretaceous, having arisen in the humid tropics. Palaeontological studies in the tropics are beginning to throw light on this critical stage of plant evolution. Though Corner's statement has yet to be demon-strated, Muller's (1968) recent palynological data from upper Cretaceous and lower Tertiary deposits in West Borneo suggest replacement of a predominantly gymno-spermous forest by successive immigration of angiosperm groups as I have suggested for Dipterocarpaceae, leading to a predominantly angiosperm flora by the Paleocene. Lack of data from other areas, however, prevents conclusions about the climate under which these early angiosperms arose.

The angiosperms thus made their first appearance in the humid tropics under con-ditions, for them, of reduced competition. At no time in the past can the humid tropical environment have been continuous and it was probably less so than now if the conti-nents were closer, for the climate of the tropical rain forest is essentially an oceanic one. It could be argued that the basic and constant differences which distinguish the families of tropical trees arose fairly rapidly under conditions of reduced competition, allopatrically but in response to similar environmental pressures. The spiny dehiscent arillate fruit found by Corner (1949) to occur in so many families may have arisen many times in this way; conversely the diverse floral structures which distinguish these families could also have arisen as a different means of exploiting the same opportunities. A parallel is found in the presence of the overwhelming majority of insectivorous and myrmecophilous tropical plants under conditions of low soil-fertility, as in the peat swamps and more especially the Heath forests; here the same environmental pressures have apparently resulted in such diverse adaptations as the leaves of *Drosera*, *Utricularia* and *Nepenthes*, the root chambers of *Myrmecodia* and *Hydnophytum*, and the swollen internodes of *Clerodendrum fistulosum* and several Macarangas. This view finds further support in Meeuse's (1962) evidence of a polyphyletic origin for the angiosperms.

When the Dipterocarpaceae first invaded the south-east Asian tropics the forest ecosystems were already complex; the large fruit, adapted to wind dispersal which would nevertheless have been effective only in the trade-wind zone, proved to have another, ecological, advantage in promoting geographical isolation and ecological diversification in the new specialized environment; and it is thus that the Dipterocarpaceae are in so many ways the apotheosis of a rain-forest tree family.

[*Editor's Note:* Figure 1 has not been reproduced here. Only those references cited in the preceding excerpt have been reproduced here.]

REFERENCES

Ashton, P. S., 1963c. *The taxonomy and ecology of the Dipterocarpaceae in Brunei State.* Ph.D. thesis, Cambridge University.

Carson, H. L., 1959, Genetic conditions which promote or retard race formation. *Cold Spring Harb. Symp. quant. Biol.* **24**:87–105.

Corner, E. J. H., 1949. The durian theory or the origin of the modern tree. *Ann. Bot., N.S.* **13**:367–414.

Dobzhansky T. H., 1950. Evolution in the tropics. *Am Scient.* **38**:209–221.

Heslop-Harrison, J., 1964. Forty years of genecology. In Cragg J. B. (ed.) *Adv. Ecol. Res.* **2**:159–247.

Meeuse, A. D. J., 1962. The multiple origin of the Angiosperms. *Advg Fronts Pl. Sci.* **1**:105–107.

Muller, J., 1968. Palynology of the Pedawan and Plateau Sandstone Formations (Cretaceous-Eocene) in Sarawak, Malaysia. *Micropaleontology,* **14**(1):1–37.

Simpson, G. G., 1953. *The major features of evolution.* New York:Columbia U.P.

Stebbins, G. L., 1942. The genetic approach to problems of rare and endemic species. *Madroño,* **6**:240–248.

Stebbins. G. L., 1950 *Variation and evolution in plants.* New York:Columbia U.P.

Stebbins, G. L., 1966. *Processes of organic evolution.* Englewood Cliffs, N.J.

Takhtajan, A., 1966. *Systema et phylogenia Magnoliophytorum.* Moscow.

Wilson, E. O. & Taylor, R. W., 1967. An estimate of the potential evolutionary increase in species density in the Polynesian ant fauna. *Evolution, Lancaster, Pa.* **21**:1–10.

Reprinted from *Science* 203:1299–1309 (1979)

Tree Dispersion, Abundance, and Diversity in a Tropical Dry Forest

That tropical trees are clumped, not spaced, alters conceptions of the organization and dynamics.

Stephen P. Hubbell

A widely held generalization about tropical tree species is that most occur at very low adult densities and are of relatively uniform dispersion, such that adult individuals of the tree species are thinly and evenly distributed in space. If true, this generalization has potentially profound consequences for the reproductive biology, population structure, and evolution of tropical tree species (*1*). In this article the adequacy of this generalization is judged with respect to a particular tropical forest, a large tract of which has been mapped in detail (*2*).

The origins of this generalization can be traced back at least to Wallace (*3*), who stated the following concerning his impressions of species densities in Malaysian forests:

If the traveller notices a particular species and wishes to find more like it, he may often turn his eyes in vain in every direction. Trees of varied forms, dimensions, and colours are around him, but he rarely sees any one of them repeated. Time after time he goes toward a tree which looks like the one he seeks, but a closer examination proves it to be distinct. He may at length, perhaps, meet with a second specimen half a mile off, or may fail altogether, til on another occasion he stumbles on one by accident.

Dobzhansky and co-workers (*4*) enumerated the species in several 1-hectare stands of Amazonian rain forest, and concluded that "the population density of a half or more of the tree species in Amazonian forests is likely to be less

The author is an associate professor in the Department of Zoology, and in the Program in Ecology and Evolutionary Biology, University of Iowa, Iowa City 52242.

than one individual per hectare." One or both parts of this generalization (low density, uniform dispersion) now appear in most ecology texts (*5*), and theories have been proposed to explain both the causes and consequences of low density and uniform dispersion of adult tropical trees. Janzen (*6*) and Connell (*7*) independently proposed theories to explain low density and spacing between adults. Janzen focused attention on the effects of host-specific herbivores that attack seeds. He noted that a high proportion of seeds falling under the parent tree are killed by such seed "predators," so called because the death of the seed is virtually assured; and he suggested that only those viable seeds transported some distance away from the parent would escape discovery and manage to germinate. The predicted result: "most adults of a given tree species appear to be more regularly distributed than if the probability of a new adult appearing at a point in the forest were proportional to the num-

ber of seeds arriving at that point" (*6*, p. 501).

Connell, in his earlier rain-forest studies, focused more attention on the dispersion and survival of young tree seedlings. In experimental studies of the fate of seedlings in an Australian rain forest, Connell, Tracy, and Webb (*8*) showed that survival was better in seedlings planted under adults of different species than under adults of the same species. They did not identify the causes of the differential mortality, but did suggest that herbivores attracted by adjacent adults would more often tend to defoliate and kill nearby seedlings rather than distant seedlings. Janzen and Connell both argued that such host-specific attack by herbivores would reduce the local density of any given species, open up habitat to invasion by additional species, and thereby maintain high species diversity (*9*).

Many explanations have been offered for the high species diversity in tropical forests, and these have been classified into equilibrium and nonequilibrium hypotheses by Connell (*10*), who now believes that high diversity is only maintained because of frequent disturbance.

Summary. Patterns of tree abundance and dispersion in a tropical deciduous (dry) forest are summarized. The generalization that tropical trees have spaced adults did not hold. All species were either clumped or randomly dispersed, with rare species more clumped than common species. Breeding system was unrelated to species abundance or dispersion, but clumping was related to mode of seed dispersal. Juvenile densities decreased approximately exponentially away from adults. Rare species gave evidence of poor reproductive performance compared with their performance when common in nearby forests. Patterns of relative species abundance in the dry forest are compared with patterns in other forests, and are explained by a simple stochastic model based on random-walk immigration and extinction set in motion by periodic community disturbance.

Potential consequences of a low-density uniform dispersion of adult trees in tropical species might include lower outcrossing success, reduction in deme size, and requirements for long-distance pollination. Thus, the generalization that adults of tropical tree species are widely spaced has also spawned a number of hypotheses about unusual breeding systems in tropical trees (*11*), or special pollinator movements over long distances (*12*). It now appears that the majority of tropical tree species is facultatively or obligately outcrossed; and the frequency of dioecy in tropical trees is very high by temperate zone standards (*13, 14*). Animals rather than wind, in most cases, are the agents of cross-pollination.

Characteristics of the Dry Forest

Many of the ideas about low density and uniform tree dispersion developed from observations in the rain forest. Therefore, the major results of this study are presented with the caveat that they may apply better to deciduous tropical forests or monsoon forests than to rain forests. Nonetheless the patterns found in the dry forest appear to be fully consistent with the available information on rain forests, with the exception that the dry forest has only a third to half the number of tree species. Conclusions about forest dynamics are necessarily tentative because they are based on the circumstantial evidence provided by one census at a single point in time.

The dry forest results were obtained from a detailed map of 13.44 hectares of forest in Guanacaste Province, Costa Rica, in which all woody plants with stem diameters at breast height (dbh) ≥ 2 centimeters were located to the nearest meter. A separate map of the population of each species was drawn by computer, with each plant marked by a letter to indicate position and dbh, and whether it was juvenile or adult. Maps were then analyzed in an attempt to answer, in part or in whole, the following questions about each of the tree species: (i) Is the dispersion of the adult tree population uniform? (ii) Are adult trees less clumped than the juveniles, or than the population as a whole? (iii) How do the densities of adult and juvenile trees change with greater distance from a given adult in the population? (iv) Is there evidence that spacing from a given adult to other trees increases from the juvenile to the adult tree classes? (v) How distant is the nearest adult, and how many adults can be expected within m meters of a given adult? (vi) What, if any, is the relationship between rarity of a tree species and its dispersion pattern or its population (size, age) structure? (vii) Does mode of seed dispersal or breeding system relate to abundance or dispersion pattern? (viii) Is there any evidence for density dependence in the per capita reproductive performance of adult trees? Questions of interest about the dry-forest community in general include: (ix) What tentative conclusions can be drawn about the equilibrium or nonequilibrium status of the forest? (x) What are the pat-

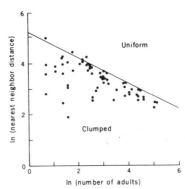

Fig. 1. Clumping in 61 tropical dry-forest tree species, as measured by mean nearest neighbor distance in meters. Diagonal line is the expected nearest neighbor distance for randomly dispersed populations [see (*37*)].

terns of relative species abundance in the dry forest, and how do they compare with the patterns in forests at other latitudes? Are the patterns of similar mechanistic origin?

Questions (i) through (iv) are of importance in evaluating the Janzen-Connell hypothesis. Question (v) is concerned with the minimum distance that a potential pollinator must fly to encounter another conspecific adult, and relates to the general question of deme size in tropical dry forest tree species. The answer to question (vi) is relevant to a discussion of the minimal density at which adult outcrossed trees of a species are capable of self-replacement. One way that a rare species can persist, in theory at least, even in the absence of long-distance pollination, would be through a higher degree of adult clumping than found in common species (*15*). Answers to questions (v) and (vi) may also suggest which of several alternative explanations of tree rarity is most probable (*16*).

The answer to question (vii) should reveal whether there are systematic differences in the adult or juvenile dispersion patterns of small-seeded species with wind or bird dispersal and large-seeded species with water or mammal dispersal. It should also reveal whether dioecious species differ systematically in density or dispersion pattern from self-compatible or hermaphroditic species. Janzen (*6*) and Bawa and Opler (*14*) hypothesized that, given similar pollination systems, the average interadult distance might be less in dioecious species because deme size would be roughly halved from that of an equally abundant hermaphroditic species. Skewed sex ratios, to the extent that they occur (*17*), should exacerbate the problem for the locally more abundant sex.

Question (viii) addresses the issue of what regulates the population size of tropical tree species. Does the number of nearby conspecific adults have a measurable influence on the reproductive performance of a given adult tree? The Janzen-Connell hypothesis would lead us to expect greater losses per adult in seeds and seedlings when adults are closer together, provided the per capita attractiveness of adult trees to herbivores is greater when the adults are in groups. Grouped adults may also compete more strongly for pollinator attention (*14, 18*) and encounter root and crown competition if they are actually adjacent to one another (*19*). Alternatively, there may be no detectable effect of adult density on per capita reproductive success (*20*).

Question (ix) relates to questions (vi) and (viii) and the reproductive success of rare trees. If the forest and its component species are in equilibrium, there should be found evidence that the rare species as well as the common species are self-replacing. In equilibrium theory, each species is presumed to be competitively superior at exploiting a particular microhabitat (*10*), with relative species abundance determined by the relative abundance of the microhabitat types. However, if the forest is in a nonequilibrium state, there should be evidence that some species are increasing in numbers while others, most likely the currently rare species, are declining.

Finally, question (x) asks whether the patterns of dominance and diversity in tropical, temperate, and boreal forests reveal any similarities that might suggest similar underlying control mechanisms. Using a simple stochastic model, I show that it is relatively easy to generate the patterns of relative species abundance in natural forest communities along a disturbance gradient. The model adds to the current theory of island biogeography a new statistical explanation for the relative abundance of species in arbitrarily defined habitat "islands" (*21*).

Study Site

The tract of mapped forest in Guanacaste Province, Costa Rica, lies approximately 8 kilometers west of Bagaces and 2 kilometers south of the Pan American Highway (*22*). The tract has been the site of studies of the pollinator community (*23*), and of the breeding systems and seasonal phenology of a number of tree species (*14, 18*). As a result of these and other studies (*24*), the pollination biology and the phenological cycles of leafing shedding, flowering, and fruiting are known for many of the deciduous forest tree species. Most species are obligately outcrossed by insects, birds, or bats. The climate of the Dry Forest Life Zone has also been described (*25*).

A total of 135 species of woody plants with stem diameters ≥ 2 cm dbh were found in the mapped area. These include 87 overstory and understory trees, 38 shrubs, and 10 vine species (*26*). About two-thirds of the tree species are deciduous through part or all of the dry season (December to May). The canopy is somewhat broken (15 to 25 meters high), and consists primarily of trees with medium-length boles and spreading crowns. Canopy cover is approximately 87 percent; the remainder consists of light gaps made by recent tree falls, and a narrow

grassland corridor through the center of the site (7 percent of the mapped area). The understory consists of a diverse array of tree seedlings and shrubs (*27*). Vines are common, but epiphytes are infrequent. Insect flower visitors, especially bees, are abundant, as are night pollinators such as bats and hawkmoths. Several types of vertebrate seed dispersers are common (*28*), and there is evidence of heavy insect and vertebrate seed predation in a number of tree species (*29*).

The mapped forest tract was a rectangular area, 420 m by 320 m, gridded into 336 quadrats 20 m on a side. All woody plants ≥ 2 cm dbh were identified to species (*30*), measured for diameter, and mapped to within ± 1 m of their true position within each quadrat (*31*). Separate data were taken on the dbh of trees and shrubs in flower or fruit to establish a lower bound on the diameter of reproductive individuals (adults) for each species (*32*). The data for each quadrat were transcribed to computer cards; programs sorted individuals to species and recomputed the coordinates of each plant from its local quadrat to its coordinates in the study area as a whole. Maps were then drawn by a Calcomp plotter for each species (*33*).

Juvenile and adult dispersion patterns were examined over a range of quadrat sizes from 4 m² to 38,416 m² (196 m on a side), by Morisita's index of dispersion, I_δ (*34*). Determining juvenile and adult densities at different distances from each adult tree in the population was done from the Calcomp maps. Average den-

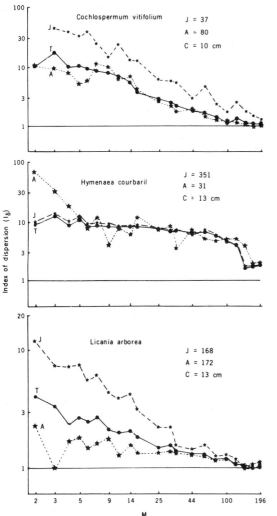

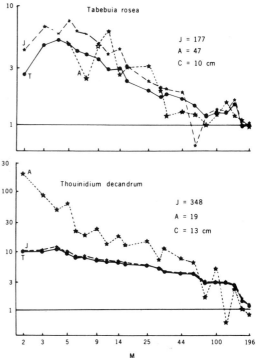

Fig. 2. Morisita's index of dispersion, I_δ, as a function of quadrat size for five sample dry-forest tree species. Numbers of the x-axis are lengths of the quadrat sides in meters. The spacing of the quadrat sizes along the x-axis is in terms of log quadrat area. Log I_δ values are plotted on the y-axis to make more visible the changes in I_δ that occur at larger quadrat sizes. The horizontal line through $I_\delta = 1$ indicates the expected value for randomly dispersed populations. *J*, juveniles; *A*, adults; *T*, total population; *C*, cutoff dbh for adults.

sities were computed for increments of 5 m out to a distance of 100 m (*35*). The maps were also analyzed to obtain the distribution of nearest neighbor distances between adults.

Tests for density dependence in per capita reproductive performance were made on 30 of the most abundant species. The gridded area was divided into 80 subplots of equal area, and the densities of adult and juvenile trees were noted for each subplot. The ratio of juveniles to adults was plotted as a function of adult density. Significant departures of slopes from zero were determined by regression analysis of variance. A necessary assumption of this test is that the adults counted in a subplot are the parents of the juveniles in the same subplot. Validity of this assumption varies with tree species and mode of seed dispersal (*36*).

Adult and Juvenile Dispersion Patterns

The adults of tropical dry-forest tree species are not uniformly dispersed in the forest, as shown in Fig. 1 for the 61 species on which information could be obtained on threshold adult diameter. The diagonal line in Fig. 1 is the expected nearest neighbor distance for randomly dispersed adults at the given mean density (*37*). Of these species, 44 (72 percent) exhibit significant adult clumping ($P < .05$, F test). The remaining 17 species (28 percent) have adult dispersion patterns which cannot be distinguished from random. No species has a significantly uniform adult dispersion.

When the dispersion of both adults and juveniles together is considered, the clumping is again pronounced. Of the 114 identified tree, shrub, and vine species having at least two individuals in the mapped area (*38*), fully 102 species exhibited significant clumping ($P < .05$, F test) in quadrats < 196 m on a side. As many as 95 species still showed significant clumping even in quadrats as small as 14 m on a side. No species showed a significantly uniform pattern of dispersion of its total population.

The pattern of change in Morisita's index of dispersion as quadrat size is increased from 2 to 196 m on a side, is shown (Fig. 2) for five species chosen at random from the 30 most common species (*39*). The dispersion indices for juveniles, adults, and total population are plotted separately. In all cases, including the species not illustrated, the dispersion index, I_δ, drops from its highest values in the smaller quadrat sizes irregularly downward toward unity as quadrat size

increases. Such I_δ patterns are typical of populations having "point sources" of relatively high population density, surrounded by more diffuse clouds of individuals diminishing in density away from the centers (*34*). Small quadrats may contain the high-density centers, thereby producing large I_δ values, whereas large quadrats tend to have lower I_δ values because they average the density of the concentrated centers with the density of the more sparsely populated surroundings.

Question (ii) asks if the adults are less clumped than the population as whole, as would be expected from the Janzen-Connell hypothesis (*6*, p. 522). At a quadrat size of 14 m on a side and for the 30 most common tree species, adults are less clumped than juveniles and than the population as a whole in 16 species, equally clumped in nine species, and more clumped in five species. The results change somewhat depending on quadrat size and the scale of pattern resolution as well. Of the five species (Fig. 2), two species (*Cochlospermum* and *Licania*) show more clumping in juveniles than in adults at small quadrat sizes, two species (*Hymenaea* and *Thouinidium*) show more clumping in adults at small quadrat sizes, and one species (*Tabebuia*) shows approximately equal clumping in adults and juveniles.

Because of such clumping many species appear to be quite rare (less than one individual per hectare) when half or more of the study area is considered, but they turn out to be quite common when the rest of the study area is included. These patchy tree distributions help to explain data on tree species densities for relatively small plots of tropical forest (1 or 2 hectares) (*40*).

Demographic Neighborhood of the Average Adult

The "demographic neighborhood" of a tree may be defined as the population of adults and juveniles of the same species occurring within a specified radius of the tree. The "average adult" may be considered for the purposes of this article as one that exhibits the expected demographic neighborhood for an adult of the given species.

Question (iii) asks how the densities of adults and juveniles change with increasing distance from the average adult in the population. If the Janzen-Connell hypothesis is correct, densities should be lower near the adult than at some intermediate distance. What constitutes an

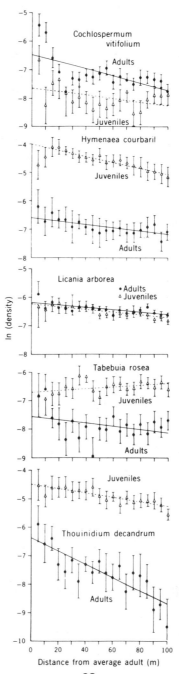

Fig. 3. Adult and juvenile densities (natural logarithm of numbers per square meter) at various distances from the average adult in the population, showing the negative slopes away from the average adult (the slope in *Tabebuia rosea* for juveniles is not different from zero) in the five sample species. The bars are 95 percent confidence limits for the means.

Table 1. Relative adult (A) and juvenile (J) densities in 5-m distance intervals away from the average adult, in the five sample species. Relative densities within adult and juvenile classes have been normalized to a sum of 100 percent for ease of visual comparison. In each distance interval, χ^2 tests were performed on the numbers of juveniles and adults counted.

Species		Relative densities at distance interval in meters:									
		0 to 4.9	5 to 9.9	10 to 14.9	15 to 19.9	20 to 24.9	25 to 29.9	30 to 34.9	35 to 39.9	40 to 44.9	45 to 49.9
Cochlospermum vitifolium	J	27.43	5.72	12.34	14.62	8.89	6.36	7.71	6.12	4.82	5.99
	A*	31.11	24.59	9.92	5.93	3.03	4.71	4.57	5.36	5.06	5.72
Hymenaea courbaril	J†	7.23	9.30	12.77	12.15	11.45	10.36	9.44	9.29	8.21	9.80
	A†	16.41	9.90	13.19	10.35	10.69	8.06	9.75	7.93	7.09	6.63
Licania arborea	J	10.17	9.18	10.93	11.60	9.64	10.13	10.33	10.76	9.42	7.84
	A†	15.44	9.47	11.79	9.63	9.35	8.53	10.02	9.83	9.75	6.19
Tabebuia rosea	J	7.74	7.12	8.14	8.35	10.35	9.99	15.90	15.68	9.06	7.67
	A*	21.24	26.61	9.81	7.92	4.44	8.82	4.90	7.12	2.54	6.60
Thouinidium decandrum	J	11.44	7.98	10.78	10.86	11.02	9.50	9.72	9.89	11.08	7.73
	A*	26.94	13.48	16.84	6.74	5.25	7.86	3.60	6.57	5.18	7.54

*Significant difference ($P < .05$). †Not significant ($P > .05$).

"intermediate distance" will depend on the tree species, but clearly it should not exceed the mean nearest neighbor distance between adults. The 30 most common species were analyzed, and the results do not support the Janzen-Connell expectation. Either adult and juvenile densities decline approximately exponentially away from the average adult, or densities remained unchanged, to and beyond the intermediate distances appropriate for the species. In 67 percent of the species, there were negative slopes for the log-transformed juvenile densities as a function of distance; and the slopes in all remaining species were not distinguishable from zero (*41*). An even greater percentage (90 percent) of these species also showed negative slopes for adult densities, and there were no positive slopes. Thus, at least for the 30 most common species, the average adult is clearly found in a clump with other adults and juveniles.

The adult and juvenile density curves for the five sample species (Fig. 3) reveal that the densities after log-transformation generally exhibited equal variance at all distances. In 15 of the 30 most common species, the highest mean juvenile density occurred in the 0 to 5-m annulus, closest to the adult. In ten additional species with horizontal density curves, mean juvenile density was no lower in the annulus closest to the adult than in other annuli. However, in the five remaining species, maximal juvenile density was achieved between 5 and 15 m from the adult. *Hymenaea* is one of these species in which there was a notable scarcity of juveniles immediately under the adult canopy. This species suffers heavy losses of seeds to predators (such as specialized insects and generalized vertebrates), but in spite of these losses, the maximum recruitment of juveniles is still quite close to the adult (15 m), only a

few meters beyond the crown perimeter (*42*).

It is possible that the distance at which the density of juveniles is maximal is not the distance with the highest probability of producing an adult. If juveniles die more often when they are growing close to an adult, there should be an increase in the mean distance between an adult and successively older cohorts of neighboring trees [question (iv)]. Alternatively, if mortality is a random thinning process regardless of distance from adult trees, there should be no significant change with increasing cohort age in the proportion of cohort individuals at a given distance. Seeds, seedlings, and saplings < 2 cm dbh—stages in which most of the mortality occurs—were not mapped. However, whatever the postulated mortality patterns in seeds and seedlings, the greatest density of surviving juveniles that remain after this mortality has taken its toll is nevertheless usually in the annulus closest to the adult. Therefore, if appreciable spacing is occurring, it must occur as a result of differential mortality among censused cohorts of juveniles ≥ 2 cm dbh.

Relative densities in juvenile and adult cohorts were computed to a distance of 50 m in the five sample species (Table 1). Two species (*Hymenaea* and *Licania*) do not show significant differences in the distributions of relative adult and juvenile densities, and for these species the null hypothesis of random thinning cannot be rejected. In the three remaining species, adult relative densities are shifted significantly closer to the average adult than are juvenile relative densities. This suggests that, contrary to prediction, the more distant juveniles suffer greater losses in these species (*43*). One explanation for such a result could be that adults are already growing in the sites most favorable to the species, with

outlying areas generally of lower microhabitat suitability. This pattern is repeated in the 30 most common tree species (including the five discussed above), of which 13 species have relative adult densities shifted closer to the average adult than relative juvenile densities. In the remaining 17 species the null hypothesis of random thinning could not be rejected. No species showed the predicted shift in relative adult densities to greater distances.

Question (v) concerns tree spacing from the point of view of pollinator movement. Although adults generally occur in clumps with other adults and juveniles, it is more specifically the absolute distance from one adult to the next that is important to pollination success. For the 30 most common species, the mean nearest neighbor distances are within 20 m in 16 species, and within 40 m in all species (*44*). These 30 species average 5.6 ± 1.3 adults within 50 m of a given adult, and 12.7 ± 2.7 adults within 100 m. A fourfold increase in area produces, on average, only a 2.3-fold increase in the number of adults, a further indication of adult clumping (Fig. 4).

The demographic neighborhood of the average adult is a composite of all the adults and juveniles in the population. Therefore, the density curves away from the average adult cannot be equated to the seed shadows of single adult trees since the curves result from the superposition of several overlapping shadows of neighboring adults. Nevertheless, the approximately negative exponential character of the composite density curves away from the average adult constitutes strong circumstantial evidence that relatively simple physical and biological mechanisms govern seed dispersal in these species.

One might expect that the effects of different seed sizes and modes of dispersal would be reflected in the slopes of these density curves [question (vii)]. In log transformation the slopes of the exponential density curves conveniently become independent of the absolute abundance and seed production of the species, permitting cross-species comparisons. Comparison of the slopes for adult and juvenile log density as a function of distance, in mammal-, wind-, and bird- or bat-dispersed species, is shown in Fig. 5. Mammal-dispersed species show the steepest slopes; shallower slopes are found in wind-dispersed species; and the shallowest slopes, on average, characterize the bird- and bat-dispersed species. All three pairwise contrasts of juvenile slopes are significantly different, and two of three adult slopes (except wind compared to bird) are significantly different ($P < .05$, Mann-Whitney U test).

Early work on dispersal by Dobzhansky and Wright (45) suggested that a bivariate normal might typically describe dispersal patterns. However, dispersal is frequently strongly leptokurtic, with a pronounced peak near the point of propagule origin (46). These results for dry-forest tree species suggest that seed dispersal as well as juvenile survival are much more leptokurtic in distribution in mammal-dispersed species than in either wind- or bird-dispersed species.

Dispersion, Abundance, and Density Dependence

In the preceding discussion I have dealt primarily with the most common third of the tree species. What is the dispersion of rare species? If individuals of rare species were dispersed at random, or spaced uniformly, nearest neighbor distances should increase with decreased density as fast or faster than the inverse square root of mean density (37); but this is not the case in adults (Fig. 1). When total population size is considered, the trend is toward increased clumping with decreased abundance (Fig. 6). A few species (outliers) do not conform to the general species sequence. I believe that these nonconforming species are probably "accidentals" or last survivors of once more abundant species (see below). Greater clumping in rare species has also been reported by Hairston (47) in old-field communities of soil arthropods.

The second part of question (vii), whether differences in breeding system explain any of the variation in dispersion, independently of population size, is answered in Fig. 6 (48). Self-compatible species are not overrepresented among rare species, nor are dioecious species more clumped for their abundance than hermaphrodite species. That most rare species are highly clumped might suggest that they are at least locally successful (reproducing themselves) when their microhabitat requirements are satisfied. However, the data on size (dbh) class suggest that per capita reproduction in rare species is considerably less per unit time than in common species. The coefficient of skewness of the dbh distribution about the midpoint diameter for each species was computed

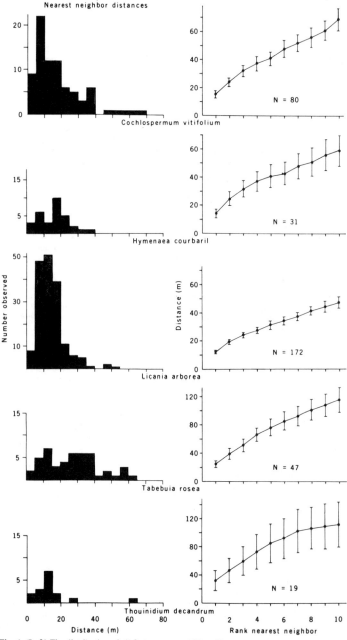

Fig. 4. (Left) The distribution of all first nearest neighbor distances for all adults, showing the closeness of the nearest adult, as well as the tendency to skew near (mode < mean). (Right) The mean distances, and their 95 percent confidence limits, to the first ten nearest neighbor adults, in the five sample species.

(49); species having an excess of large adults show positive or zero skewing, and species with an excess of juveniles show negative skewing.

Rare species exhibit positive or zero skewing about the midpoint dbh, whereas common species exhibit negative skewing (Fig. 7). No biology is needed to explain negative skewing in common species because it would be physically impossible to fit so many trees into the available space if all were large adults. However, skewing in rare species is not constrained in either direction, yet no rare species is negatively skewed. This pattern does not prove that the rare species are failing to reproduce themselves—for example, successful reproduction might be extremely episodic *(16)*—but it is clear evidence for a much slower per capita rate of juvenile establishment in rare species than in common species *(50)*.

Some of the rare species in the forest may be last survivors of species once more common in earlier successional stages, and some may be accidentals that became established through good fortune outside the habitat to which they are optimally adapted. All seven of the outlier species (Fig. 6) may be such accidentals because they are represented solely by very large, randomly scattered adults *(51)*. These and other rare species are common on sites elsewhere in Guanacaste Province *(52)*, and, where common, they exhibit strong negative dbh class skewing, as would be expected. Unless one is prepared to accept radically different life history strategies in adjacent plant populations of the same species, these results point to reproductive failure in the rare species in the tract of forest described here *(53)*. Thus, the available circumstantial evidence suggests that the forest is in a nonequilibrium state (question ix).

Rare species might persist longer in a given forest stand than otherwise if the abundance of common tree species is limited, short of complete space monopoly, by density-responsive herbivory or seed predation, or by other density-dependent processes. There is no argument that seed predation by specialized herbivores is intense and results in considerable thinning. Even if seed and seedling predation is a random thinning process in relation to distance from the average adult, it is still quite possible that seed predation can lower the per capita reproductive performance of whole clumps of adults compared to more isolated adults.

Density dependence can be detected in the tree species [question (viii)]. Of

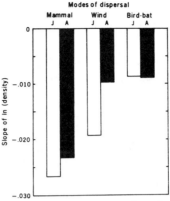

Fig. 5. Relation between the mode of seed dispersal and the steepness of the slope with which the natural log of the density (numbers per square meter) in adult (*A*) and juveniles (*J*) decreases with distance (meters) from the average adult in the population, showing the steeper slopes found in the heavier seeded, mammal-dispersed species. Sample sizes: 9 mammal-dispersed species; 12 wind-dispersed species; 9 bird- or bat-dispersed species.

the 30 most common species, 17 species (57 percent) exhibited significantly negative slopes in the regression of juveniles per adult on adult density. All 13 remaining species also showed a negative slope, but the null hypothesis of zero slope could not be rejected ($P > .05$). Density dependence was detected more frequently in species producing a large number of small seeds (generally the wind- and bird-dispersed species) as compared to species producing a smaller number of large seeds (mammal-dispersed species). Two-thirds of the wind- and bird-dispersed species exhibited density dependence, whereas only one-third of the mammal-dispersed species did *(54)*.

What is the source of this apparent density dependence? If it is primarily density-responsive seed and seedling predation, one might expect that the species attacked most heavily would in general be those exhibiting the strongest density dependence. However, in general, the large-seeded species are more frequently attacked by host-specific seed predators (commonly bruchid weevils) than are the smaller seeded species. An alternative hypothesis is that the density dependence is occurring via reduced per capita seed output in crowded adults, perhaps because of competition for pollinator attention or root and crown competition. It is also possible that some of these intraspecific effects are more apparent than real. The apparently greater frequency of density dependence among wind- and bird- dispersed tree species may be spurious *(36)*. Moreover, if the adults of several species are positively associated in space, the shade and root competition they collectively produce would result in "diffuse" competition against the seedlings of all species growing beneath them *(55)*.

Dominance-Diversity Relationships

Although it has long been clear that tropical land plant communities are far richer in species number than their temperate or boreal counterparts, sufficiently large data sets from which to make quantitative comparisons of relative species abundance have become available only in the last 25 years. When the species of a plant community are arranged in a sequence of importance (using a measure such as standing crop, basal area, or annual net production) from most to least important, they form a smooth progression without major discontinuities from the common to the rare

Fig. 6. Relation between total abundance (juveniles and adults in 13.44 hectares) and dispersion pattern for 87 tree and 8 large shrub species, showing the increased clumping in rare species. The *y*-axis is log I_δ for a quadrat size of 14 m on a side. Solid circles are outcrossed hermaphroditic species; open circles are self-compatible hermaphroditic species; and stars are dioecious species. The seven outlier species (inside dotted line) are not significantly clumped and are represented solely by very large adults.

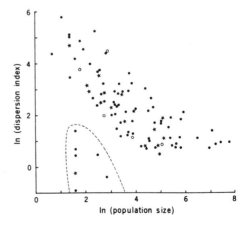

species. Different plant communities produce characteristic "dominance-diversity" curves when the importance values are log-transformed and plotted against the rank of the species in importance. Simple communities with few species generally yield almost straight ("geometric") lines on these semilog plots, whereas species-rich communities characteristically exhibit S-shaped ("lognormal") progressions (56).

Figure 8 compares the dominance-diversity curve for the tropical dry forest at 10°N (this study) with the curves for an equatorial (Amazonian) forest (57), for a rich temperate forest, and for a species-poor temperate montane forest similar to boreal forest (56). A number of factors distort the quantitative differences between the temperate and tropical dominance-diversity curves (58), but the qualitative pattern is clear: tropical forests exhibit the same general lognormal curve characteristic of rich temperate forests, but the distributions differ in location (mean) and scale (variance).

The rank-1 species in the dry forest has an importance value of 11 percent, compared with only 4.7 percent in the Amazonian forest. At 35°N, the rank-1 species in rich temperate forest has an importance value of 36 percent, which increases to about 65 percent in the montane spruce-fir forest. If the latter dominance-diversity pattern is comparable to patterns in the boreal coniferous forest at 50° to 60°N, then, as a general rule, the importance value of the dominant species in neotropical and nearctic forests increases by approximately one percentage point for every latitudinal degree of northward movement, starting from a base of a few percent at the equator. Factors such as topographic diversity, elevation, the frequency and magnitude

of disturbance, and physical harshness will likely cause local deviations from this general pattern (59).

Relative Species Abundance in Nonequilibrium Communities

The qualitative similarity between the dominance-diversity curves for temperate and tropical forests suggests that similar processes control the relative abundance of tree species in the two regions (question x). May (60) has cautioned against reading too much significance into lognormal species abundance patterns, noting that lognormal distributions may be expected when many random variables compound multiplicatively, given the "nature of the equations of population growth" and the central limit theorem. Recently, Caswell (61) built several "neutral" models of community organization and relative species abundance, based on neutral-allele models in population genetics; but lognormal relative abundance patterns were not obtained from any of them.

Another line of reasoning, however, does generate lognormal relative abundance patterns under one set of circumstances, as well as geometric patterns under other circumstances. The model is essentially a dynamic version of MacArthur's "broken stick" hypothesis (62), and is based on a nonequilibrium interpretation of community organization. Suppose that forests are saturated with trees, each of which individually controls a unit of canopy space in the forest and resists invasion by other trees until it is damaged or killed. Let the forest be saturated when it has K individual trees, regardless of species. Now suppose that the forest is disturbed by a wind storm,

landslide, or the like, and some trees are killed. Let D trees be killed, and assume that this mortality is randomly distributed across species, with the expectation that the losses of each species are strictly proportional to its current relative abundance (63). Next let D new trees grow up, exactly replacing the D "vacancies" in the canopy created by the disturbance, so that the community is restored to its predisturbance saturation until the next disturbance comes along (64). Let the expected proportion of the replacement trees contributed by each species be given by the proportional abundance of the species in the community after the disturbance (65). Finally, repeat this cycle of disturbance and resaturation over and over again.

In the absence of immigration of new species into the community, or of the recolonization of species formerly present but lost through local extinction, this simple stochastic model leads in the long run to complete dominance by one species. In the short run, however, the model leads to lognormal relative abundance patterns, and to geometric patterns in the intermediate run. The magnitude of the disturbance mortality, D, relative to community size, K, controls the rate at which the species diversity is reduced by local extinction: the larger D is relative to K, the shorter the time until extinction of any given species, and the faster the relative abundance patterns assume an approximately geometric distribution (66).

The random differentiation of relative species abundance in a 40-species community closed to immigration is illustrated in Fig. 9 (67). Given equal abundances at the start, after 25 disturbances the species form the set of approximately lognormal dominance-diversity curves;

Fig. 7 (left). Coefficient of skewness of dbh distribution about midpoint dbh, as a function of total species abundance. Rare species are positively skewed toward large dbh, suggesting infrequent or highly episodic reproduction. Fig. 8 (right). Comparison of the dominance-diversity curves for two tropical forests and two temperate forests. Importance values for the temperate forests are based on annual net production; importance values for the dry forest in Costa Rica are from basal area (cross-sectional area of all stems of a given species); importance values in the Amazonian forest are based on above-ground biomass [see (58) for additional comments].

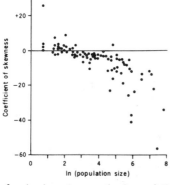

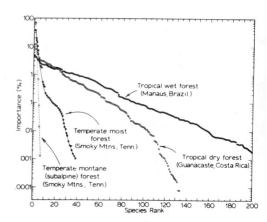

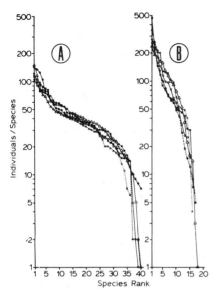

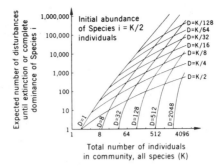

Fig. 9 (left). Randomly generated patterns of relative species abundance in a 40-species community closed to immigration; $K = 1600$, $D = 400$. (A) Dominance-diversity curves after 25 disturbances, showing an approximately lognormal pattern. (B) Dominance-diversity curves after 250 disturbances, showing an approximately geometric pattern. Five independent simulations are plotted to show the high predictability of the patterns for given values of K and D. Fig. 10 (right). Expected number of disturbances of size D in a community of K individuals until a species starting with $K/2$ individuals either goes extinct or becomes completely dominant, in a community closed to immigration. The actual expectations are points for discrete (integral) values of D and K. These points have been connected for fixed D, or D as a fixed fraction of K, for graphical clarity.

whereas after 250 disturbances with no immigration, the species have formed the set of approximately geometric curves. These transient distributions are not perfectly lognormal or geometric, but exhibit a "tailoff" phenomenon in the abundances of rare species. Rare species are somewhat less common than would be expected from the symmetric lognormal distribution, resulting from the continual attrition of rare species through local extinction. The large data set for the dry forest reveals this same rare-species tailoff in the dominance-diversity curve (Fig. 8).

If total community size is expanded (K larger), the number of disturbances to extinction of any given species gets rapidly larger for any fixed disturbance size, D (Fig. 10) (68). If D is small and K is moderate-sized to large, time to extinction or complete dominance can be very long. For example, if eight trees are killed with each disturbance out of a community total of 512, it will on average take about 90,000 such disturbances to remove or "fix" a species which begins with 256 individuals (69).

When immigration of new or old species is allowed, however, eventual complete dominance by one species is precluded; and a stochastic equilibrium is established between species immigration and local species extinction. The resulting relative abundance patterns can be either more nearly lognormal or more nearly geometric, depending on the relative importance of immigration in replenishing species diversity reduced by dis-

turbance. If local extinctions outpace immigrations, local relative abundance patterns become increasingly geometric with time. As the rarer species drop out, the remaining species fill the gaps and grow more abundant, and thereby become less prone to extinction per unit time for a given level of disturbance. This causes the local extinction rate to drop into balance with the immigration rate. In contrast, if species immigrations outpace local extinctions, species must accumulate in the community, and local relative abundance patterns become more lognormal with time. Since more and more species are being packed into a finite space of size K, however, some species inevitably random walk to such rarity that they have a high risk of local extinction per time as compared with common species (70). A point is reached when sufficient numbers of species are rare and locally going extinct per unit time that the immigration rate is balanced, and the number of species in the community stops increasing.

Observed latitudinal changes in dominance-diversity patterns of local forest communities are predicted by the model provided that there are (i) fewer species in the source pool of potential immigrants, and (ii) more frequent or severe disturbances (or both) in boreal and temperate forests than in tropical forests. Whether these conditions are met is not yet clear. Persistently lower species richness in boreal and temperate forests may be due to elevated rates of extinction during the Pleistocene (71). In

historical times boreal and temperate forests have been disturbed frequently by fires, plagues, and strong winds (72). On somewhat longer time scales, paleoecological evidence reveals that interglacial periods have been too brief for equilibria to be achieved in temperate deciduous forests before glacial periods set in again (73).

Disturbance in equatorial forests tends to be a localized, rather than a large-scale phenomenon, with blowdowns generally involving only a few trees at a time (74). Indeed, the observed clumped dispersion pattern of dry-forest tree species is just what one would expect if the forest is essentially a palimpsest of small, regenerating light gaps of different ages. However, stability of tropical climates over Pleistocene times is not certain, in view of mounting evidence of long dry periods during full-glacial and mid-Holocene times in equatorial regions (75).

Obviously this model is an oversimplified representation of the dynamics of natural communities (76), but it does provide a number of important lessons. First, we may expect to observe substantial differentiation of the relative abundance of species in natural communities as a result of purely random-walk processes—a kind of "community drift" phenomenon. Second, we cannot necessarily conclude that, just because a species is of rank-1 importance in a community, its current success is due to competitive dominance or "niche pre-emption" (56), stemming from some superior adaptation to the local environment. Because such a simple model generates the basic patterns of relative species abundance in natural plant communities, it would perhaps be preferable to use departures from the lognormal or geometric distributions as evidence for competitive dominance. Finally, whether or not forest communities are at equilibrium, our understanding of community organization would profit from more study of processes of disturbance, immigration, and local extinction, in conjunction with the more traditional studies of the biotic interactions of species (such as competition, niche differentiation, and seed predation).

References and Notes

1. P. S. Ashton, *Biol. J. Linn. Soc.* **1**, 155 (1969).
2. A companion paper (S. P. Hubbell, J. E. Klahn, G. Stevens, R. Ferguson, in preparation) describes the forest site in detail, and discusses the patterns for each tree species.
3. A. R. Wallace, *Tropical Nature and Other Essays* (Macmillan, London, 1878), p. 65.
4. G. A. Black, Th. Dobzhansky, C. Pavan, *Bot. Gaz.* **111**, 413 (1950); J. M. Pires, Th. Dobzhansky, G. A. Black, *ibid.* **114**, 467 (1953).
5. P. A. Colinvaux, *Introduction to Ecology* (Wiley, New York, 1973), p. 477ff.; J. M. Emlen, *Ecology: An Evolutionary Approach* (Addison-Wesley, New York, 1973), p. 400; C. J. Krebs, *Ecology: The Experimental Analysis of Distribution and Abundance* (Harper & Row, New York, 1972), p. 520; R. H. MacArthur and J. H. Connell, *The Biology of Populations* (Wiley, New York, 1967), p. 37; R. H. MacArthur, *Geographical Ecology* (Harper & Row, New York, 1972), p. 191; E. R. Pianka, *Evolutionary Ecology* (Harper & Row, New York, ed. 2, 1978), p. 296; P. Price, *Insect Ecology* (Wiley Interscience, New York, 1975), p. 49; J. L. Richardson, *Dimensions of Ecology* (Williams & Wilkins, Baltimore, 1977), p. 219; R. E. Ricklefs, *Ecology* (Chiron, Portland, 1973), p. 721. Most texts present several of the competing theories to explain the low density of tropical tree species.
6. D. H. Janzen, *Am. Nat.* **104**, 501 (1970).
7. J. H. Connell, in *Dynamics of Populations*, P. J. den Boer and G. R. Gradwell, Eds. (PUDOC, Wageningen, Netherlands, 1970), pp. 298–312.
8. J. H. Connell, J. G. Tracy, O. O. Webb, unpublished result cited in (7).
9. D. H. Janzen, in *Taxonomy and Ecology*, V. H. Heywood, Ed. (Systematics Association, special volume No. 5), (Academic Press, New York, 1973), chap. 10; *Annu. Rev. Ecol. Syst.* **2**, 465 (1971).
10. J. H. Connell, *Science* **199**, 1302 (1978).
11. H. G. Baker, *Cold Spring Harbor Symp. Quant. Biol.* **24**, 177 (1959); A. A. Fedorov, *J. Ecol.* **54**, 1 (1966); A. Kaur, C. O. Ha, K. Jong, V. E. Sands, H. T. Chan, E. Soepadmo, P. S. Ashton, *Nature (London)* **271**, 440 (1978).
12. D. H. Janzen, *Science* **171**, 203 (1971); G. W. Frankie, in *Tropical Trees: Variation, Breeding, and Conservation*, J. Burley and B. T. Styles, Eds. (Linnean Society Symposium Series No. 2, 1976), pp. 151–159.
13. K. S. Bawa, *Evolution* **28**, 85 (1974).
14. ___ and P. A. Opler, *ibid.* **29**, 167 (1975).
15. I do not mean to suggest that clumping is any sort of evolved adaptation to rarity.
16. For example, if a rare species occurs only as widely scattered, very large adults, we may suspect that the population is a relict of an earlier successional episode when the species was once more abundant. Alternatively, if a rare species is locally abundant with a high proportion of juveniles, it may be self-replacing and may be rare only because its microhabitat is local and rare. This sort of evidence is only circumstantial, at best, since it is always possible to construct a life table for any observed age distribution consistent with population growth, constancy, or decline. However, the reasonableness of some life tables necessary to balance some age distributions may be questionable. For example, to balance a population consisting only of extremely old individuals requires (i) highly episodic reproduction so that there is very little overlap of generations, (ii) extremely delayed age at maturity, (iii) high survival from seed to adult, or else extremely high adult fecundities, and (iv) synchronous fruiting, at long intervals, corresponding to (i). [Also see (53).]
17. K. S. Bawa and P. A. Opler (14), and M. N. Melampy and H. F. Howe [*Evolution* **32**, 867 (1978)] have reported that, although sex ratios may be skewed from 1:1, the staminate and pistillate plants are distributed at random with respect to one another in the species they studied. Determining what "sex ratio" means in plants can be difficult at times, particularly if there is temporal variation in the number of staminate and pistillate flowers offered by individual plants. See also R. W. Cruden and S. W. Hermann-Parker [*ibid.* **31**, 863 (1977)] and K. S. Bawa (*ibid.*, p. 52).
18. G. W. Frankie, H. G. Baker, P. A. Opler, *J. Ecol.* **62**, 881 (1974).
19. Information on the effects of root and crown competition and thinning on the growth of plantation stands is available in references on tropical trees [T. C. Whitmore, *Tropical Rainforests of the Far East* (Clarendon, Oxford, 1975)].
20. Potentially positive effects of density on reproductive success are also possible, within certain density ranges. For example, if pollinators are limiting seed set, a group of adult trees might exhibit greater per capita seed set than isolated adults if per capita rates of pollinator visitation were greater in the group.
21. The mathematical model is similar in aspects to verbal models developed by J. H. Connell (10) and J. Terborgh [*Am. Nat.* **107**, 956 (1973)].
22. The site is a nearly level plain at 100 m elevation, 10°32′N, 85°18′W. Soils are uniform, pale orange-brown silty clays derived from low bluffs of rhyolitic tuff, and exposed basaltic flows 200 to 600 m north of the site. Soils are 1.4 to 2.2 m deep, underlain by basaltic flows. Because the forest is seasonally dry for 6 months, a shallow layer of leaf litter (A-0) and humus (A-1) persists all year. A small seasonal creek flows through the site in a channel cut to the basaltic bedrock.
23. E. R. Heithaus, *Ann. Mo. Bot. Gard.* **61**, 675 (1974); L. K. Johnson and S. P. Hubbell, *Ecology* **56**, 1398 (1975); S. P. Hubbell and L. K. Johnson, *ibid.* **58**, 949 (1977); *ibid*, in press.
24. D. H. Janzen, *Evolution* **21**, 620 (1967); R. Daubenmire, *J. Ecol.* **60**, 147 (1972).
25. L. R. Holdridge, *Life Zone Ecology* (Tropical Science Center, San José, Costa Rica, 1967); J. O. Sawyer and A. A. Lindsey, *Vegetation of the Life Zones of Costa Rica* (Indiana Academy of Sciences, Indianapolis, 1971); C. E. Schnell, Ed., *O. T. S. Handbook* (Organization for Tropical Studies, San José, Costa Rica, 1971).
26. This total represents approximately a third of all woody species reaching a size of 2 cm dbh or larger in the Dry Forest Life Zone of Costa Rica.
27. Cattle enter the forest on occasion and cause some damage to the shrub understory by trampling and browsing. This disturbance is discussed in relation to the dispersion patterns found (2); cattle disturbance cannot have produced the clumped dispersion patterns of adult trees, nor the concentrations of juveniles around adults found in nearly all species (2).
28. Potential mammalian seed dispersers common in the forest include deer, howler monkeys, variegated squirrels, and agoutis. Visits by white-faced monkeys, spider monkeys, tapirs, peccaries, and pacas have also been recorded. Other mammals include armadillos, coatimundis, and kinkajous. Frugivorous birds are also common.
29. It is difficult to find intact seeds of species such as *Hymenaea courbaril*, *Enterolobium cyclocarpum*, *Cassia* spp., and many others that have not been attacked by weevils.
30. Many of the unknown plants were identified in the field by P. A. Opler and confirmed by voucher specimens sent to W. Burger (Field Museum, Chicago) or R. Leisner (Missouri Botanical Gardens, St. Louis). A few specimens could not be identified to species because they lacked reproductive structures.
31. Individual maps for each quadrat were drawn on graph paper (scale 2 m/cm) in the field. Plants within 5 m of the quadrat perimeter were mapped first; plants inside the inner square (10 by 10 m) were added afterward, the perimeter plants being the reference points. Independent checks of the accuracy of mapping showed that two people could locate most plants to within 1 m of each other, and of the plant's true position. Mapping took two people 3 months. Gross forest structure, by dbh class, is as follows: 2.0 to 4.9 cm, 9788 stems; 5.0 to 9.9 cm, 2606 stems; 10.0 to 14.9 cm, 1359 stems; 15.0 to 19.9 cm, 650 stems; 20.0 to 29.9 cm, 768 stems; 30.0 to 39.9 cm, 411 stems; 40.0 to 49.9 cm, 117 stems; 50.0 to 99.9 cm, 163 stems; > 100 cm, 27 stems.
32. It was not possible to check the reproductive capacity of every tree or species. Information was obtained on 61 tree species. The somewhat arbitrary rule of setting the lower limit for adult size at the smallest-sized individual of the species in flower or fruit was used. In general, this rule means that we probably have virtually all of the adults in the "adult" class. By the same token, however, some nonreproductive subadults are probably included with the adults. For purposes of this study, it was better to risk overestimating the number of adults since, in general, there were many more juveniles than adults.
33. Maps were drawn to a scale of 1:450. Locations of individual plants were marked by a letter. The letter A represented plants < 2.5 cm dbh; B represented plants between 2.5 and 5.0 cm dbh; and thereafter, letters represented 5 cm dbh increments.
34. M. Morisita, *Mem. Fac. Sci. Kyushi Univ. Ser. E* **2**, 215 (1959). Morisita's dispersion index is a ratio of the observed probability of drawing two individuals randomly without replacement from the same quadrat (over q quadrats), to the expected probability of the same event for individuals randomly dispersed over the quadrats. The index is unity when individuals are randomly dispersed, regardless of quadrat size or the mean density of individuals per quadrat. Values greater than unity indicate clumping, and values between 0 and 1 indicate uniformity. An F statistic can be computed to test for significant departure of the index (symbolized by I_δ) from unity (randomness).
35. Circular, clear plastic overlays with concentric rings drawn to scale every 5 m were made by Thermofax copier. These overlays were centered in turn over each adult tree on the map, and the numbers of adults and juveniles were counted in each successive 5-m annulus, out to 100 m. Density was computed by dividing the number of counts by the area of the annulus in question.
36. Small-seeded species with wind, bird, or bat dispersal should exhibit greater dispersal distance than large-seeded species dispersed by ground mammals. The greater dispersal distance expected in small-seeded species should tend (i) to obscure the local difference in seed production by both crowded and scattered adults and (ii) to increase the apparent per capita reproduction of scattered versus crowded adults. Consider, for example, the limiting case in which regional averaging of juvenile production is so complete that all quadrats have an equal density of juveniles. Then quadrats that have a small number of adults will show a high ratio of juveniles to adults, whereas quadrats that have many adults will show a low ratio, regardless of any density dependence.
37. The expected nearest neighbor distance, d, is given by $d = 1/(2\sqrt{\rho})$, where ρ is the mean density in number of adults per square meter [P. Clark and F. C. Evans, *Ecology* **35**, 445 (1954)].
38. Of the 21 remaining species of trees, shrubs, or vines represented by one individual, eight species could not be identified from our vegetative samples.
39. *Cochlospermum vitifolium* Spreng. (Cochlospermaceae) is a deciduous species found in large light gaps. It grows rapidly and has wood of low density. Its large dry-season flowers (see cover) are pollinated by big anthophorid bees, and its seeds are wind-dispersed. *Hymenaea courbaril* L. (Leguminosae) is a slow-growing, evergreen species commonly found in riparian areas, and it has dense wood. It has fairly large flowers which are bat-pollinated. Its seeds are large and encased in a thick, woody pod, are attacked on the tree before dispersal by bruchid weevils, and are food for a number of mammals [D. H. Janzen, *Science* **189**, 145 (1975)]. *Licania arborea* Seem. (Rosaceae) is a slow-growing, evergreen, high-canopy species of the mature dry forest, with very dense wood. It has small, bee-pollinated flowers, and its seeds are dispersed by bats, birds, and monkeys. *Tabebuia rosea* DC (Bignoniaceae) is a tree of intermediate stature characteristic of forest edge and large light gaps. It is relatively fast growing, with wood of intermediate density. It is deciduous and produces a showy display of large flowers in the dry season. Pollination is by large anthophorid bees. Its pods are straplike, containing numerous small, flat, winged, wind-dispersed seeds. *Thouinidium decandrum* Radlk. (Sapindaceae) is a medium-sized, deciduous tree of mature forest, with a moderate growth rate and medium to dense wood; its seeds are wind-dispersed.

40. P. S. Ashton (*1*) and M. E. D. Poore [*J. Ecol.* **56**, 143 (1968)] also report clumping in rain-forest tree species in Malaysian dipterocarp forests. Their large rain forest maps cannot be used very effectively to test the Janzen-Connell hypothesis since only trees with a circumference ≥ 30 cm were mapped, and they did not distinguish between adult and juvenile trees.

41. Because the demographic neighborhoods of adjacent adults are not completely independent (some juveniles are counted in the neighborhoods of more than one adult), compensation for this partial nonindependence should be made by a downward adjustment of the degrees of freedom of the regression analysis of variance. Accordingly, the average redundancy of juvenile counts was determined by the ratio of apparent number of juveniles in the neighborhoods of all adults to the actual number of juveniles in the union of all adult neighborhoods. The number of degrees of freedom was reduced by dividing the number of adults by the mean juvenile redundancy. This procedure results in a conservative estimate of the true degrees of freedom (R. Lenth, Department of Statistics, Univ. of Iowa, personal communication), and corresponds to the case of completely dependent demographic neighborhoods. For example, suppose that there are only three coincident adults in the population, which consequently have completely identical demographic neighborhoods in which every juvenile is counted three times. Therefore, mean juvenile redundancy is 3, and the number of independent data sets is found by dividing the number of adults by 3, giving one degree of freedom. A similar procedure was followed in testing the slopes of the adult density curves, with mean adult redundancy being used to adjust the degrees of freedom. However, even without the downward adjustment of degrees of freedom, the significance tests are conservative since individual trees counting in the neighborhoods of more than one adult should always bias the slopes in the positive, not in the negative, direction. The slopes found were all negative or zero.

42. Fifteen meters away from the parent tree might be enough to permit the seeds to escape from discovery by seed predators. However, D. E. Wilson and D. H. Janzen [*Ecology* **53**, 954 (1972)] found that in *Scheelea* palm there was no reduction in the percentage of seeds attacked under the palm and at a distance of 8 m, when seed density was held constant.

43. This analysis, of course, is analogous to constructing a vertical life table, which assumes that the population size has been approximately stationary for some time. While the conclusions are not necessarily invalid if the population is increasing, their validity cannot be confirmed.

44. In the 30 most common tree species, immediately adjacent adults with touching crowns are not infrequent.

45. Th. Dobzhansky and S. Wright, *Genetics* **28**, 304 (1943).

46. J. A. Endler, *Geographic Variation, Speciation, and Clines* (Princeton Univ. Press, Princeton, N.J., 1977).

47. N. G. Hairston, *Ecology* **40**, 404 (1959).

48. The self-compatible species may do little if any selfing (R. Cruden, personal communication). K. S. Bawa (personal communication) has shown that most of the hermaphroditic species in the dry forest he has studied are self-incompatible. Therefore, in Fig. 6, hermaphroditic species for which self-compatibility data are lacking are pooled with the obligately outcrossed hermaphrodites.

49. I chose the midpoint of the dbh range as the pivotal size in order to be conservative in my estimate of the number of species showing positive skewness. This means that, when positive skewing is detected, it is actually very pronounced.

50. Similar patterns of skewing have been reported in rare species by Connell (*10*) and by D. H.

Knight [in *Tropical Ecological Systems* (Ecology Series No. 11), F. B. Golley and E. Medina, Eds. (Springer-Verlag, New York, 1975), pp. 53–59]. Knight studied late second-growth stands of semi-evergreen forest on Barro Colorado Island, Panama.

51. In 7 years of work at the study site, I have seen no juveniles produced by these species in spite of repeated flowering (seed set was not observed).

52. Of the 59 species in the forest studied, which occur with an average density (ignoring dispersion) of greater than one individual per hectare, I know of at least 42 species that occur locally in much higher densities elsewhere in Guanacaste; and I expect the same is true for most of the remaining species.

53. The most parsimonious explanation for the data on rare species in the dry-forest stand described here is general reproductive failure. However, undoubtedly there are some rare species in the forest that are replacing themselves, and there are probably some species that are everywhere rare. G. S. Hartshorn [in *Tropical Trees as Living Systems*, P. B. Tomlinson and M. H. Zimmerman, Eds. (Cambridge Univ. Press, New York, 1978)] has suggested that "nonregenerating" rare species may simply require particular types of light gaps in order to regenerate, and that rarity is due to the infrequency of creation of such gaps.

54. Of 21 wind- or bird-dispersed species, 14 showed density dependence, but of nine mammal-dispersed species, only three showed density dependence.

55. Diffuse competition is a term coined by R. H. MacArthur [*Geographical Ecology* (Harper & Row, New York, 1972)]. It refers to the sum of competition from all interspecific competitors in the community acting on a given species. I have not yet analyzed the association of species to check this possibility.

56. R. H. Whittaker, *Science* **147**, 250 (1965).

57. Data of H. Klinge, as was reported by E. F. Brunig [*Amazoniana* **4**, 293 (1973)].

58. The curves for the Smoky Mountains are derived from analysis of single, 0.1-hectare stands in the cove and spruce-fir forests, respectively. Also, they represent all vascular plants, not just woody plants; and the importance values are based on annual net primary production. The respective dry-forest and rain-forest curves, however, are based on larger quadrats (13.44 and 1.0 hectares), but only woody plants; and the respective importance values are based on basal area and above-ground biomass. The quantitative effects of these differing methods are difficult to assess, but fortunately the effects are partially canceling (larger plots mean more species, but eliminating nonwoody species means fewer species). If there is a greater percentage of nonwoody plants in temperate forests, the difference between the species richness of temperate and tropical forests may be somewhat underestimated in Fig. 8.

59. M. P. Johnson and P. H. Raven, *Evol. Biol.* **4**, 127 (1970); S. J. McNaughton and L. L. Wolf, *Science* **167**, 131 (1970).

60. R. M. May, in *Theoretical Ecology: Principles and Applications*, R. M. May, Ed. (Saunders, Philadelphia, 1976).

61. H. Caswell, *Ecol. Monogr.* **46**, 327 (1977).

62. R. MacArthur, *Am. Nat.* **94**, 25 (1960); G. Sugihara (unpublished result) has also developed a "sequential breakage" model that generates lognormal patterns.

63. Losses are governed by the hypergeometric distribution. Thus, for the *i*th species the probability of losing *j* individuals, *j* ≤ *D* ≤ N_{it}, where N_{it} is the population of species *i* at the current time *t*, is given by

$$\binom{N_{it}}{j}\binom{K - N_{it}}{D - j}\Big/\binom{K}{D}$$

64. A disturbance can be as small as the death of a single tree.

65. Recruitment to fill the *D* disturbance vacancies is governed by the binomial distribution. Let N'_{it} be the number of the *i*th species at time *t* after disturbance. For the *i*th species, the probability of contributing *m* replacement individuals, *m* ≤ *D*, is:

$$\binom{D}{m}\left(\frac{N'_{it}}{K - D}\right)^m\left(\frac{K - N'_{it} - D}{K - D}\right)^{D - m}$$

66. It is easy to prove that rare species are more likely to go extinct per unit time than are common species: If $N_{it} > D$, there is no chance that species *i* will go extinct in the next disturbance; but if $N_{it} \le D$, there is a nonzero chance of extinction in the next disturbance. Therefore, the mean time to extinction of a species *j* with $N_{jt} > D$ must exceed that of a species *i* with $N_{it} \le D$ by at least the mean time it takes for N_j to decrease to *D*. If a "rare" species is one for which $N \le D$, then increasing *D* will increase the number of rare species going extinct per disturbance over successive disturbances.

67. In this example, I chose a large *D* simply to speed up the process of random-walk extinction. These transient distributions obey the "canonical hypothesis" of F. W. Preston [*Ecology* **43**, 185 and 410 (1962)]. It has been shown (*21*) that the canonical lognormal is the result of imposing a fixed ceiling, *K*, on the total number of individuals of all species in the community. This result has also been discovered independently by G. Sugihara (personal communication). The Monte Carlo simulations were performed at the University of Iowa Computer Center on an IBM 360/70.

68. Because the species can random walk up or down in abundance, either outcome (extinction or complete dominance) is possible. I chose to illustrate a species with *K*/2 because it represents the abundance at which a species is equally likely to go to either outcome. It is also the abundance with the longest mean transient time, both outcomes considered.

69. The model is a Markovian random walk of the abundance of the *i*th species between 0 and *K*. Mean transient times from a starting abundance of *K*/2 individuals were found from the fundamental matrix determined for particular values of *D* and *K*.

70. There is an added risk of extinction for rare species if they are more clumped than common species, such that a single disturbance might kill all individuals in a given local area.

71. The extinction of many temperate tree species is well documented in Europe (*73*).

72. For example, see M. L. Heinselman, *Quat. Res.* **3**, 329 (1973); J. D. Henry and J. M. A. Swan, *Ecology* **55**, 772 (1974).

73. M. B. Davis, *Geosci. Man.* **13**, 13 (1976).

74. R. Foster, personal communication; S. P. Hubbell, personal observations.

75. B. S. Vuilleumier, *Science* **173**, 771 (1971); J. E. Damuth and R. W. Fairbridge, *Bull. Geol. Soc. Am.* **81**, 189 (1970); J. Haffer, *Science* **165**, 131 (1969); B. S. Simpson and J. Haffer, *Annu. Rev. Ecol. Syst.* **9**, 497 (1978).

76. The model in its simplest form as presented here corresponds to the "equal chance hypothesis" discussed by Connell (*10*), provided that per capita chances of reproduction or death are made the same for all species. For greater realism, species differences in dispersal, fecundity, competitive ability, and resistance to environmental stresses need to be treated.

77. Help from the following people was vital to the completion of this study; Jeffrey Klahn, George Stevens, Paul Opler, Richard Ferguson, Ronald Leisner, William Burger, Daniel Janzen, Joseph Connell, Leslie Johnson, Robin Foster, and Douglas Futuyma. Others have also made contributions which are appreciated. The study was supported by the National Science Foundation.

Editor's Comments
on Papers 8 Through 11

HOW TO DESIGN NATURE RESERVES FOR
THE TROPICS

Another line of research that has its origins in the question of tropical species richness is that of co-evolution. Robinson (Paper 8) presents examples of the many co-evolutionary schemes and suggests that the high incidence of mutualism in the tropics is a feature that distinguishes tropical ecology from other types of ecology.

Gilbert (Paper 9) shows how mutualism plays a crucial role in the survival of tropical ecosystems and the implications mutualism has for ecosystem management. Some key organisms in mutualistic relationships are restricted to a particular microhabitat, while some animals required for reproduction and dispersal of plants often depend on several microhabitats. In order to accommodate all the species necessary for the survival of a tropical ecosystem, reserves should be managed to maintain a variety of microhabitats.

Gomez-Pompa et al. (Paper 10) point out that the more complex mutualistic interactions in tropical forests, as well as the short dormant period of rain forest seeds, means that tropical gene pools are more difficult to preserve than gene pools of temperate forests. They state that "a gene pool of primary trees can be maintained along roads, near houses and the like, for temperate areas, but not for the tropical rain forest."

Greater interdependence of all species in tropical ecosystems means that tropical nature reserves must be large in comparison to temperate reserves. For example, a reserve to save 90 percent of the species in a tropical forest must be bigger than a reserve to save 90 percent of the species in a temperate forest.

There is little disagreement among ecologists that in tropical areas a bigger reserve will be more effective than a small one. The question seems to be that if large reserves are not possible, can there be compromises such as several small reserves (Simberloff and Abele, 1976; Terborgh, 1976). Diamond (Paper 11) applies the theory of island biogeography to designing a series of small reserves so as to maximize species preservation when a single large reserve is not feasible. His models indicate, for example, that small reserves close together would be more effective than the same number of reserves widely spaced. Diamond's paper is notable in that it bridges the gap between theory as developed by tropical ecologists and practical applications as needed by tropical land managers.

REFERENCES

Simberloff, D. S., and L. G. Abele, 1976, Island Biogeography Theory and Conservation Practice, *Science* **191**:285–286.

Terborgh, J., 1976, Island Biogeography and Conservation: Strategy and Limitations, *Science* **193**:1029–1030.

8

IS TROPICAL BIOLOGY REAL?

Michael H. Robinson

Smithsonian Tropical Research Institute, P. O. Box 2072, Balboa, Canal Zone, Panama

Abstract: The term *tropical biology* can be used as a mere descriptive, geographical or administrative convenience. However, it is argued that the biology of whole organisms in rich tropical environments may in fact be *qualitatively* distinct. It is here suggested that the distinction lies in the great and disproportionate complexity of inter-specific interactions between tropical organisms. Such complexity should also affect the ecological, behavioral and structural adaptations that mediate interspecific interactions. Intraspecific interactions, on the other hand, may well have a similar level of complexity irrespective of climatic regime.

It is suggested that the interspecific complexity that is characteristic of tropical biology is not merely a proportional product of increased species richness but rather is the result of an evolutionary process that has been called *biological accommodation*. Studies of interspecific relationships are relatively few compared to the preponderant studies of social (intraspecific) interactions. In addition tropical studies are much fewer than temperate region studies. A number of selected studies on interspecific interactions of tropical organisms are reviewed herein. This review suggests that the potentially unique aspects of tropical biology may best be revealed by further studies of interspecific relationships rather than by concentration on "sociobiology".

Résumé : Le terme *biologie tropicale* peut étre utiliser simplement comme une convenance descriptive, géographique ou administrative. Cependant, c'est argumenté que la biologie des organismes qui se trouvent dans de riches ambiances tropicales peut étre en realité *qualitativement* distincte. Mais dans ce travail on a suggeré que la distinction reste dans la grande complexité disproportion néede lainteraction interspécifique entre les organismes tropicaux. Telle complexité peut aussi etre demontrée dans les adaptations écologiques et structurelles et dans les adaptations du comportement ayant rapport direct aux interactions interspécifique. De l'autre côte, des interactions intraspécifiques peut bien avoir un niveau similaire de complexité indépendant du régime climatique.

On a aussi suggeré dansce travail, que la complexité interspécifique qui est une des characteristiques de la biologie tropicale n'est pas simplement le produit proportionnel de l' accroissement de la richesse des espèces mais plutôt le résultat du procés evolutionniste qui s'appelle "*biological accommodation*". Les études des rélations interspécifiques ontrelativement peuencomparison avec les études prèponderantes sur des interactions sociales (intraspécifique). Il y a moins études sur les zones tropicales que sur les zones moderées. Ici, on a seleccionè certaines études au sujet des interactions interspécifique des organismes tropicaux et on les a examinées. Ce revue suggere que les aspects pratiquement unique de la biologie tropicale-peut se révèler miexu par des études plus profondes des relations interspécifique au lieu de par la concentration sur la "*sociobiology*".

INTRODUCTION

In this article I want to raise the question of whether or not "tropical biology" is a phenomenologically distinct subject. Of course, even if the answer is no it would still be useful to file the results of tropical studies separately. This would be an administrative and bibliographic convenience but would not be part of a natural taxonomy of sciences. I believe that the answer is yes, that tropical organisms from complex habitats are part of a unique system and that this uniqueness will be reflected in the results of organismic studies but *not necessarily* in studies carried out below the level of the whole organism. Ecology, ethology and ecological physiology should reveal the distinctive aspects of tropical biology. My own interests are those of an ethologist working, in the tropics, on the borderline between ethology and ecology. There are many definitions of ecology but a recent one by Ricklefs (1973) seems to me to be both comprehensive and succinct: "ecology now denotes the study of the natural environment, particularly the interrelationships between organisms and their surroundings". The relationships between *animals* and their environment is mediated by behaviour ; the study of behaviour is thus directly related to ecology. Animal behaviour is studied by ethologists, experimental psychologists and even by anthropologists. Ethology is firmly rooted in biology and has a primary concern for the functional and evolutionary aspects of behaviour. Morris (1969) emphasised the naturalistic element of ethology when he said that ethologists "were all asking the same question : not what can we *make* animals do, but what do they do ?" Research into naturally occurring behaviour puts emphasis on field work rather than on laboratory studies. It also leads to an interest in responses to naturally occurring environmental factors. An important element of the environment of any animal is the existence of other organisms of the same and different species. The pre-eminence of the biotic element of the environment is, in my view, the unique feature of tropical biology.

THE UNIQUENESS OF TROPICAL BIOLOGY

There is nothing new in the view that interrelationships between organisms is the overriding feature of complex tropical ecosystems. A number of biologists have been intuitively (and inductively) drawn to this conclusion. Some of these interpretations of tropical complexity are reviewed below. The overview of a number of ecologists concerned with explaining latitudinal gradients in species diversity (reviewed in Robinson *et al.* 1974) has involved overt or implied assumptions about a qualitative difference between the emphasis of selective agents between the tropics and temperate regions. This was perhaps first, and most clearly expressed by Dobzhansky (1950) : "Where physical conditions are easy interrelationships between competing and symbiotic species become the paramount adaptive problem". Later the designation of climatic regions where communities were "physically controlled" as opposed to others where communities were "biologically accommodated" (Sanders 1968, Slobodkin & Sanders 1969) gave new expression to this old view. In equating the tropics with a biologically accommodated community, or an area where "physical conditions are easy", we would be guilty of gross oversimplification if we applied a blanket definition of tropics. Certainly not all the land lying between latitudes $23\frac{1}{2}$ N and $23\frac{1}{2}$ S, or within a certain isotherm, is necessarily tropical in its biology as defined here. For present purposes I would exclude grasslands, deserts, the rocky and sandy littoral regions and confine attention to tropical forests and coral reefs. I suspect that the exclusions should also apply to tropical lacustrine ecosystems and most freshwater habitats (except, perhaps, the giant river systems of Amazon and Congo ?).

Within the tropics as so defined the evidence for a unique degree of interrelational complexity is fragmentary. Whether we define complexity in terms of complexity of networks of relationships with other organisms or by the number of steps in a single straight-line relationship, there is little exact or quantified direct evidence. This is to some extent the product of the comparatively low volume of tropical studies and the restricted nature of inter-region comparisons. The world distribution of biologists unfortunately reflects the world distribution of economic wealth. This means that areas supremely rich in plants and animals have comparatively few endemic (or migratory) biologists. When a specialist is fortunate enough to make broad studies (in the geographic sense) we get tantalising insights.

A case in point is Karr's (1975) study of the ecology of forest birds in the neotropics, Africa and Asia. This revealed substantial differences in species diversity and trophic specialization between the regions. What we do not know is whether trophic specialists not represented in the birds of Africa (say) are missing because the resource is missing or because the niche is filled by another organisms not recognised (or studied) by Karr. This kind of example could be multiplied in a parade of our fundamental ignorance. Nonetheless there seems to be an overwhelming impression among tropical biologists that complexity of biotic relationships is the essence of the system. If this is true, then in terms of behaviour *we should look for complexity in interspecific relationships.* An overwhelming proportion of the relatively few ethological studies carried out in the tropics has been of the social behaviour of animals.

Social behaviour seems to be susceptible to pressures that result in the simplification of signals and their restriction in number (Moynihan 1970a, Smith 1969). There is little reason to assume that the amount of social information transmitted between animals of the same species should reflect environmental complexity. Selection for unambiguous displays that convey information quickly, accurately, and perhaps urgently (and are often accompanied by a high degree of redundancy), in short, selection for communication, has important consequences for ethologists. Where the displays are in sensory modalities that are well-developed in humans they are inevitably more amenable to "translation" by us than are behaviours that are not specifically selected for communication purposes. (Such displays may be intrinsically more fascinating to most scientists than are other elements of behaviour). Given a preponderance of studies of social behaviour, not only in the tropics but elsewhere, studies of interspecific behaviour have been massively overshadowed and largely overlooked.

I propose here to review a number of studies that reveal interspecific complexity in tropical organisms. This review will set the scene for a concluding re-examination of our original hypothesis of tropical uniqueness and a statement of its implications which may be wider than is at first apparent. The examples below are drawn from studies mainly intermediate between ecology and ethology, from the field that could be called eco-ethology.

STUDIES THAT REVEAL TROPICAL INTERSPECIFIC COMPLEXITY

The interspecific relationships that have been most intensely studied are those involved in symbioses and predator/prey interactions. Competitive relationships have, on the whole, been little studied by ethologists.

An outstanding observational/experimental study of relationships between several species of birds, insects and mites by N. G. Smith (1968, pers comm.) revealed a system of bewildering complexity. Over a wide area of tropical America, oropendolas and caciques nest in colonies. They build complex nests that hang from the branches of tall trees- and they are frequently parasitized by "cuckoos" (tropical cowbirds). Study showed that the cuckoos have two basic types of behaviour. They are either stealthy invaders that lay a single mimetic egg in a host nest or bold intruders that lay several non-mimetic eggs in host nests. They are thus "matchers" or "dumpers". The hosts are also polyethic, they either reject the cuckoo's egg shortly after finding it in their nest or leave it to remain there. Hosts are thus classified into "discriminators" and "non-discriminators". There are even more variables than this. Some colonies nest in trees containing the nests of bees or wasps, some do not. When the birds nest in trees with large social insect nests in them, they cluster their own nests around the bee or wasp nests. Colonies associated with vigorous hymenopteran colonies turn out to be discriminators and few cuckoos are raised in their nests. Colonies that are not associated with bees and wasps are frequently very heavily parasitized and raise large numbers of cuckoos year after year. So far it is a complicated story but it becomes more so. Careful study showed that a major cause of host mortality was the botfly *Philornis* whose larva fed upon the tissues of the young oropendolas. Adult flies simply enter the nest and oviposit on the chicks. In colonies with wasps and bees, host chicks were rarely parasitized and flypapers hung in the trees caught few flies. Remains of flies were, however, collected beneath the wasp nests. It thus seems that the air-space around the nesting birds was kept free of botflies by the social insects (as predators in the case of the wasps). Association with social insects was clearly beneficial to the birds. Such insects were probably also powerful deterrents to arboreal vertebrate predators that might otherwise attack the fledgelings. (Myers, 1935, noted many instances of associations between nesting tropical birds and aculeate hymenoptera, including associations between caciques and wasps.)

In the case of the non-discriminators there is a further complication. These are not associated with bees and wasps and are subjected to *Philornis* attacks. Flypapers set in their colonies catch innumerable botflies. Despite this, examination of the nestlings revealed a striking fact. If a cuckoo chick was present the host chick was unlikely to be parasitized by botflies ! Only 8% of the chicks with cuckoo nestmates had botfly infections, whereas 90.1% of chicks in nests without cuckoos (and without wasp protection) were botfly victims. This was a highly suggestive correlation. Investigation of the behaviour of the young cuckoo completed the solution of the puzzle. The young cowbird needed 5 to 7 days less incubation time than the host, it emerged covered in down rather than naked and its eyes were open within 48 hours of hatching, compared to the host chicks' 6-9 days. At five days old the cuckoo was preening itself and its nestmates, and snapped "aggressively", and in a directed manner, at intruding objects. Stomachs of cowbirds of this age contained botfly adults, eggs and larvae. Clearly the young cuckoo was capable of defending itself and (incidentally ?) its nestmate against botfly attack. This is a remarkable form of symbiosis ; non-discriminators "allow" cuckoos to be present and benefit from brood parasitism. In a four-year analysis the probability of a host chick fledgling, under conditions of botfly attack, was about three times greater if its nestmate included a cuckoo !

This outline account of the situation concentrates on only part of the story. The extensive study revealed the developmental background to the division into discriminators and non-discriminators and that mites are important causes of fledgling mortality. The steps in the evolution of the relationships outlined above are, to me, mind-boggling in their probable complexity without considering any of the further complications.

Situations like the oropendola/cowbird/botfly interaction may not be exceptional in tropical regions. There are indications that nesting birds may be involved in other complex defensive symbioses. In Ghana (West Africa) there appears to be a fairly complex association between nesting Heuglin's weavers (*Ploceus heuglini* Neumann and red ants (*Oecophylla longinoda* Latreilli). According to Grimes (1973) the ants run all over the birds' nests and nestlings without harming them but immediately attack the nestlings if they fall to the ground. *Oecophylla* is a nasty ant to disturb and forms the model for a fairly extensive mimicry complex involving mantids, spiders and other arthropods (Edmunds 1974 : 115-120). Other species of tropical birds are known to nest in close proximity to the nests of raptorial birds. The oropendola story may only be exceptional because it is one where the details have been worked out by patient investigation and inspired experimentation.

Studies of relationships of birds, and arthropods, to the doryline Army Ants have revealed a wide spectrum of complex interspecific relationships. These ants are numerous and conspicuous in the Neotropics. Their Old World tropical counterparts in Africa, the Driver Ants, have not been anything like as extensively studied. The studies of the ecology and behaviour of the ants themselves, pioneered by Schnierla (see bibliography in Wilson 1971) show that they are extremely interesting animals. They are nomadic large-scale predators of insects ranging from the north temperate region to the south temperate (of the Americas). The tropical species are the most conspicuous and *the only ones to have specialist followers in the form of birds.* Two species that raid in above-ground swarms have been particularly well studied. These are *Eciton burcheli* and *Labidus predator* (F. Smith). *Eciton burchelli* (Westwood) was studied in depth by Schnierla (1956, 1957 for instance). The foraging behaviour is strikingly impressive as up to 200,000 ants advance on a broad front across the forest floor. The advancing ants march in a swarm that is frequently at least six metres wide. The ants pass over and under fallen leaves, probe through debris and climb up trees as they move onwards. Insects are either captured or make frenetic escapes. Waiting for the escapers that are flushed out of their hiding places are the other hunters. Tachinid and conopid flies follow the ant swarms and parasitize the cockroaches and gryllids flushed out by the army (Rettenmeyer 1961). A whole assemblage of birds species also uses the ants as beaters. Willis has made intensive and extensive studies of neotropical antbirds, perhaps the most detailed of which is his study of the bicolored antbird (Willis 1967).

As many as fifteen bicolored antbirds (*Gymnopithys bicolor* Lawrence) may accompany a given swarm of ants. They quickly set up a dominance hierarchy around such swarms. As they forage the birds move with the ants perching less than a metre above the swarm. The birds often perch on vertical saplings in the understorey and from these they sally down to feed. The bird usually snaps up a flushed insect and quickly returns to a perch, very occasionally they may fly up to catch prey that are on the wing. They do this very clumsily. Above the bicolored antbirds on higher vegetation other antbirds wait for larger or strongly flying insects. Willis (1967) records that birds actively foraging at good sites over an ant column darted after prey every 42.6 seconds ! In time individual birds become sated and start ignoring prey of suboptimal size, in time concentrating more and more on maintenance activities such as preening. Eventually the bird leaves the immediate vicinity of the ant swarm and simply "loafs" at a distance of 10-15 metres from the column. The birds seem to have an extraordinarily easy life with regards to food-finding. Willis (1967) remarks "when a bicolored antbird follows a swarm of ants, it easily captures one insect after another." Finding the ant swarms, in the first place, may present some difficulties. It has been suggested that the birds find the swarms by attending to the calls of their conspecifics. Certainly, to me, ant swarms always seem to be locatable by the excited birdnoises that accompany them. However there is little direct evidence that the birds locate the ant swarms in this way and Willis failed to attract birds to playback songs in 51 out of 76 experiments (Willis 1967). In any case the *first* birds to arrive at a swarm would have to find it by other means. They could respond to visual stimuli from the raid front itself (the sight of the advancing column and the fleeing insects are conspicuous to field workers). The birds could also encounter trails rather than swarms and would then have to trace these to the raid front. This procedure could involve complex learning. The front is in the direction opposite to that in which the ants carry their prey ; do the birds note this or do they merely search randomly in both directions ?

It is possible to regard the ant-bird phenomenon as another response to tropical informational complexity. Professional ant-followers presumably evolved from ancestors that were visually-hunting wide-ranging foragers. Many of the nonprofessional antbirds still hunt independently by careful searching but also, at times, join ant swarms where they behave just like the professionals. Willis (1972) describes exactly this sort of behaviour in his descriptions of ant-tanagers.

Complex relationships between a wide range of arthropods and the army ants have already been mentioned. Wilson (1972) reviews the field in considerable detail. Research by Akre and Rettenmeyer, in Panama, has revealed a whole range of bizarre examples of *ecitophiles* (associates of the genus *Eciton*). One of their most interesting findings is that a wide range of ecitophiles can follow the trails of their host species. In some cases the guests are more sensitive to the trail pheromones than the hosts themselves (Akre and Rettenmeyer 1968). Beetles, flies, thysanurans and diplopods all showed an ability to follow trails. The highly specialized ecitophilous staphylinids showed a broad spectrum of behavioral adaptations to their association with army ants. They run with the ants when the latter emigrate and run in the centres of emigration columns. They groom army ant workers and live within the bivouacs of the ants (Wilson 1972). The degree of "social mimicry" involved in these associations seems to me to be far in excess of anything implied in Moynihan's (1968) conception. The evolutionary steps to a situation where the symbionts accurately interpret the hosts' trail pheromones and elements of its social behaviour are almost inconceivable.

Another area of extreme interspecific complexity is revealed by Moynihan's (1960, 1962, 1963, 1968, 1970b, 1976) studies of mixed species bird flocks and associations between New World primates, other mammals and birds. Of the birds Moynihan (1962 :10) remarks "Mixed species flocks are found almost everywhere, in almost all environments ; but they seem to be most common and varied and probably attain *the greatest structural complexity in certain regions of the humid tropics*" (my italics). The subject is a vast one and the complexity revealed by Moynihan's extensive studies is paralleled by the results of other studies in other tropical regions (for instance see Winterbottom 1943, 1949 for Rhodesia, Davis 1946 for Brazil). The details of even a single mixed-species flock-type are too complex to do more than summarise here. The mixed blue and green tanager and honeycreeper flocks that Moynihan studied in lowland Panama are not mere fortuitous or unstructured aggregations. They are (Moynihan 1962) "rather complex societies. Each of the more common species of the alliance plays a characteristic social role, more or less distinctly different from that of every other species in the mixed flocks of the alliance. These roles are the results of complex interactions between each species and at least one (usually several) other species. Several of the species tend to react differently to several other species. Most of the more common species have evolved special adaptations, of plumage and/or behaviour, to facilitate the performance of their characteristic roles in mixed flocks." Eighteen species of birds may be regularly involved in these flocks and at least ten more species constitute a pool of casual "joiners".

It is impossible to mention here all the types of interspecific interactions described by Moynihan. They are complex and complicated by the fact that a particular species behaves in a different manner according to which species it is reacting to at a given time. There are *nuclear* species and *attendant* species. Nuclear species are those species whose behaviour stimulates the formulation of mixed species flocks, or contributes to their subsequent cohesiveness, or both. Attendant species contribute little or nothing other than their presence. A further point : there are *active* nuclear species and *passive* nuclear species. Active nuclear species join and follow other species but seldom attract others, passive nuclear species are not themselves "joiners" but stimulate others to join them. Superimposed on this classification is a further one into "regular" and "occasional" species associated with mixed flocks. Thus a species can be passive regular nuclear, active regular nuclear or even active occasional nuclear to cite but three cases. Mixed species flocks of resident birds may be joined by migrants to further complicate the issue.

Complex "societies" of many different species could clearly not have arisen unless there were some strong adaptive advantage to the association. There seems to be little rigorous evidence supporting the two adaptive roles usually suggested for mixed species bird flocks. Moynihan (1962) rejects the suggestion by Rand (1954) that flocks result from gregariousness as a functional explanation (it *is* in fact a causal explanation). The two commonly proposed functional explanations for mixed species flocking, that the flocks are an adaptation to food-finding or to protection, are not mutually exclusive. Many of the first reviews of mixed species flocking (Rand 1954) emphasized that fact that the birds are insectivorous. In this case it is easy to develop an *a priori* argument for a food-finding function. Elsewhere (Robinson, in press) I have stressed the extreme nature of the problems faced by the visually-hunting tropical-forest insectivore. These problems are problems of informational complexity resulting from the extreme species diversity of tropical insect faunas and the rich complexity of their anti-predator adaptations. The informational complexity could be overcome (at least in part) by multiplying the eyes, and brains, involved in foraging.

Moynihan suggests that similar advantages do not accrue to nectivores or frugivores in the particular case of the tanager/honeycreeper flocks (1962). He bases his view on the presumption that such birds are usually familiar with the actual and potential sources of food and nectar within or near their territories or ranges. This may well be true but the possibility that food-finding may be a complex process for frugivores and nectivores, involving the necessity for complex learning, should not be overlooked. (I have suggested (Robinson, in press) elsewhere that the adaptive strategies of plants should be to conserve energy while assuring maximum dispersal of fruits or pollen. Such strategies have almost certainly led to the equivalent of Batesian mimicry in flowers and fruits. In this case the mimics would be energetically worthless but would have rewarding models.)

Protection in numbers may well be a crucial function of mixed flocks. Multiplying the number of predator detectors could well be as important as the deterrent effect of a larger counterattacking force than a single species could produce. In summary, mixed-species flocking probably serves different functions from flock to flock and even in the same flock at different times and places. In addition, different species within each flock probably obtain different benefits at different times. Tropical complexity ?

It is noteworthy that the phenomenon of mixed-species flocking led Moynihan (1968) to propose the ethologically novel and stimulating theory of "social mimicry". This hypothesis, to my view, could only have been conceived by a tropical ethologist. Social mimicry includes all the resemblances, behavioral and morphological, that are special adaptations to facilitate social relations *between species*. The evolutionary implications of the possibility that social mimicry is a widespread phenomenon are vast in scope. They are exciting, but beyond the theme of this paper.

Studies of anti-predator adaptations and the specializations of predators reveal two sides of a major set of interspecific reactions. They have been massive reviews of anti-predator adaptations that draw largely on tropical examples (Cott 1942, Edmunds 1974, Wickler 1968). Reviews of predatory behaviour also (considering the world distribution of ethologists) draw disproportionately on tropical studies (for instance, Curio 1977). My own studies have been concerned with both parties to the predator/prey interaction and I am convinced that there is a clear-cut case for the overwhelming complexity of the tropics in this field of study. Complexity of behavioural and morphological adaptations, that often function synergistically, can be fully illustrated from studies of cryptic and plant-part mimicking orthopteroid insects. (This is merely selecting one small area out of a large literature.) Detailed treatments of this subject are to be found in Robinson (1969a, 1969b, 1970a, 1973) and Edmunds (1972, 1974, 1976). Here I simply present a condensed review of some significant findings.

There are at least three basic strategies of *primary defence* available to animals (Robinson 1970b). Two of these are involved in the cases I will describe here. One strategy is to suppress (as far as possible) all signals that could reveal the presence of the potential prey to the predator. Information suppression of this kind is achieved in an enormous variety of ways but the basic policy is that of camouflage or crypsis (perhaps better called *eucrypsis*, see Robinson 1969a). The potential prey animal may alternatively adopt the strategy of signalling false information to the predator, a technique of "suggesting" that it is either inedible or obnoxious. This strategy is that of Batesian mimics and plant-part mimics. The Batesian mimics resemble obnoxious or distasteful models ; the plant-part mimics resemble sticks or leaves that are normally simple inedible to insectivorous predators. Batesian mimics are active by day but cryptic and plant-mimicking insects are nocturnal. By day such insects assume resting postures that either enhance their concealment (cryptic postures) or enhance their resemblance to the object mimicked (mimetic postures).

Cryptic postures, in combination with camouflaged form, function to suppress information about outline, profile and structure that might otherwise permit the predator to detect the presence of the insect. (Clearly such devices are never perfect, the predator/prey interaction is a dynamic struggle.) Figure 1 illustrates some cryptic postures of phasmids and tittigoniids, they show remarkable convergences that are paralleled in other insect orders, in several families of spiders and even in some vertebrates. To mention the details of one of these syndromes of convergence the *Prisopus berosus* Westwood posture can be used as a starting point (Fig. 1a). The insect is dorso-ventrally flattened, has a concave undersurface, has flattened limbs and the edges of the body and limbs are hair-fringed. In addition the animal has pigmented dorsal surfaces and unpigmented ventral surfaces (in the case of the appendages these are pigmented on their outer surfaces and unpigmented within). At rest the animal assumes a posture in which its ventral surfaces are tightly appressed to the substrate and the flattened legs assume the postures shown in Fig. 1a. This posture conceals the discrete nature of the first legs and elongates the animal's apparent profile. The legs enclose the protracted antennae and conceal them. The third legs are extended posteriorly and opposed to the sides of the abdomen, where they merge into the body,

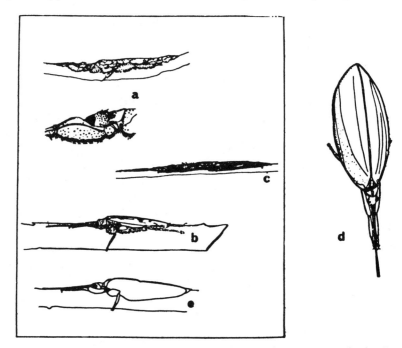

Fig. 1. In rectangle : a. *Prisopus berosus*, full animal in cryptic posture and detail of head. b. *Acanthodis curvidens*, tettigoniid in convergent posture. c. the phasmid *Acanthometriotes crassus* in a cryptic posture. All after Robinson 1969a, 1969 b. e. The tettigoniid *Sathrophyllia femorata* karry after Chopard 1938, compare with b. Outside the rectangle : d. the tettigoniid *Acauloplacella immunis* in a flattened cryptic posture, from Robinson 1973. See text for detailed explanations.

all margins are obscured by hair fringes.   Only the second legs are extended laterally and they are concealed by their coloration, flattening, hair fringes and posture. Almost identical structural and postural devices are found in the tettigoniid *Acanthodis curvidens* (Stål, and the phasmid *Acanthometriotes crassus* Hebard (Fig. 1b, c).   Chopard (1938) describes a tettigoniid from Java which is strikingly similar in cryptic posture to *Acanthodis* and *Prisopus*.   Flattening and profile concealment are achieved in a totally different way by the tettigoniid *Acauloplacella immunis* Brunner in New Guinea (Robinson 1973).   Here (Fig. 1d) the insect has an extremely specialized posture for the tegmina which contrasts very markedly with the normal attitude of these parts during feeding and locomotion.   Essentially this insect has gone a stage further in profile concealment than *Scorpiorinus fragilis* (Hebard) a tettigoniid that also rests on leaf surfaces in a prostrate attitude (Robinson 1969, Fig. 3).   Other examples of prostrate cryptic postures are illustrated in the references cited above (Edmunds 1972, 1974, 1976), has some nice examples among Ghanaian preying mantids).   The trend can clearly lead in two directions, towards flattening or elongation.   I have argued elsewhere (1969a) that elongation could be a preadaptation to stick-mimicry and flattening a pre-adaptation to leaf-mimicry.   This hypothesis provides an adaptive explanation for the survival value of intermediate stages in the evolution of specialized plant-part mimicry.   Some of these can be seen in Central American phasmids (Fig. 2).   Note that concealment of typical insect structures (paired legs, antennae, heads, segmentation, etc.) is necessary for signal suppression (eucrypsis) and then can become part of the false-signals used by plant-part mimics.   Thus legs that are protracted ahead of the body and enclose (and conceal) the antennae apparently enhance the size of the elongate insect and incidentally render it more stick-like (Fig. 1, 2).   The height of specialization in mimetic postures occurs in some phasmatids that have leg concealment postures that produce a stick with branches effect.   These occur in *Pterinoxylus spinulosus* Redtenbacher (Robinson 1968) shown in Figure 2.   West African mantids of the genus *Stenovates* also produce "branched" mimetic postures (Robinson 1966, Edmunds 1972, 1974).

The complexity of cryptic and mimetic postures from a small sample of tropical insects is a reflection of the informational complification that is part of the predator/prey interaction.   In secondary defences the picture is equally plain.   Tropical insects as a totality have in general more lines of defence, more specialization within each line of defence and more defensive diversity than their temperate region counterparts as a totality.   Kettlewell (1959), in a survey of the defensive adaptations of Brazilian insects, makes this point very well.

A further example of complex interspecific interactions has recently been elucidated by Waage and Montgomery (1976).   This concerns the relationships between sloths of two species and a pyralid moth *Cryptoses choloepi* Dyar.   It has long been known that moths live in the long fur of both three-toed (*Bradypus infuscatus* Wagler) and two-toed (*Choloepus hoffmanni*, Peters) sloths.   Adult moths can be found walking and fluttering around the long hairs on the sloth.   For some time there has been general speculation about what the moths were actually doing there. Populations of moths can reach 132 individuals on a single three-toed sloth, two-toed sloths usually have fewer.   No eggs, larvae or pupae were found on any of the sloths that were examined, despite the sometimes enormous adult populations.   Study

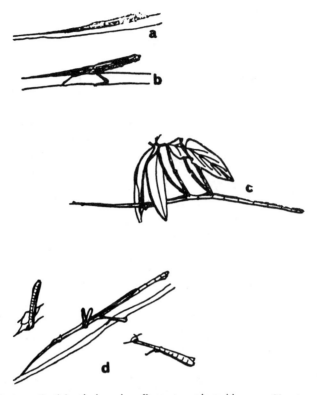

Fig. 2. The evolution of stick-mimicry in Panama phasmids. a. Simple prostrate attitude with elongation in *Metriotes diocles* Westwood b. Partially prostrate attitude but with abdomen angled off the substrate and entirely pigmented in *Isagoras dentipes* Redtenbacher c. Complete stick attitude in *Bacteria ploiaria* Westwood. d. Advanced stick-mimicry in *Pterinoxylus spinulosus* details show leg-concealment postures of intermediate and posterior legs. a–c after Robinson 1969 b; d. from Robinson 1968.

showed that the larvae of sloth moths will not feed on sloth hair or the leaves of the sloth's favourite food plant. They will however, feed on sloth dung and complete their life cycle within dung piles. Sloths descend to the forest floor to defaecate, at about weekly intervals (Montgomery and Sunquist 1975). At this stage the female moths riding in the fur-forest of the sloth descend to the dung to oviposit. The moths thus (at least) use the sloths as resting places and transport agents from one oviposition site to the next. They may mate in the fur and could feed when on the sloth's body. Potential sources of food are the lachrymal and perineal secretions of the host, and alga-enriched rainwater trapped in the sloth's coat. If the latter were the case, yet another species would be involved in the relationship.

An elegant experimental/observational study by W. Eberhard in Colombia revealed another system as complex in its way as the cuckoo/oropendola situation. Erberhard studied a subsocial pentatomid bug *Antiteuchus tripterus* Ruckes that defended its eggs and first instar nymphs against two species of parasitic scelionid wasps. The bugs lay their eggs in regular patterns and the females stand over the clutch, covering it with their bodies. The regular pattern of egg deposition and the highly stereotyped egg-guarding behaviour of the females make it possible to correlate the success of the parasites with the egg position (in the clutch) and the defensive activity of the parent. The female bugs stood guard over their clutches (90% of which consisted of 28 eggs) by day and by night, seldom if ever feeding during the incubation period, or during the development of the first instar nymphs. When standing in defence of the eggs the parent indulges in a number of fairly complex behaviours some of which seem to be oriented at intruders, while other defences are simply released by them (all details from Eberhard 1975). The behaviours include kicking out the second and third legs, antenna-waving and scraping the edge of the egg-mass with the body. Behaviours indulged in by bugs defending eggs differed significantly from those of parents defending young nymphs, at least as far as the experimental presentation of dummy wasps was concerned. Parents defending eggs scraped more than those defending nymphs and parents defending nymphs kicked more than those defending eggs. Two species of scelionid wasps regularly parasitized the eggs, *Phanuropsis semiflaviventris* Girault and *Trissolcus bodkini* Girault. Eberhard is convinced that the wasps are attracted to the gravid bugs, in the first case, by their appearance and position on leaves, but the wasps then require further cues to allow them to locate oviposition sites. When bugs that were guarding eggs were displaced (experimentally) wasps already in attack positions moved after the bugs rather than remaining with the eggs. That vision alone was not sufficient to elicit wasp attacks is suggested by the fact that when dead bugs were "staked-out" as attractants only one in fact was found with a wasp nearby and this left.

The parasitizing tactics of the two wasp species were different. Before sallying into attack, *T. bodkini* would remain for some time out of range of the bug's antennae, facing towards the eggs, and then rush in and turn just at the edge of the clutch. The bug responded to the rushes by scraping, kicking, shuddering and antenna waving. The behaviour sometimes drove the wasps away. Attacks by *P. semiflaviventris* females were very distinct. These wasps patrolled close to the edge of the egg-mass, from where they would turn to back in to the clutch, extending their abdomen to about twice its normal length as they did so. When the wasp contacted an egg with her extended abdomen she would lay an egg and then move to the adjacent one, and so on.

Studies of the patterns of parasitised eggs in hatched clutches can be used to determine the efficacy of the bug's defences against the two attacking species (or conversely to gauge the effects of the two different wasp attack strategies). This Eberhard did. The results were striking. They showed that since the bug always faces the same way with respect to the clutch it pays a penalty, the back row of eggs is always heavily parasitized by both species of wasps. However the lateral egg rows are defended fairly successfully against the sneak raider *T. bodkini* and much less successfully against *P. semiflaviventris*. The story is further complicated by the fact that egg clutches vary in their compactness and there seems little doubt that bugs are better able to defend smooth-sided compact clutches than irregular disorderly groups with more "edges". There is a premium on neatness!

Parental care has been recorded for fourteen species of Hemiptera, in this particular case, the only detailed study of the effects of parental care, the final conclusion is interesting. Eberhard removed guarding female bugs from egg clutches and scored survival. After a period of from 1-2 weeks nearly all the experimental egg groups had disappeared altogether ! These results show that parental care is highly effective in defending the egg against generalised egg-predators of which Eberhard identified at least eight (mainly ants). However removal of the female in antless trees results in *decreased* parasitization. This paradoxical result is probably due to the fact that the wasps use the presence of the bug as the initial cue in finding the eggs ! All this (and other details and complexities omitted here) lead Eberhard to the interesting speculation that there has been a continuing dynamic interaction between the bug and its egg predators and parasites. In the course of this interaction parental care provided such an increase in protection that the bugs were able to economise in egg shell thickness and are now trapped into maintaining a behavioral defence (guarding) despite the advances in the wasps' technique that threatened this kind of defence with obsolescence.

The use, by marine ethologists, of the Aqualung (SCUBA) has led to some exciting discoveries about the behaviour of marine organisms of all kinds. Symbioses reach extraordinary levels in the sea, probably because of the greater prevalence of sessile animals. As in the terrestrial case there seems little doubt that complex symbioses are more common in tropical seas than in temperate ones. Coral reefs in particular may rival rainforests in their overall intricacy of interspecific interactivity. Striking examples of complex interspecific relationships are found in symbioses between fishes that form the cleaner fish syndrome, and also in relations between fishes and anemones. Both these general relationships are worth describing in some detail.

Relationships between smaller organisms and large tropical sea-anemones quite clearly benefit the anemone associate. The anemone is armed with formidable batteries of thousands of sting-capsules on its antennae and acontia on other parts. In general it is something to be avoided. The benefits to the anemone of harboring a variety of ectosymbionts are not always readily apparent. This is certainly true in the case of the damselfish/anemone symbioses (see later). However associations between shrimps and anemones may have more easily calculable balance sheets. W. Smith (1977) studied the relationship between a snapping shrimp *Alpheus armatus* H. Milne Edwards and an anemone *Bartholemea annulata* (LeSuer). Smith found that 77% of all the *Bartholomea* in his study area (at Galeta, Panama Canal Zone) had at least one symbiotic shrimp but never more than two. In the field the anemone was seen to be approached by a coral predator, the fireworm *Hermodice carunculata* (Pallas), on three occasions. In each case the approaching worm was rebuffed by snapping shrimps. In laboratory experiments the fireworms proved capable of killing anemones, but only in the absence of shrimps. If shrimps were present the worm would attack but as soon as it contacted the anemone the shrimp would rush toward it "snapping its chela until the worm moved away from the area" (Smith 1977). The laboratory data were supported by field censusing showing that mortality rates were significantly higher in anemones that lacked symbionts during the survey period. (Interestingly Glynn (1976, 1977) has convincing observations on the defense of *Pocillopora* coral colonies by the crustacean symbionts *Trapezia furruginea* Latreille and *Alpheus lottini* Gúerin. The shrimp attacked *Acanthaster* by pinching and snapping at the asteroid's advancing arm, an attack reminiscent of the defence of *Bartholomea* described above. Crabs defended the coral by attacking the *Acanthaster*, seizing tube feet and ambulacral spines and vigorously jerking the sea star up and down.

In most cases the coral predator was driven off. Only very persistent *Acanthaster* manage to gain a foothold on their prey when faced with such defenders.)

A number of pomacentrid species of the genus *Amphiprion* associate with tropical sea-anemones. These fishes are invariably brightly coloured and in general quite small. There are magnificent colour pictures of the fishes (and anemones) in Allan (1975). The host anemones are giants of the family Stoichactiidae. Both the coelenterates and their symbionts are Indo-Pacific in distribution. There has been considerable ethological research into the relationship (reviewed in Mariscal 1970, 1971, for instance). Essentially the situation is one where the fish is so intimate an associate of the large anemone that it appears to bathe within its tentacles, it is apparently unaffected by the thousands of stinging capsules that are regularly used to kill the anemone's normal prey—small fishes. There is clearly a great advantage to the fish in being able to move into the protective hemisphere that comprises the anemone's tentacular crown. Clearly it can only achieve this virtual sanctuary if it becomes somehow protected against the host's stings. Controversy has raged about how this is accomplished. The fish could either develop an immunity to the host nematocysts *or* the host could somehow come to recognise and tolerate the presence of the fish so that it no longer tried to sting it.

Observations show that the fish goes through a complex series of actions before it settles in as an intimate symbiont of the anemone. This behavioral series has been called "acclimation" (Mariscal 1972). Before it takes place the fish is stung by the anemone, afterwards it is no longer stung by the same anemone (Mariscal 1972).

Acclimation behaviour consists of the fish progressively increasing its contact with the anemone's tentacles. It starts by cautiously approaching the anemone and simply nosing or nibbling at the tentacles. The tentacles stick to the fish but bend away from it. After a while the fish moves down the oral disc and allows itself to sink slowly down until its ventral and caudal fins touch the tentacles. At contact the fish swims quickly out of reach behaving as though it has been stung. Eventually the fish is able to bury itself in the tentacles without (apparently) being stung. At this stage acclimation is considered to be complete ! In experiments conducted by Mariscal (1971) separation of an acclimated fish from its host resulted in partial deacclimation if the duration of separation were over one hour. After twenty hours of separation the deacclimation was complete in all cases, fish that were then reintroduced to anemones had to start the acclimation process from scratch. There is now little doubt that the acclimation process induces changes, or produces changes, in the surface properties of the fish's skin. Thus although complex interspecific signalling may not be involved there is a high level of complexity in the interspecific relationship. The relationship between assemblages of marine fishes and their cleaners are cases of complex interspecific communication but are not exclusively confined to the tropics. Feder (1966) made generalisations about differences between temperate and tropical cleaning organisms that have been contested (see Losey 1971, Hobson 1969) though not disapproved. These generalisations are (Losey 1966) :

1. Temperate forms are usually less conspicuous than tropical forms.

2. *Tropical forms obtain a larger proportion of their food by cleaning* and put on elaborate displays to attract fishes to be cleaned.

3.   Colder-water forms are usually highly gregarious or schooling while warm-water forms are solitary, paired or slightly gregarious.

4.   Known cleaners of temperate waters seem to be more numerous as individuals than those of the tropics, though the number of species is less.

If point 2 above is correct then tropical cleaning symbionts are probably good examples of increased interspecific complexity in the tropics.

Fishes in general and marine fishes in particular tend to be hosts to a wide variety of ectoparasites, many of them crustaceans that are highly adapted to clinging tight to the fish's skin.   Fish can do little themselves to dislodge these ectoparasites except rub against the substrate.   They do this regularly but the widespread existence of cleaning symbioses suggests that utilizing other species as groomers may be more effective than simply rubbing.   (It is possible that the proximal causation of a fish presenting itself to cleaning symbionts is that it gets satisfaction—at a tactile level—from their attentions, in evolutionary terms this would be a causal explanation of a behaviour with the function of ensuring parasite removal.)   Whatever the causation there is no doubt that in many areas fishes of many species aggregate at special points where cleaners wait to remove ectoparasites.   Such "cleaning stations" are common features of coral reef areas where I have dived and may be "traditional".   In the more complicated cases the fish that arrived seeking cleaners, posture at the cleaning station in a distinctive way.   Postures vary widely from species to species (Losey 1971) but the *poses* are maintained for considerable periods (which increase with cleaner deprivation) and elicit quite predictable responses from the cleaners.   Cleaners respond to a host posing with staged responses.   After approach, which may involve a very distinct swimming pattern ("dancing"), the cleaner then may "inspect" the posing fish and proceed to clean it.   The behaviour of inspecting often elicits posing in fish that are at the cleaning station but are not soliciting cleaning in a direct manner. There is little doubt that signalling is mutual.

An even more complex evolutionary development of the cleaning situation has been described by Wickler (1961, 1963, 1968).   This is a case of mimicry in which a common and strikingly marked Indo-Pacific cleaner fish *Labroides dimidiatus* Cuvier & Valenciennes (a wrasse) is mimicked by a sabre-toothed blenny *Aspidontus taeniatus* Quoy & Gaimard.   The blenny approaches the passively soliciting host and bites chunks out of it!   In the guise of the cleaner the blenny can approach large fishes which it would otherwise only be able to attack by sneak-raiding from ambush (relatives of *Aspidontus* that are not mimics in fact do just this).   The mimic resembles the bona fide cleaner in colour, shape, patterning and stereotyped approach behaviour. The behavioral mimicry is interesting.   Wickler has traced its derivation from conflict behaviour in intraspecific situations that has become ritualised as an inter-specific display.

The implications of this mimicry situation are worth exploring.   The benefits of cleaning symbiosis must be considerable if the tendency to pose has not been overcome by selection to eliminate the depredations of mimics.   One way to counter the mimic would be for the host species to learn to discriminate between the cleaner and its mimic.   This in fact happens, older fishes can distinguish the mimic from the model.   As fish become more experienced their discrimination sharpens so that *Aspidontus* mainly succeeds with younger, inexperienced, fishes.   In this dynamic relationship selection should be strongly favouring enhancement of mimicry on one side and sharper discrimination on the other.

CONCLUSION

The cases of interspecific complexity in inter-relationships reviewed above are based on a very non-random sampling of the ethological, ecological and natural history literature. It must be remembered that behavioral research in the tropics comprises only a small fraction of behavioral research as a whole and that this fraction is overwhelmingly dominated by studies of social behaviour and autethological studies. The era of fashionable primate studies, partly justified by the hope that insight into human behavioral problems could be gained from studying living primates other than man, is largely past but the latest trend in biological research "fashions" is "sociobiology". (The study of fashions in research, perhaps themselves a product of intant communication between widespread institutes, would be a fascinating topic of intrinsic interest. Such a study might eventually save biology from fretfully changing directions and emphasis with such an inefficient and wasteful periodicity. An associated study of changes in the fashionable "in-language" of biological sub-disciplines would be revealing in its sociological and psychological implications.) If the half-life of "fashions" is diminishing, as it seems to be, it is possible that tropical studies may equilibrate to a more balanced approach before too long. If so I would suggest that the material reviewed above gives hope that studies of interspecific behaviours are likely to reveal much that is exciting and of overall interest. My own, perhaps overly pessimistic view, is that we will probably not understand tropical ecosystems, in time to save them. This task probably requires a huge integrated research programme on a world scale. I think that it should start from the point of recognising interactive interspecific complexity.

My colleagues at the Smithsonian Tropical Research Institute, and particularly Dr. M. H. Moynihan (to whom I am greatly indebted, as always, for advice and criticism) point out that my presentation of case histories of complexity does not prove the hypothesis of tropical uniqueness. The admitted complexity could, they argue, be the simple proportional consequence of the greater number of species present in the tropical habitats considered here. Thus tropical biological phenomena would be part of a continuum that starts in the regions with low species diversity. Fig. 3 shows a graphical model of this point of view and a model of the hypothesis that the tropical situation is qualitatively distinct and disproportionately different. The difference between the two models is (theoretically, at least) testable, the hypothesis of qualitative difference is, in principle, falsifiable. At any level where a non-tropical system and a tropical one have equal species numbers the tropical one should have more complexity of interactions (the exclusions made on page being in operation). It is not very probable that comparable situations can be found but perhaps the data can be collected to permit extrapolation of the two curves shown in Fig. 3. Certainly the factor or factors plotted against species number on Fig. 3, could be chosen from a multitude of different aspects of interactivity. One could plot number of species against number of feeding specialisations, complexity of parasitization, complexity of food chains, number of defensive strategies, and many other aspects of biotic interactions that would all relate to the diversity of adaptations to "competing and symbiotic species". In practice this means that biologists studying very diverse aspects of whole-animal biology could collect their data in such a way as to establish points on the curves of the two models in Fig. 3.

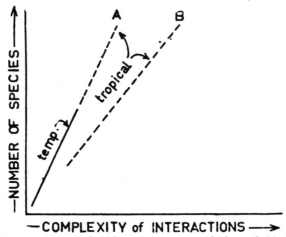

Fig. 3. Models of the possible relationships between complexity of biotic interactions and species numbers. A. The tropics and temperate regions lie on the same curve. B. The tropics are distinct from the temperate region....the view supported in this article.

There are implications in the concept of biological accommodation that are worth exploring in relation to a possible uniqueness of tropical biology. One of these "buried corrolaries" is that one would expect tropical organisms to have fewer adaptations to cope with climatic variables than do organisms from regions where physical conditions are difficult. This may well be the case, physiological ecologists are providing the basis for testing this point. Another more apparently esoteric possibility is that animals may have a finite capacity to store "survival information" genetically. If so then a big investment in biotic survival information could be associated with a small *capacity* for storing physical survival information and vice versa. This situation is shown in the models in Figure 4. Of course these could be representative of the *situation* that obtains in tropical and temperate organisms even if there were no such limiting *mechanism*. The exigencies of life in the two zones would then simply necessitate a different apportioning of survival information between the two categories, biotic and physical. For this reason the models illustrate the theoretical basis of our basic assumption that tropical biology is qualitatively different. It is different because biotic interactions are paramount in the tropics and animals *therefore* have to store a preponderance of genetic information enabling them to cope with such interactions. It would be interesting if comparisons between similar taxa in the tropics and temperate regions could give estimates of the information storage capacity of their genetic material. Unfortunately it would be most difficult to separate function from causation if a similar genetic "memory" capacity were found to be the case. These implications do, however, generate certain predictions about the capacity for animals that are well-adapted to complex tropical habitats to invade temperate regions. Basically they lead to the conclusion that many tropical animals should be competively superior to temperate zone animals but totally without the capacity to cope with

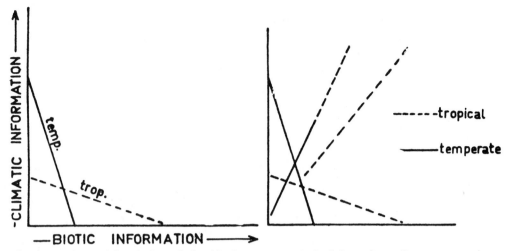

Fig. 4. (Left) Models of the relationship between survival information of two categories : biotic and climatic related to two climatic regimes. 4A. (Right) shows the models of Fig. 3 superimposed on this graph.

physical rigors. Invaders therefore should occur only in the ranks of those tropical organisms that are still subjected to major physical stresses. This, in considering terrestrial animals, would exclude most forest species but would perhaps include animals adapted to exploit areas of forest regeneration (in fire and landslide successions), forest edge organisms and some grassland and savanna species. It is interesting to apply these criteria to Moynihan's (1971) list of recent successful tropical invaders of the temperate region. Endotherms may be less liberated from climates than is generally considered to be the case. Migratory birds that have to cope with biotic and physical interactions in a number of very distinctly different climatic regimes should be of great interest. Could their behaviour at the two extremes of their migrations provide insights into the theory of tropical uniqueness ? Do they carry more stored information than non-migratory species ?

In the above review and discussion I have completely ignored inter-relationships between plants and animals. This is an area of tropical biology that is undoubtedly rich in research potential and that has hardly been touched by modern investigative techniques. Janzen's (1975) review of some salient points in the ecology of plants in the tropics dwells heavily on what is really the ethology of plant/animal interactions (Chapters 2, 3 and 4 are almost wholly devoted to this). Again and again the examples cited can be read to support the general hypothesis of inter-relational complexity developed here. The fields of pollination, fruit and seed dispersal and relations with herbivores will certainly yield evidence of complexity at many levels. Smythe (1971, in press) studied the ecology and behaviour of forest-floor rodents in the Neotropics. His studies led him to suggest that these animals have strongly influenced the fruiting strategies of a range of forest trees and have, in turn, been influenced by these strategies. He documents a complex case of coevolution that rivals many of the inter-relational studies reviewed above. There is a general tendency to think that plant "behaviour" occurs, in nearly all cases, with a comparatively extensive time delay between stimulus and response. This is almost certainly the case where behaviour involves a growth-based movement of the plant. Turgor-change can power more rapid movements as in *Mimosa* species, but it seems entirely possible to me that plants could make rapid defensive responses to herbivores that are entirely biochemical and therefore covert. If the general hypothesis of maximised interactivity in the tropics is correct, the place to look for sophisticated plant responses might well be in the rainforest.

One way of coping with complex biotic interactions is to build the capacity for plastic behaviour into the animal and for the individual to store, process and adaptively reorganize information during its own lifetime. This is the path of learning rather than instinct. I have suggested that *complex* tropical habitats provide conditions that should be favourable to the evolution of intelligence in some animals (Robinson, in press). The niche for the intelligent organism may not be a particularly big one but given time, and stability, it should eventually be filled. (That it apparently hasn't been in some tropical regions is worth further consideration elsewhere). In the conditions that favoured the evolution of intelligence its capacity should be tied up full-time in the exploitation of complex resources not available to the non-intelligent. Perhaps only when the intelligent animal found itself in an environment where biotic interactions were reduced in complexity would intelligence lead to abstract thought. (Certainly humans are often incapable of simultaneously attending to complex sensory imput and performing mental tasks.) A release from biotic pressures (coping with informational complexity =complex sensory imput processing) would almost certainly result from a movement out of tropical forests onto the parklands, grasslands or other strikingly less complex habitats outside the forest. It is possible that our capacity to think about whether there is really a tropical biology could be a consequence of the fact that there is.

REFERENCES

Akre, R. D. and C. W. Rettenmeyer. 1968. Trail following by guests of army ants (Hymenoptera : Formicidae : Ecitonini). *J. Kansas Ent. Soc.* **41** : 745-782.

Allen, G. R. 1975. *The anemone fishes : their classification and biology.* T. F. H. Publications, New Jersey, U. S. A.

Chopard, L. 1938. *La Biologie de Orthoptères.* Paul Lechevalier, Paris.

Cott, H. B. 1942. *Adaptive coloration in animals.* Methuen & Co. Ltd., London.

Curio, E. 1977. *The ethology of predation.* Springer, New York.

Davis, D. E. 1946. A seasonal analysis of mixed flocks of birds in Brazil. *Ecology*, 27 : 168-181.

Dobzhansky, T. 1950. Evolution in the tropics. *Amer Scientist*, 38 : 209-221.

Eberhard, W. G. 1975. The ecology and behavior of a subsocial pentatomid bug and two scelionid wasps : strategy and counterstrategy in a host and its parasites. *Smithson. Contr. Zool.* **205** : 1-39.

Edmunds, M. 1972. Defensive behaviour of Ghanian praying mantids. *Zool. J. Linn. Soc. Lond.* **50** :339-396.

Edmunds, M. 1974. *Defence in animals.* Longmans, London.

Edmunds, M. 1976. The defensive behaviour of Ghanaian praying mantids with a discussion of territoriality. *Zool. J. Linn. Soc. Lond.* **58** : 1-37.

Feder, H. M. 1966. Cleaning symbiosis in the marine environment. *In* S. M. Henry (ed.). *Symbiosis.* Academic Press, New York and London.

Glyan, P. W. 1976. Some physical and biological determinants of coral community structure in the Eastern Pacific. *Ecol. Monogr.* **46** : 431-456.

Glynn, P. W. 1977. Interactions between *Acanthaster* and *Hymenocera* in filed and laboratory. *Proceeding of the Third International Coral Reef Symposium*, Miami.

Grimes, L. G. 1973. The breeding of Heuglin's masked weaver and its nesting association with the red weaver ant. *Ostrich*, **44** : 170-175.

Hobson, E. S. 1969. Comments on generalizations about cleaning symbiosis in fishes. *Pacific Science*, **23** : 35-39.

Janzen, D. H. 1975. *Ecology of plants in the tropics*. Edward Arnold, London,

Karr, J. 1975. Production, energy pathways and community diversity in forest birds. pp. 161-176. *In* Golley, F. B. and E. Medina (eds). *Tropical ecological systems*. Springer Verlag, Berlin.

Kettlewell, H. B. 1959. Brazilian insect adaptations. *Endeavour*, **18** : 200-210.

Losey, G. S. 1971. Communication between fishes in cleaning symbiosis. *In* T. C. Cheng (ed.) *Aspects of the biology of symbiosis*. Butterworths, London.

Mariscal, R. N. 1970 The symbiotic behaviour between fishes and sea anemones. *In* H. E. Winn and B Olla (eds.) *Behavior of marine animals—recent advances*. Plenum Publishing Co., New York

Mariscal, R. N. 1971 Experimental studies of the protection of anemone fishes from sea anemones. *In* T C Cheng (eds.) *Aspects of the biology of symbiosis*. Butterworths, London.

Montgomery, G. G. and M. E. Sunquist. 1975 Impact of sloths on neotropical forest energy flow and nutrient cycling. pp. 69-78. *In* Golley, F. B. and E. Medina (eds.) *Tropical ecological systems*. Springer Verlag, New York.

Morris, D. 1969. *Patterns of reproductive behaviour*. Cape, London.

Moynihan, M. H. 1960. Some adaptations which help to promote gregariousness. *Proc. 12th Int. Ornithol. Congress*, 523-541.

Moynihan, M. H. 1962. The organization and probable evolution of some mixed species flocks of neotropical birds. *Smithson. Misc. Coll.* **143** : 1-140.

Moynihan, M. H. 1963. Inter-specific relations between some Andean birds. *Ibis*, **105** : 327-339.

Moynihan, M. H. 1968. Social mimicry, character convergence versus character displacement. *Evolution*, **22** : 315-331.

Moynihan, M. H. 1970a. The control, suppression, decay, disappearance and replacement of displays. *J. Theoret. Biol.* **29** : 85-112.

Moynihan, M. H. 1970b. Some behavior patterns of platyrrhine monkeys. II. *Saguinus geoffroyi* and some other tamarins. *Smithson. Contr. Zool.* **28** : 1-77.

Moynihan, M. H. 1971. Successes and failures of tropical mammals and birds. *American Naturalist*, **105** : 371-383.

Moynihan, M, H. 1976. *The New World primates*. Princeton Univ. Press, U. S. A.

Myers, J. G. 1935. Nesting associations of birds with social insects. *Trans. R. Ent. Soc. Lond.* **83** : 11-22.

Rand, A. L. 1954. Social feeding behavior of birds. *Fieldiana Zoology*, **36** : 1-171.

Rettenmeyer, C. W. 1961. Observations on the biology and taxonomy of flies found over swarm raids of army ants (Diptera : Techinidae : Conopiade). *Kansas Univ. Science Bull.* **42** : 993-1066.

Rickleffs, R. E. 1973. *Ecology*. Thomas Nelson, London.

Robinson, M. H. 1966. Anti-predator adaptations of stick-and leaf-mimicking insects. D. Phil. thesis. Bodleian Library, Oxford.

Robinson, M. H. 1968. The defensive behavior of *Pterinoxylus spinulosus* Redtenbacher, a winged stick insect from Panama (Phasmatodea). *Psyche* **75** : 195-207.

Robinson, M. H. 1969a. Defenses against visually hunting predators. *In* T. Dobzhansky, M. K. Hecht & W. C. Steere (eds). *Evolutionary Biology III*.

Robinson, M. H. 1969b. The defensive behaviour of some orthopteroid insects from Panama. *Trans. R. ent. Soc. London.* **121** : 281-303.

Robinson, M. H. 1970a. Insect anti-predator adaptations and the behavior of predatory primates. *In* Act. IV Congr. Latin. Zool. **2** : 811-836.

Robinson, M. H. 1970b. Animals that mimic parts of plants. *Morris Arboretum Bull.* **21** : 51-58.

Robinson, M. H. 1973. The evolution of cryptic postures in insects, with special reference to some New Guinea tettigoniids (Orthoptera). *Psyche*, **80** : 159-165.

Robinson, M. H. (In press). Informational complexity in tropical rain forest habitats and the origins of intelligence. *Proc. Congr. Trop. Ecol.*, Panama.

Robinson, M. H., Y. D. Lubin and B. Robinson. 1974. Phenology, natural history and species diversity of web-building spiders on three transects at Wau New Guinea. *Pacific Insects.* **16** : 117-163.

Schnierla, T. C. 1956. A preliminary survey of colony division and related processes in two species of terrestrial army ants. *Insectes Sociaux*, **3** : 49-69.

Schnierla, T. C. 1957. A comparison of species and genera in the ant family Dorylian.: with respect to functional pattern. *Insectes Sociaux*, **4** : 259-298.

Smith, N. G. 1968. The advantages of being parasitized. *Nature, Lond.* **219** : 690-694.

Smith, N. G. 1969. Provoked release of mobbing—a hunting technique of *Microstur* falcons. *Ibis.* **111** : 241-243.

Smith, W. J. 1969. Messages of vertebrate communication. *Science*, **165** : 145-150.

Smith, W. L. 1977. Beneficial behavior of a symbiotic shrimp to its host anemone. *Bull. Marine Science*, **27** : 343-344.

Smythe, N. 1970. Relationships between fruiting seasons and seed dispersal methods in a neotropical forest. *Amer. Naturalist*, **104** : 25-35.

Smythe, N. (In press). The natural history of the Central American agouti (*Dasyprocta punctata*). *Smithson. Contr. to Zool.*

Waage, J. K. and G. G. Montgomery. 1976. *Cryptoses choloepi* : a coprophagous moth that lives on a sloth. *Science*, **193** : 157-158.

Wickler, W. 1961. Uber das Verhalten der Blenniiden *Ranula* und *Aspidontus* (Pisces : Blenniidae). *Z. Tierpsychol.* **18** : 421-440.

Wickler, W. 1963. Zum Problem der Signalbildung am Beispiel der Verhaltensmimickry zwischen *Aspidontus* und *Labroides*. *Z. Tierpsychol.* **20** : 657-679.

Wickler, W. 1968. *Mimicry.* Weidenfeld and Nicolson, Ltd., London.

Willis, E. O. 1967. The behavior of bicolored antbirds. *Univ. of California Publications in Zoology*, **79** : 1-132.

Willis, E. O. 1972. Taxonomy, ecology and behavior of the sooty ant-tanager (*Habia gutturalis*) and other ant-tanagers. *Amer. Mus. Novitates*, **2480** : 1-38.

Wilson, E. O. 1971. *The insect societies.* Belknap Press, Harvard, U. S. A.

Winterbottom, J. M. 1943. On woodland bird parties in Northern Rhodesia. *Ibis*, **85** : 437-442.

Winterbottom, J. M. 1949. Mixed woodland bird parties in the tropics with special reference to Northern Rhodesia. *Auk*, **66** : 258-263.

[*Note from the author:* This article has been reprinted from uncorrected proofs, which were not seen by the author.]

9

Reprinted from pp. 11–33 of *Conservation Biology: An Evolutionary-Ecological Perspective*, M. E. Soulé and B. A. Wilcox, eds., Sinauer Associates, Sunderland, Mass., 1980 395 p.

FOOD WEB ORGANIZATION AND THE CONSERVATION OF NEOTROPICAL DIVERSITY

Lawrence E. Gilbert

Population biologists have been more concerned with the design of future natural areas than with their management (Sullivan and Shaffer, 1975; Simberloff and Abele, 1975; Diamond, 1976; Diamond and May, 1976b; Terborgh, 1976; Whitcomb et al., 1976). In this chapter I apply biological insights to problems of reserve management rather than to those of design, and stress the interface between plants and animals rather than particular vertebrate taxa. Rationale for these points of emphasis follow:

1. Biologists will probably not be given a significant role in land-use planning anywhere in the world. Ultimately most conservation programs will be the management of the scraps of nature left untouched by various governments.

2. Though biogeographers may design some natural areas, these "experiments" may or may not support their theories. While sound design reduces the need for management, in many cases management will have to deal with mistakes in design and unforeseen complications.

3. The history of land use in much of the world precludes any opportunity for large natural reserves. In countries like Britain, such constraints lead to an emphasis on problems of management rather than of design (Duffey and Watt, 1971). Thus, conservation biologists there have participated little, if at all, in the controversy over reserve design (e.g., Whitcomb et al., 1976).

113

4. Although conspicuous vertebrates receive most of the attention from conservationists, the bulk of terrestrial species are insects. My own research emphasis, and that of this paper, is on plants and herbivorous insects—which together probably constitute over half of all existing species (Gilbert, 1979).

If maximum diversity is what we wish to maintain in a reserve, the ecology of plants and their associated faunas should be the basis of sound management. Many birds and larger vertebrates are less sensitive to changes in plant species composition than to changes in habitat structure, productivity and land area. For example, consider two reserves in which the equilibrium diversity of insectivorous birds is deliberately maintained. The two might develop great differences in overall community diversity if insects and plants are not equally well monitored. Moreover, the assumption that reserves designed to maintain larger mammals will automatically be large enough to perpetuate smaller organisms without management is only true if natural disturbance rates or "patch dynamics" (page 27) are considered in the initial design.

In addition to these theoretical justifications, a focus on insects and plants provides biologists with powerful economic arguments for natural habitat conservation (Myers, 1976; Oldfield, 1976): (1) Research on natural populations of insects in intact ecosystems provides a theoretical basis for developing biological controls for insect pests and weeds; (2) Natural area preserves provide permanent reservoirs of potential control agents; (3) Research on patterns of natural resistance of wild plants to insects suggests directions for crop breeding programs; (4) Natural areas contain wild relatives of cultivated plants and are a permanent source of genetic resources in crop breeding (Browning, 1975).

In the last decade there has been an upsurge of interest in animal-plant interaction in the neotropics. While only a fraction of the known taxa have been examined ecologically, sufficient long-term studies of representative animal-plant systems are available to reveal some consistent patterns of food web organization relevant to managing biological diversity in neotropical forest preserves.

NEOTROPICAL FOOD WEBS AND THE MAINTENANCE OF DIVERSITY

The relatively stable and mild conditions of tropical rain forests, combined with high productivity and complex vertical structure, have often been used to explain the high species richness in such habitats (see MacArthur, 1972 for a review). While these factors may be valid explanations for observed differences between tropical and temperate regions, they are too general to be of much use to a manager faced with an eroding fauna and flora on a particular reserve. On a local level we need to understand the specific components of diversity within a habitat so that controlling elements may be discovered and manipulated if necessary.

From the perspective of insects and other animals which specialize on particular host plants, it is possible to identify four strictly biotic features, probably common to many terrestrial ecosystems, which are of key importance to understanding and managing the bulk of neotropical species diversity. The following synthesis emerges from the work of numerous independent researchers and builds on a previous paper (Gilbert, 1977). The four key organizational features of neotropical forests will be discussed in this order: (1) the chemical mosaic and coevolved food webs; (2) mobile links; (3) keystone mutualists; and (4) the ant mosaic.

The Chemical Mosaic and Coevolved Food Webs

As Janzen (1977) pointed out, the relatively uniform green color of vegetation disguises its true nature from the perspective of herbivores. Neotropical forests, for example, are exceedingly complex mosaics with respect to interspecific variety in secondary compounds (alkaloids, terpenes, etc.), chemicals thought to serve a defensive function in plants (see review by Levin, 1976). Consequently, most insect herbivores have evolved sensory and digestive specializations which restrict their diet to a relatively small, chemically similar fraction of the available plant species. For example, some beetles feed only on the seeds of a single tree (Janzen, 1977), presumably because of the extreme digestive specialization required to deal with highly toxic seeds.

An important feature of the chemical mosaic is the fact that plants of the same genus or family often share many compounds as well as those insects specialized to deal with such compounds. A large fraction of herbivorous insects, along with their own parasitoids, are thus organized into many separate food webs. These webs occur side by side, but are based on different, chemically and taxonomically delimited compartments of primary production.

Reference to such systems as "coevolved food webs" (Gilbert, 1977) reflects my assumption that patterns of the chemical mosaic (and thus of insect feeding) have been and will continue to be generated largely by a process of coevolution between plants and their parasites. It is, of course, possible that insects diversified against an already existing mosaic of plant chemistry and have no selective impact on the chemical mosaic. Root's (1973) "component community" or Cohen's (1977) "source food web" may be preferable to those wishing to use a more neutral terminology in discussing ecosystem substructure.

Regardless of how ecosystem substructure evolved, the important fact for conservation biology is that much of the total species diversity in a neotropical forest is to be found in many parallel, host-restricted food webs which are similar in trophic organization but different taxonomically. If we understand how diversity is maintained within a representative sample of the hundreds of such sub-systems, we may find general rules for managing the many species that we do not yet know of or about which we know little.

Two partially analyzed neotropical food webs will serve as examples for the major points involving reserve management. These food webs are delimited at the level of two plant families, the Passifloraceae and the Solanaceae, and are components of all neotropical forest habitats. Butterflies with aposematic (warning) coloration are the most visible manifestation of the herbivore community supported by these plant groups. Many of the general insights relevant to this paper derive from population and community studies of these insects. Only brief highlights of available information on the two food webs follow.

Passifloraceae. The passion flower family is represented by about 500 species in the American tropics, most being vines of the complex genus *Passiflora*. Broad geographical information on host relations of the major herbivores of this family, heliconiine butterflies (Figure 1A–C), and evidence for coevolution with these insects are summarized by Benson, Brown and Gilbert (1976). Other important herbivores of *Passiflora* include flea beetles, coreid bugs and dioptid moths, but these insects are ecologically and taxonomically less well known at present. The flea beetles resemble heliconiines in local diversity and host relationships (Gilbert and Smiley, 1978).

Only a small fraction of the known neotropical species of Passifloraceae occurs in any given local flora. Moreover, the local diversity of this plant family is strikingly constant across neotropical rainforest sites (Gilbert, 1975), ranging between 10 and 15 species. It appears that this consistency reflects combinations of microhabitats and pollinators which will support different *Passiflora* species locally. Gentry (1976a) noted similar patterns in neotropical Bignoniaceae. Since the numbers of heliconiines and other host-specialist insects correlate with the number of host species available (Gilbert and Smiley, 1978), we clearly need a better understanding of such geographically constant diversity within plant taxa to manage the diversity of associated food webs in reserves.

Studies of the passion flower food web at the Organization for Tropical Studies (OTS—La Selva field station in Costa Rica) by Smiley and Gilbert indicate that the Passifloraceae, like many other woody tropical plant families, is represented locally by species differing in growth form, successional stage occupied and pollination relationships (Figure 4F). In the flood plain forest at La Selva, for example, only three species become large canopy-emergent vines. One species is a small forest understory tree

FIGURE 1. Life histories of Heliconiine and ithomiine butterflies. A: *Heliconius ismenius* ovipositing on seedling of *Passiflora serratifolia* in forest understory, Sierra de los Tuxtlas, Vercruz, Mexico. B: *H. cydno* collecting pollen from *Anguria* (Cucurbitaceae) in Atlantic lowland forest, Costa Rica. C: Gregarious larvae of *H. doris* on canopy vine, *P. ambigua*, in Atlantic lowlands, Costa Rica. D: *Ithomia heraldica* male displays scent hairs while perched in a sun-fleck near the forest floor in Costa Rica. E: Males of several ithomiine species visiting *Eupatorium* (Compositae) for nectar and (presumably) sex pheromone precursors in successional vegetation, Corcovado Park, Costa Rica.

and six species can be found in various successional stages along with seedlings and suckers of canopy species. The full complement of ten Passifloraceae thus depends upon a mix of successional stages.

The insects specialized on *Passiflora* are further restricted with respect to the microhabitat occupied. Smiley (1978) found that within an area, the most closely related *Heliconius* species occupy different microhabitats and participate in different mimicry associations with more distantly related congeners. Within microhabitats, forest understory for example, Smiley found that mimetic pairs of heliconiines are of more distantly related species and utilize different host plants. Thus it appears that habitat heterogeneity has contributed to the evolutionary radiation of *Heliconius* into several color pattern complexes, each of which is closely tied to particular microhabitats.

Several other aspects of this food web are worth mentioning. First, each *Passiflora* population is typically composed of rare, scattered individuals which produce flowers and fruit infrequently and asynchronously. Consequently, these plants provide little incentive for the evolution of host-restricted pollinators. Instead they rely on pollinators (bees, moths, bats, hummingbirds) and seed dispersal agents (birds, bats) which visit rare, scattered plants of many taxonomic groups.* I call such organisms "mobile links" because through their foraging movements, they are of mutual concern to the reproduction of many different unrelated plants which, in turn, support otherwise independent food webs.

Another category of mobile link species should be mentioned briefly in reference to *Passiflora*. Most passion flower vines possess extrafloral nectaries which attract predators and parasitoids (ants, wasps, microhymenopterans) and a host of other insects. Some of these visitors are known to be significant as mortality factors to the insects which attack the plants. While highly specialized parasitoids are considered part of the coevolved food web (Gilbert, 1977), predators such as ants and wasps forage over many kinds of plants possessing leaf and petiole nectaries, thereby forming another kind of connection between host-based food webs (Figure 2A, C–E).

Another notable aspect of the passion vine system concerns the butterfly genus *Heliconius,* the most numerous and diverse genus of the heliconiines. Not only do these insects have the problem of rare and scattered larval host-plants, but most are specialists on new shoots of *Passiflora,* a resource more rare in space and time than the plants themselves. In spite of this, local populations can be remarkably constant at low density for years (Ehrlich and Gilbert, 1973). How do *Heliconius* populations persist in spite of low density and fluctuating larval food supplies? Part of the answer involves the cucurbit vines, *Gurania* and *Anguria*. Male flowers of these vines provide *Heliconius* with adult resources (pollen) known to increase both the fecundity and longevity of adults (Dunlap-Pianka et al., 1977). Adult reproductive longevity reduces the probability of local extinction by reducing the impact of fluxes in the larval food supply (Gilbert 1975; 1977). Although *Heliconius* adults can be opportunistic in plant species visited for pollen and nectar, it is the relatively constant supply of pollen from *Anguria* in most neotropical

---

*D. H. Janzen first applied the term "traplining" to emphasize that the foraging of such animals is similar to that of trappers running strings of traps. The terminology suggested here emphasizes their ecological role.

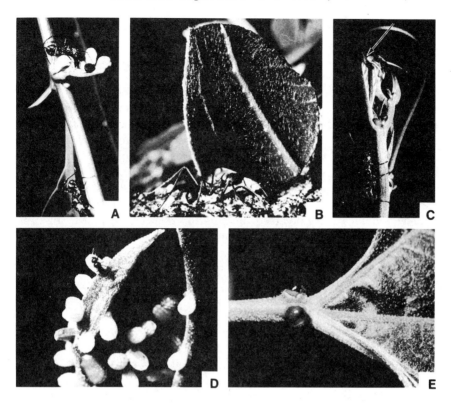

FIGURE 2. Ants, parasitoids and plant nectaries. A: *Ectatomma* tending a membracid (Homoptera) on inflorescence of *Solanum* in forest understory, Atlantic lowlands, Costa Rica. B: Leaf cutter ant *(Atta)* carrying leaf in dry forest of Tamaulipas, Mexico. C: *Ectatomma* approaches newly hatched *Heliconius* larva (arrow) on shoot of *Passiflora vitafolia* at Corcovado Park, Costa Rica. Note empty egg shell near larva and nectar glands on petioles of young leaves. D: Female chalcid parasitoid (probably Scelionidae) ovipositing on *H. sara* eggs which are always placed on *P. auriculata* in the Atlantic lowlands of Costa Rica. Eggs are less than 1 mm. tall. E: Nectar produced by petiole glands of *P. incarnata,* Braziora County, Texas.

rainforests that insures the persistance of the rarer members of the *Heliconius* community. It should be noted that *Heliconius* species which are largely separate in microhabitat preference and larval hosts can, nevertheless, be seen to overlap on the same pollen sources along forest edges and in the canopy.

Solonaceae. The tomato and eggplant family, like Passifloraceae, is represented by shrubs, vines, trees, herbs and epiphytes although shrubs and small trees are probably the dominant growth form of the family. In the neotropics, the large fauna of insects specialized on the Solanaceae includes grasshoppers, hemipterans, homopterans, chrysomelid beetles (including flea beetles) and various lepidopterans (including ithomiine butterflies). Like the heliconiines, the ithomiines form an important group of conspicuous neotropical butterflies with warning coloration (Figure 1D, E). A local community of Solanaceae might include 45 to 50 species, most of which would be of the genus *Solanum*. *Solanum* species can be specialized to any part of the forest and they comprise the major larval resources for ithomiines (although numerous other solanaceous genera are utilized).

The Solanaceae are diverse in pollination relationships, utilizing bees (most *Solanum),* bats *(Markea),* hummingbirds *(Juanulloa),* and lepidopterans *(Cestrum).* Birds and bats are known to be seed dispersal agents for certain species. Thus, as was the case in the Passifloraceae, most members of the Solanaceae rely heavily on mobile links for their continued presence in the ephemeral, patchy microhabitats to which they are restricted.

Research on ithomiine ecology by several field workers over the last decade, most notably that of Drummond (1976) and Haber (1978), provides details of larval and adult biology. Like heliconiines, ithomiine species are restricted as to suitable host plant and microhabitat. Like *Heliconius,* ithomiines depend on adult resources important to their reproductive biology. Adults regularly use fresh bird droppings and nectar plants as food, and still other plants (Boraginaceae, Compositae) which provide precursors for male courtship pheromones.*

Unlike heliconiines, ithomiines have localized courtship areas in the forest where males of several species may gather to wait for females. Adults of species with larval host plants in the forest understory can be seen visiting floral resources in early successional areas; species with larval hosts in early successional areas visit the forest understory for bird droppings and court and mate there. Thus a juxtaposition of microhabitats is important to many ithomiine species.

Even in sympatry, no species appear to be in both the Passifloraceae- and Solanaceae-based food webs (except possibly for a few mobile links). However, the diversity of each system seems to be based on similar factors. First, in both cases full diversity at the herbivore level depends upon maximum species diversity of host plants locally. In both cases, maximum plant diversity requires a balance of healthy successional stages. Second, in both cases maximum diversity of the insects depends upon important resources (adult host plants, courtship sites, and so on) other

*In the closely related danaines, pheromones necessary for mating (Pliske and Eisner, 1969) are known to be derived from the compounds (Schneider et al., 1975) which attract male ithomiines (Pliske, 1975).

than larval host plants. These secondary resources may be found in successional stages other than that occupied by the larval host. Thus, once again, a balance of successional patches often appears necessary to prevent the loss of those insect species which depend upon taxonomically and ecologically different plants in larval and adult stages. This phenomenon is not restricted to the tropics (Wicklund, 1977).

Where tropical forest has been replaced by pastures, both food webs described above are still present. However, they conspicuously lack those species dependent upon spatial heterogeneity of the habitat. Such species also appear to decline as successional stages give way to climax forest. However, since shrubs are less likely than vines to survive shading as the forest grows up, the Solanaceae food web would be expected to decline in diversity more rapidly than the Passifloraceae web. This appears to be the case at Barro Colorado Island, Panama. During the time between isolation of the island by flooding in 1914 and a census of the butterflies in the early 1930's, the ithomiines on the island dropped from about 20 species (a conservative estimate for intact forests) to 11 species (Huntington, 1932). By 1969 only six species could be found in a five month study by Emmel and Leck (1970). This drop is correlated with a decline in Solanaceae diversity as succession returned disturbed areas to forest (T. Croat, personal communication). Meanwhile, heliconiines actually increased from 12 to 13 species between 1930 and 1969. The ability of *Heliconius* to persist as small populations may also help account for the relative stability of the number of heliconiine species (Gilbert and Smiley, 1978).

Based on community samples taken by Brown (1972) heliconiines are two to five percent and ithomiines five to 11 percent of total butterfly species in neotropical rainforest. Together, they are the principal models for several major butterfly mimicry complexes, so that the loss of such species could have a drastic impact on the many edible and mimetic species in an area. Thus, a management scheme which maintains successional patchiness would help preserve a minimum of seven to 16 percent of the butterfly fauna. Furthermore, since many other plant taxa show patterns of ecological and morphological diversification similar to those of Passifloraceae and Solanaceae, it is likely that a large but unknown fraction of the plant and insect diversity would be maintained by management aimed specifically at these two food webs.

Mobile Links

Animals that are significant factors in the persistence of several plant species which, in turn, support otherwise separate food webs can be called mobile links (Figure 3). In addition to being rare and dispersed, most neotropical forest plants are obligate outcrossers (Bawa, 1974) and rely on hummingbirds (Feinsinger, 1976), bees (Delgado Salinas and Sousa Sánchez, 1977), hawk moths and bats (Heithaus et al., 1975) for pollination. Likewise, efficient seed dispersal is required for the colonization of habitat patches suitable for germination and seedling establishment.

FIGURE 3. Some mobile links and plants they visit. A: Long-billed hermit, *Phaethornis superciliosus* (Osa, Costa Rica), an important pollinator in neotropical rainforests, exemplifies a mobile link. B: *Heliconia longa* (Osa, Costa Rica), frequently visited by hermit hummingbirds. Some *Heliconia* qualify as keystone mutualists. C: Blue honey creeper, *Cyanerpes cyaneus,* feeding on fruit of *Miconia* (Melastomaceae), Atlantic lowlands, Costa Rica. *Miconia* species are key resources of many small frugivorous birds. D: Toucan, *Ramphastos swainsonii,* in Atlantic lowland rainforest. This species relies on fruit of the canopy tree *Casearia* during parts of the year. At other times it and its congeners are dispersal agents for other tree species. E: Bat-dispersal cluster of *Piper* (Piperaceae) seeds. *Piper* fruits are a year-round staple for some bats. F: "Monkey pot" fruit of *Lecythis costaricensis* (Lecythidaceae), canopy tree related to brazil nut, has bat-dispersal seeds. Operculum drops from pod of mature nut. Food bodies are at end of nut (arrow) away from opening, so that bat must grab seed on its hard end, fly to a better perch and rotate it before eating food body. Seed is then dropped, completing the dispersal sequence.

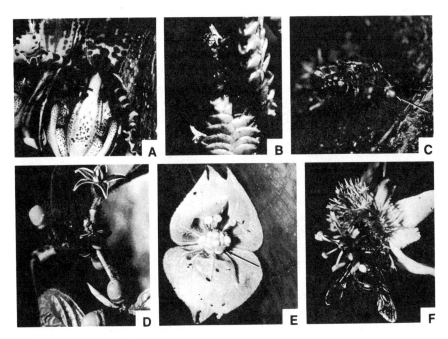

FIGURE 4.   Euglossine bees *(Euplusia, Euglossa, Eulema)*, a Carpenter bee and plants they visit. A: Male *Euplusia simillima* visiting *Stanhopea tigrina* for volatile chemicals in a Mexican cloud forest. B: *Euglossa* female collecting nectar from *Calathea insignis* (Marantaceae) in successional area, Atlantic lowlands, Costa Rica. C: *Euglossa* male attracted to cineole placed on tree in Costa Rica rain forest. Arrow points to orchid pollinium stuck to head. D: *Eulaema migrifacks* female which was collecting pollen (note pollen load on hind leg) on *Solanum sanctae-clarae,* a purple-flowered epiphytic shrub in Atlantic lowlands of Costa Rica. E: Flower of *Dalechampia aff. tillifolia* (Euphorbiaceae), a common successional vine in the Sierra de los Tuxtlas. Resin gland is indicated with arrow. F: Carpenter bee *(Xylocopa)* visiting *Passiflora pulchella* at Santa Rosa Park, Costa Rica.

Birds and bats are the major neotropical seed dispersers (McKey, 1975). Ants may well be important to seed movement as they appear to be in other regions (see review in Gilbert, 1979). Certain ants may also patrol and defend a variety of nectar-producing plants (Bentley, 1977). Seedling establishment as well as reproduction by adult plants may be enhanced by ant protection.

Among the most important and best researched groups of neotropical link organisms are the euglossine or orchid bees (Figure 4, A–E). It is worth elaborating the biology of these spectacular insects because they can be involved in a major way with hundreds of plant species in one area. A single euglossine species may link plant species from all stages and strata of a forest into a system of indirect mutualism.

Dodson (1966) has given a general account of euglossine-plant relationships. Euglossine females gather pollen from successional plants such as *Solanum* and *Cassia*. In some cases they are simply part of a complex of bees which visit the plants—for example, *Cassia* (Delgado Salinas and Sousa Sánchez, 1977). In other cases (for example, some *Miconia* and *Solanum* species) they may be the primary pollinators. Females are known to travel great distances in foraging and are thus important to the reproduction of low density plants (Janzen, 1971).

Resin for nest construction may be obtained opportunistically from damaged trees or, more predictably, in successional patches where certain *Dalechampia* (Euphorbiaceae) vine species provide resin as a floral reward for pollination (Armbruster and Webster, in press). Nectar is obtained from a variety of successional (Apocynaceae, Marantaceae, Rubiaceae and others), understory (Rubiaceae, Marantaceae) and canopy (Orchidaceae, Bignoniaceae, Bromeliaceae) plants by both sexes. A single euglossine species may pollinate a dozen species of nectar plants representing every level and successional stage available in a forest.

Like male ithomiine butterflies, euglossine males collect volatile chemical compounds for use in courtship behavior (Dodson, 1975). Compounds collected differ for each of the 30 to 50 species of euglossines which might occur together. Some orchids rely on a single euglossine species for pollination and provide volatile compounds such as cineole to visiting males. Mechanical methods (Dressler, 1968) and chemical divergence enforce sexual isolation of many genetically similar sympatric orchid species (Hills et. al., 1972). Other plant groups have also evolved male euglossine pollination systems. In Veracruz, Mexico, Armbruster and Webster (1979) found that while one *Dalechampia* species provides resin for and is pollinated by females, a second species produces volatile compounds and is male pollinated. Many epiphytic *Spathiphyllum* and *Anthurium* species (Araceae) have likewise specialized on male euglossines for pollination. It has been suggested that for the aroids (Araceae), as for orchids, a proliferation of species has been made possible by the availability of a diverse euglossine fauna (Williams and Dressler, 1976).

Many euglossine species rely on early successional plants for larval resources and at the same time are important, even necessary, pollinators of plants restricted to later successional stages of forest. Canopy orchids and aroids are thus indirectly dependent upon the presence of early succession patches. Such specialization evolved under conditions of continuous forest where appropriate successional stages were always available within the flight range of euglossines. In isolated habitat preserves, however, certain successional plant species might be lost during periods of low habitat disturbance and euglossine species critical to canopy plants may die off as commuting distances between diverse resources become too great.

From these considerations one might logically predict the loss of many orchids and aroids from Barro Colorado Island now that successional changes have reduced the diversity and productivity of early succession plants like some Solanaceae. Several years ago I attempted to test this prediction but learned from T. Croat, then conducting a floral survey of the island, that the orchid list had increased since the last inventory. In addition to the obvious problems (the earlier survey was probably not sufficiently intensive and orchids are long lived), it should also be pointed out that euglossines have been seen crossing Gatun Lake to the island (Dressler, 1968). Thus, my prediction may be realized in the next 30 years as adjacent mainland forest in central Panama is removed.

Rather than test such predictions with our few habitat reserves, I believe we should devise schemes to monitor and manage the diversity of such important link organisms as orchid bees. This would involve keeping track of various successional stages and some of the plants in them. In lowland areas (where land slips and other forms of natural disturbance occur at a lower rate per hectare than in steeper areas) managed disturbance will be necessary, particularly as the preserve becomes smaller and more isolated.

Representative link species should be a special focus for autecological research in conservation biology. From available information on bats, hummingbirds and bees, it appears that management rules designed for the few species we know something about will apply to many other, ecologically similar species. It is safe to say that maintaining successional variety is one general rule clearly mandated by existing information, though the details are lacking.

Keystone Mutualists

Keystone mutualists are those organisms, typically plants, which provide critical support to large complexes of mobile links. The loss of a keystone mutualist would, with some time delay, cause a loss of mobile links, followed by losses of link-dependent plants through a breakdown in reproduction and dispersal. Finally, host-specialist insect diversity should decline with the reduction in diversity of host communities. This is the sort of linked extinction predicted by Futuyma (1973).

The canopy tree *Casearia corymbosa* (Flacourtiaceae) fits the concept of keystone mutualist. Howe (1977) has shown that although only one bird, the masked tityra *(Tityra semifasciata),* is a highly effective seed dispersal agent at La Selva, 21 other fruit feeders use the tree. Howe suggests that *Casearia* is a "pivotal species" because it supports several obligate frugivores which depend on it almost entirely during the two to six week annual scarcity. Howe predicts that the loss of *Casearia* at La Selva would lead to the disappearance of *T. semifasciata* (with consequences for other trees whose seed it disperses), and *Ramphastos* toucans (with effects on *Virola* and *Protium). R. swainsonii* is shown in Figure 3D. Although *Casearia* is a canopy tree at La Selva, it supports birds which may be significant to successional as well as primary forest tree species. Perhaps the most important fruit source for smaller frugivores, *Miconia* (Melastomaceae) is shown in Figure 3C.

Two detailed hummingbird studies reveal similar keystone mutualists. Again at La Selva, Stiles (1975) found that most hummingbirds depend on *Heliconia* (Figure 3B) during the early wet season. Hermits (Figure 3A), on the other hand, rely on *Heliconia* for nectar at all times of the year. The various sympatric *Heliconia* species have displaced flowering times so that at least one or two species are always in bloom. However, as Stiles notes, flowering for some species depends on light-gap formation. In a small reserve it might, therefore, be necessary to manage light-gaps around *Heliconia* clumps to insure the survival of hermits, an important group of link species. Incidently, a study is in progress on the insect fauna of *Heliconia* (Strong, 1977).

In a mountainous locale near Monteverde, Costa Rica, Feinsinger (1976) found that *Hamelia patens* (Rubiaceae) sustains hummingbird populations during periods when other foods are scarce. *H. patens* flowers year-round but peaks from May to October when other flowers are rare.

Bats account for a large fraction of mammalian diversity in the neotropics. Like other link species, many bats require a variety of resources including insects, nectar, pollen and fruit (Figure 3E, F). Suitable roosting sites such as large hollow trees, may also be habitat restricted (Figure 5G). Heithaus et al. (1975) identified major floral sources *(Ceiba,*

FIGURE 5. **Neotropical rain forest habitats. A: Successional stages along airstrip near Rincon, Osa, Costa Rica. Grasses predominate in frequently cut area. B: Forest understory, Corcovado Park, Costa Rica. C: Natural disturbance along small stream constantly generates successional habitats. Osa, Costa Rica. D: Epiphyte load of large canopy tree at La Selva, Atlantic lowlands of Costa Rica. Hundreds of plant species occur in the area shown in this photograph. E: River canyon on slopes of Vólcan Barba, Costa Rica, where land slips along steep banks maintain successional variety. Human exploitation is almost impossible here. F: Heavily grazed pasture which was recently forest in the Atlantic lowlands of Costa Rica. Few elements of natural successional vegetation can be found in such areas. G: Large hollow tree at La Selva, Costa Rica, being checked out for roosting bats.**

Ochroma) and fruit sources *(Piper, Ficus, Solanum)* which would qualify as keystone mutualists for bats. It is clear from observed bat-plant inter-actions that few of the plant species which depend on bats for pollination or seed dispersal are sufficiently constant (in reproductive activity) or

127

abundant to support even a single bat species. In contrast, each keystone mutualist may support an entire bat assemblage for part of the season.

Another kind of keystone mutualist is a plant which produces large quantities of extrafloral nectar or accessible floral nectar. Smiley and Gilbert (unpublished) aimed time-lapse cameras at the leaf nectary of a large *Passiflora ambigua* vine at La Selva. In addition to *Pseudomyrmex* ants visiting the nectar glands, dozens of Hymenoptera, Hemiptera, and Diptera not otherwise involved with *P. ambigua* exploited the nectar resource. In temperate zone forests similar relationships between parasitoids of tree-feeding insects and certain nectar plants have been studied. Hassan (1967) found 148 species and 1,535 individual hymenopterous parasitoids associated with flowering herbs and shrubs within north German forests. *Daucus* (Umbelliferae) qualifies as a keystone mutualist in attracting 68 species during that study.

The study of Syme (1977) underscores the significance of Hassan's data to forest stability. Syme showed that *Asclepias* and *Daucus* nectar increases the reproductive longevity of parasitoids (which would, in turn, decrease chances for herbivore outbreak). Anderson (1976) reviewed the role of egg parasitoids in reducing rates of defoliation of forest trees. Kulman (1971) summarized experimental and observational data relating defoliation to growth and mortality of trees. Tying these observations together, it is obvious that the energetic support of the most minute Hymenoptera has demographic consequences for trees (through reduction of defoliation and other damage) and thus is of concern in forest reserve management. Key plants for the parasitoids may occur in different microhabitats from the forest "pest" insects attacked. A wise policy in temperate zone and tropical forests, therefore, would be the management of successional communities so as to maintain plants which support these tiny "control" agents (Gilbert, 1977).

Ant Mosaic

Ants defend many plants against herbivores (Bentley, 1977), but some carry about and tend plant-feeding homopterans and lepidopterans. Other ants, such as *Atta* (Figure 2B) in the neotropics, are major defoliators of forest shrubs and trees (Lugo et al., 1973). Work by Leston and others on the pests of tree crops in the old world tropics has revealed a three dimensional "ant mosaic" which results from the nonoverlap of territories of a few dominant ant species. Different assemblages of subdominant ants coexist with each dominant species so that the mosaic is a patchwork of ant communities. A consequence of the ant mosaic important to agriculture and conservation is the fact that the immunity of a plant to disease and defoliating animals is a function of its position in the ant mosaic.

Recently Leston (1978) reported the existence of a similar ant mosaic in neotropical forests. He demonstrated that whether a tree was cut by *Atta* depended upon the dominant territorial species associated with the tree. Trees occupied by *Ectatomma* (Figure 2A, C) were cut since this ant does not defend trees against *Atta*. Conversely, the presence of other species prevented leaf removal by the leaf cutters. An unexplained aspect of *Atta* behavior has been the apparent "conservation" of favored tree species near its nests (see review in Gilbert, 1979). Leston's observations may help explain such intraspecific differences in susceptibility to *Atta* attack.

It is probable that the composition of the ant fauna in a tree will also affect its suitability to foliage-feeding monkeys such as howlers *(Alouatta)*. Anyone having been stung by neotropical *Paraponera* would not doubt their ability to deter other primates. For plants highly coevolved with ants (an extreme case of the ant mosaic), the defensive role of ants against vertebrates has been thoroughly studied (see review by Bentley, 1977).

Knowledge of the ant mosaic could be important to the management of small neotropical forest reserves. For example, if the number of reproductive individuals of a keystone mutualist species in an area is small, defoliation by *Atta* or howlers of all or part of the population could be sufficient to immediately eliminate important mobile links, with subsequent loss of species dependent upon the links. Just as the ant mosaic is manipulated to improve cocoa production in Africa (Majer, 1976), it could be manipulated when necessary to insure the flowering and fruiting of keystone mutualists in undersized reserves.

PLANT/ANIMAL RELATIONSHIPS AND PATCH DYNAMICS

The species diversity of animals which depend upon one or a few species of plants for critical resources is related to the population size and age structure of the appropriate plants in an area. In England, Ward and Lakhani (1977) found that 79 to 87 percent of between-site variation in species diversity of juniper feeding insects was due to variations in the numbers of host bushes available at each site. They point out that the continuous availability of reproductive aged plants is crucial for the fruit feeding fraction of the juniper fauna. Among tropical birds, fruit and flower feeders are known to be "extinction prone" (Chapter 7) presumably because the area required to assure the continuous availability of such resources is much greater than that required for the continuous availability of insect food.

Since the persistence of specialized herbivores undoubtedly depends upon the density of host plants, one must consider the factors that influence plant populations (for example, Duffey, 1977). In the case of neotropical forests where extreme microhabitat specialization has been seen in plant species, a key factor to consider is the relative abundance and continuity of various kinds of microhabitats (Figure 5). In an important paper on this subject, Pickett and Thompson (1978) suggest that successional patches of various sizes and stages within an isolated biological reserve should be considered the only source for colonizing future openings in the forest. They apply island biogeography theory to systems of microhabitat islands within a reserve, and suggest that an important factor to consider in the design of nature reserves is the size distribution, spacing and turnover rates of microhabitat patches.

The factors that influence such "patch dynamics" are of concern to both the design and management of nature reserves. These include topographic, geologic and climatic features of a region (see Chapter 5 for a detailed discussion). For example, the frequency of landslides per unit area increases from level to inclined terrain and from earthquake-free to earthquake-prone regions. In contrast, the rate of succession back to mature forest may remain relatively constant over these same gradients. Therefore, since the relative area of any particular successional stage is a dynamic balance of disturbance and succession, areas with intrinsically higher rates of disturbance (Figure 5E) will be more likely to have all stages of succession constantly available within a unit area.

Key elements of food web organization are related to habitat patchiness in Figure 6. This figure includes many of the animal/plant interactions explained in this chapter. The interactions shown in the figure are based on real cases but are oversimplified to enhance understanding of the major points in the chapter. Plant species are represented by solid hexagons (the sizes of which indicate abundance) and are organized into six chemically distinct higher taxa (shown as six "islands" in the figure). Each "island" possesses three different microhabitats: successional (light gray), forest understory (medium gray) and forest canopy (dark gray). Host specialist herbivores and their specialized parasites are shown as small solid triangles and small solid boxes, respectively. Insects are arranged above their favored hosts and microhabitat. Generalist herbivores are represented by a large triangle at the center of the figure, typified by the most habitat and host-plant generalized herbivore in the system: leaf-cutter ants *(Atta)*. Mobile link groups important in pollination, seed dispersal or plant protection are represented by the six large circles (labeled A to F). These are connected by thin lines to plants with which the mobile links interact. Interactions between mobile links and plant groups not represented in the figure are indicated by outward-radiating lines. Plants critical to the support of particular groups of mobile links, keystone mutualists, are connected to appropriate link taxa with heavy arrows. Specification of plant groups represented in Figure 6 (Roman numerals I through VI) and a brief discussion of the ecological details shown is worth emphasis.

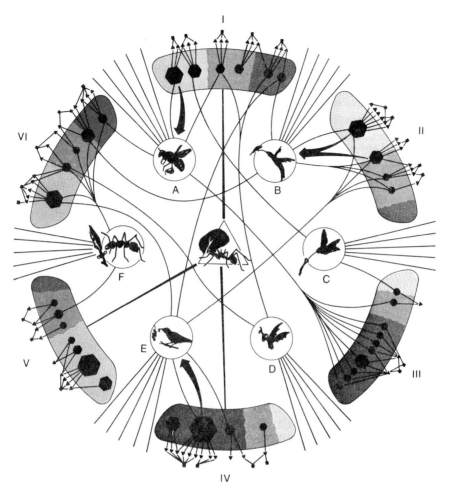

FIGURE 6.   Idealized scheme to show how the bulk of neotropical diversity is organized by chemical mosaic and mutualism (see text).

I. The *Solanum* family has representatives in all microhabitats. Note that one successional species is a keystone mutualist for orchid bees (A). One canopy member, *Juanulloa,* is pollinated by hummingbirds (B), but dispersed by frugivorous birds (E). Bats (D) pollinate one canopy species *(Markea),* and disperse an understory species *(Solanum).*

II. *Heliconia* has representatives in successional habitats and in the forest understory. The two species on the left are key resources for hermit hummingbirds. None are represented in the canopy.

III. Orchids are represented primarily in the canopy and most exist at relatively low densities as shown. One species shown is pollinated by hawk moths (C), the remainder by male euglossine bees (A). Orchids have wind-dispersed seeds. Also, many are ant-defended but this is not shown.

IV. Canopy-emergent trees (family not specified) can be key resources for complexes of frugivorous birds. Juvenile stages of one species are shown in the understory. As a rule, vines and trees which reproduce in the forest canopy are nevertheless represented as juveniles in other microhabitats where leaves are available to a slightly different set of herbivores. This has been largely omitted for simplicity.

V. Grasses are poorly represented in the canopy and have relatively few mutualistic interactions. One species in the diagram has seeds dispersed by ants (F). A few (not shown) are insect pollinated.

VI. Passion flower vines are characterized by ant-visited extrafloral nectaries. *P. vitafolia*, a canopy-emergent, presents flowers in the understory which are visited by hummingbirds. Herbivores "partition" *P. vitafolia* by microhabitat as shown. *Passiflora* are largely free of *Atta* herbivory (possibly because of ant defense).

In virtually all cases the pollination, seed dispersal and defense of a plant species are carried out by different animals as indicated in plant groups I and II. For simplicity, most such multiple interactions are not indicated. Note also that most mobile links shown visit plants in several microhabitats.

As stressed in Figure 6, coevolved food webs, mobile links and keystone mutualists all depend directly or indirectly on local microhabitat heterogeneity (expressed as a balance of successional stages). This heterogeneity may be lost in two important ways. First, if a reserve is too small relative to the size of the largest disturbance patch type, elements of mature forest may all be lost at once with no ready source for recolonization (Pickett and Thompson, 1978). Second, if the reserve is too small relative to the natural disturbance rate, certain patch types might disappear temporarily through succession. While the same type patch might be returned by a later disturbance, there may be no colonization source for the plants and animals previously associated with it if the reserve is isolated.

In areas where the first problem is of concern the only practical answer seems to be to design reserves with the largest possible disturbance in mind. The second problem is more likely to occur in flat lowland forest. In this latter case, managing the disturbance rate seems a necessity in reserves too small to insure a continuum of all the important successional microhabitats. I would include the Barro Colorado Island reserves and, if it becomes further isolated, the OTS-La Selva reserve in this category. Because of important links among species which characterize different successional stages, even the temporary loss of one such stage will have a ripple effect throughout the entire system. From the perspective of plants and their closely associated animals, the prevalent idea that early successional areas need not be considered in forest reserve management is a serious misconception that is not restricted to the tropics (Wright, 1974).

CONSERVATION BIOLOGY, AGRICULTURE, AND FORESTRY

Considering both the biological complexities and the sociopolitical problems associated with the development and management of neotropical nature reserves, I believe that conservation biology should be concerned with two major questions involving the interaction of biological reserves with agriculture. First, what forms of agriculture are most compatible with the management of nature reserves? And second, what kinds of practical benefits might nature reserves have for surrounding agroecosystems?

The value of agricultural practices might be twofold. First, if primitive agricultures which maintain high crop diversity (Soemarwoto et al., 1975) or active interaction between natural habitats and crops (Wilken, 1977) are encouraged in a buffer zone around biological reserves, sources for colonization of successional patches within the reserve would be increased. Second, the primitive agricultural zone would buffer the area against harsher forms of land use and diminish political pressures for alternative uses of the reserve itself. However, it is likely that such traditional agriculture will, eventually, require subsidies from conservation organizations.

The values of biological reserves to agriculture have not been adequately addressed by conservationists. In the long run such areas are a valuable source of new cultivars, medicines and industrial products and biological control agents. In addition, new ideas to improve forestry and agriculture may originate from work in reserves. Unfortunately, at present, these values are less obvious in areas where high energy input monocultures dominate agricultural practices. It is already becoming clear that current practices cannot provide stable solutions to our food, fiber and wood needs; as energy supplies decline, this failure will be even more obvious. Conservation biology must anticipate the future needs of forestry and agriculture and develop clear connections between the preservation and management of biological diversity and these "practical" aspects of land use.

An example of the sort of situation which requires an integration of forestry, agriculture and conservation is the problem of *Passiflora mollissima* in Hawaii. *P. mollissima,* a native of the Andes, was introduced along with about 15 other *Passiflora* species. One, *P. edulis,* is a valuable crop below 1,500 ft. *P. mollissima* has become a serious pest, (the "kudzu"* of Hawaii), blanketing about 84,000 acreas of forest, killing trees and causing economic devastation to the local forest industry. The spread of *P. mollissima* threatens endemic forest species and should be of concern to national park officials and others responsible for the remnants of the Hawaiian flora. Meanwhile, the solution to the dilemma may exist in neotropical reserves where intact food webs based on *Passiflora* suggest ways to control *mollissima* without injury to the *P. edulis* industry (Gilbert, 1977; Waage, Smiley and Gilbert, in preparation). Unfortunately, the fragmented structure of our academic, research and funding institutions tends to isolate the disciplines of basic ecology, agriculture, conservation and environmental protection. At best, there is little overlap between these areas; at worst, there is active antagonism. It is incumbent upon those concerned with the long-term biological diversity of the earth to ignore the traditional boundaries of scientific subdisciplines and to seek, instead, more holistic solutions.

SUMMARY

The structure of neotropical forests as determined by animal/plant interactions is highly organized (Figure 6). The system consists of many parallel, structurally similar but taxonomically different, food webs based on particular groups of plants. Mutualism plays a crucial role in the maintenance of diversity in the system. Mobile links are animals required by many plants for reproduction and dispersal. Keystone mutualists are plants which support link organisms and indirectly support the food webs which depend upon mobile links for all or part of their species richness. Finally, because ants control rates of herbivory on many plants, the existence of a mosaic of different dominant, territorial species of ant creates a subtle but important form of patchiness.

These interactions occur against a patchwork of microhabitats defined by the size of disturbance and successional status. Many key organisms in the system are restricted to one microhabitat while others, especially mobile links, depend on the constant availability of several. The need for management of disturbance rates in neotropical reserves increases toward areas with less topographic relief and smaller size.

I suggest the following research priorities for neotropical conservation biology: (1) Further analysis of plant specific food webs; (2) Autecology of link species, keystone mutualists and dominant ant species; (3) Classification of microhabitats and study of patch dynamics; (4) Integration of conservation, agriculture and forestry.

Finally, it should be pointed out that such research should be an integral part of managing the diversity of neotropical forests.

*Kudzu is an exotic legume vine *(Pueraria)*. Introduced into the Southeastern U.S. for erosion control, it is now a serious forest pest.

SUGGESTED READINGS

Baker, H. G., 1973, Evolutionary relationships between plants and animals in American and African tropical forests, in *Tropical Forest Ecosystems in Africa and South America,* Meggers et al. (eds.), Smithsonian Inst. Press, Washington D.C., pp. 145–159. A comparison of new and old world tropical forests from the perspective of plant reproductive biology.

Brown, K. S., 1978, Heterogeneidade: fator fundamental na teoria e práctica de conservacão de ambientes tropicais, in *I Encontro Nacional Sobre a Preservacão da Fauna e Recursos Faunisticos,* (Brasilia, 1977). I.B.D.F., Brasila. The first Portuguese language discussion of some of the ideas presented in this paper.

Futuyma, D., 1973, Community structure and stability in a constant environment, *Amer. Nat.,* 107, 443–446. An important theoretical discussion of tropical diversity maintenance.

Gilbert, L. E., 1979, Development of theory in the analysis of insect-plant relationships, in *Analysis of Ecological Systems,* D. Horn, R. Mitchell and G. Stairs (eds.), Ohio State Univ. Press, Columbus. A general review of insect-plant interactions from a perspective of ecological theory.

Gilbert, L. E. and P. Raven (eds.), 1975, *Coevolution of Animals and Plants,* Univ. of Texas Press, Austin. A diverse collection of papers dealing with animal-plant interactions.

Janzen, D. H., 1968, Host plants as islands in evolutionary and contemporary time, *Amer. Nat.,* 102, 592–595. A classic paper on the evolutionary ecology of plant-based insect faunas.

Mound, D. and N. Waloff, (eds), 1978, *Diversity of Insect Faunas,* Blackwell Scientific Publications, London. The most recent symposium on insect diversity.

Pickett, S. T. A. and J. N. Thompson, 1978, Patch dynamics and the design of nature reserves, *Biol. Conserv.,* 13, 27–37. The key paper on plant community dynamics and conservation.

Price, P. W., 1977, General concepts on the evolutionary biology of parasites, *Evolution,* 31, 405–420. A good introduction to the biology of host-restricted animals.

10

The Tropical Rain Forest: A Nonrenewable Resource

A. Gómez-Pompa, C. Vázquez-Yanes, S. Guevara

There is a popular opinion that the tropical rain forests because of their exuberant growth, their great number of species, and their wide distribution will never disappear from the face of the earth.

On the other hand, it has often been stated that the tropical rain forests (tall evergreen forests in tropical warm and humid regions) around the world must be protected and conserved for the future generations (*1*). It has also been stated that it is most important that knowledge about the structure, diversity, and function of these ecosystems has priority in future biological research (*2*). Unfortunately, either these voices have not been heard or their arguments have not been convincing enough

The authors are in the department of botany at the Institute of Biology, National University of Mexico, Mexico 20, D.F.

to promote action in this direction.

It is the purpose of this article to provide a new argument that we think is of utmost importance: the incapacity of the rain forest throughout most of its extent to regenerate under present land-use practices.

Even though the scientific evidence to prove this assertion is incomplete, we think that it is important enough to state and that if we wait for a generation to provide abundant evidence, there probably will not be rain forests left to prove it.

During the last few million years of their evolution, the rain forests of the world have produced their own regeneration system through the process of secondary forest succession. This regeneration system evolved in the many clearings that occurred naturally as a result of river floods, storms, trees

that die of age, and the like. The genetic pool available for recolonization was great, and a number of populations and species with characteristics that were advantageous in the rapid colonization of such breaks in the continuity of the primary rain forest were selected. These plants were fast-growing heliophytes, with seeds that have dormancy and long viability, and efficient dispersal mechanisms (*3, 4*). These sets of species played a fundamental role in the complex process of regeneration of the rain forest, and it is astonishing that very little is known about their biology, their behavior in the succession, and their evolution, even though they are the key to understanding the process of secondary succession, which is one of the most important ecological phenomena. The few works on the subject point out that there are certain repetitive patterns that can be predicted and that the species involved are fundamentally different from the primary species (*5–7*).

It is still uncertain how most of the primary species of the rain forest reproduce themselves and how the forest is regenerated, but from the evidence available it seems that there is a very complex system working at different times and in different directions, depending on the local situation and the plants involved (*3*).

One of the most important aspects

of natural regeneration is that on the floor of the primary rain forest there are always seedlings of young plants of many of the primary tree species. Under the effects of disturbance these seedlings will continue growing at an increased rate (8), and, at the same time, the secondary species start growth from dormant seeds in the soil. After several years the primary species will have grown taller than the secondary ones, and the major step in the regeneration has been accomplished.

There are taking place at the same time other processes, such as the colonization of trees and shrubs by epiphytic plants, about which even less is known, as well as the growth and establishment of climbing plants that has occurred probably since the early stages and of which many species will grow to the upper canopy of the rain forest. In all these cases the key plants are the seedlings of tall primary trees that will take over the upper canopy of the old successional series.

The regeneration of these primary species by the seedlings or young plants inside the forest is not the only possible way. Another important means of regeneration comes from seeds in the soil (9). In tropical rain forests this type of regeneration seems to be very effective mainly for species that happen to be in fruit during the disturbance of the area, because apparently there is a very short dormancy of the seeds of most of the primary species and the entire life of seeds of tropical tree species is in many cases very short (8, 10–12).

It seems appropriate to mention that we realize, of course, that the many, and quite different, primary tree species in the rain forests of the world may behave very differently in their germination responses and life-span. The available evidence indicates, nevertheless, that many primary tree species have large seeds (8, 13) with either short dormancy or none at all (10). The biological implications of this phenomenon are barely known (14), but it seems that the general trend is toward rapid germination, which is usually advantageous to the survival of the species. If one considers the predators of all types (fungi, bacteria, animals of various types) that are present in tropical warm and humid conditions, it seems reasonable to attribute survival power to the species, the seeds of which can germinate quickly and the seedlings of which can remain

alive for a long time in a slow-growing condition (8). The scarcity of seeds of primary trees stored in soils from rain forests has been demonstrated in one area in Mexico (9), but many more studies in this direction are needed, even though this fact has been noted earlier (8, 11).

Another possibility for the establishment of primary tree species in the early stages of regeneration is by long-distance dispersal by birds and by other animals such as monkeys, rodents, and others.

Very little is known about fruit and seed dispersal of tropical forest species (12, 15), but it seems that long-distance dispersal has played, and is playing, an important role in areas where human disturbance has not reached a critical level. The phenomenon of long-distance dispersal of tropical trees has almost never attracted the attention of researchers, but it may be extremely important in understanding the evolution of local populations and the adaptation to local ecological conditions (16), even though, from the point of view of regeneration, it may have little importance. Still another means of reestablishment of primary species in the early stages of succession is vegetative reproduction by means of rhizomes, bulbs, and roots that may remain alive after the destruction of the original forest and become active soon after the disruption.

Man and the Tropical Rain Forest

All that has been considered to this point concerns natural regeneration caused by natural catastrophes. The regenerative system of the rain forest seems to be very well adapted to the activities of primitive man. The use of small pieces of land for agriculture and their abandonment after the decrease of crop production (shifting agriculture) is similar to the occasional destruction of the forest by natural causes (17, 18). This type of activity can still be seen in many tropical areas where a mosaic pattern can be found, with large pieces of primary rain forest and patches of disturbed forest of different ages from the time of their abandonment. Several studies of these successional series are available (9, 17, 19–21), and in most cases they tend to agree that shifting agriculture has been a natural way to use the regenerative properties of the rain forest

for the benefit of man. How this operates is not well known, but we can extrapolate our knowledge of the natural regeneration of the rain forests and compare it with the data available. After the abandonment of the land by the primitive farmer, regeneration starts with the available seeds and other propagules in the soil. At present the seeds known to remain viable in the soil are mainly those of secondary species (9). After cultivation of an area, the possibility of any seedlings or young plants of the trees of the closest primary forest persisting is almost nil, and, because of the time involved in crop production, most of the future propagules of primary trees have to come in by natural dispersal (such as animals, water, gravity, or air).

There are several problems involved in connection with the speed of regeneration under these circumstances, but, in general, one can say that under the shifting cultivation system, the genetic pool of primary trees is retained, and from this pool comes the raw material for the successional processes. Of course, this is true in all cases where demographic pressure has not forced an intensive shifting agriculture with short periods of recovery, as in some tropical regions (20, 22).

We think that it is evident that the importance of retaining pieces of the original forest as the only way to reconstruct future forests cannot be overstressed. Whether the system of shifting agriculture is responsible for the extinction or simplification of some of the present tropical ecosystems is not completely certain, but it does seem clear that it prevents a mass extinction of species.

We cannot overstress the importance of the space factor in these considerations, because it makes an enormous difference if we use or destroy thousands of square kilometers or if we destroy one or two.

The Tropical Rain Forest in the Green Revolution Era

In recent times the trend in many tropical areas has been to look for ways to make permanent use of the land, in contrast with the old way of shifting cultivation. Permanent use can be accomplished with the help of the new technology and chemicals that have proved to be successful in many tropical areas.

Fig. 1. This destruction of a rain forest in southern Mexico is an example of what is occurring in large areas throughout the world.

These methods have opened greater possibilities for making available large extents of land for agricultural crops; the new trends can be seen in almost any tropical area today (Fig. 1). We shall not consider the problems of such methods and their possible consequences (21, 23, 24). Instead, we analyze this trend in relation to the natural regeneration process.

Under an intensive and extensive use of the land, sources of seeds of primary tree species for regeneration becomes less and less available because of the dispersal characteristics of those species and because of the scarcity of individuals of most of the tree species (25). The only species available that are preadapted for continuous disturbance are secondary species or primary species with some of the characteristics of the secondary ones (26). This group of species has characteristics that enable them to thrive in such conditions; they produce large numbers of seeds, which have means of long-distance dispersal and dormancy; these seeds accumulate and stay alive in the soil (that is, they have a long life-span). The process often called "savannization" and "desertization" (27, 28) of the tropical humid regions can very well be explained by these characteristics.

Also, plants preadapted for disturbance such as ones from drier environments with built-in adaptations to re- main alive in a dormant condition for long periods of time may invade these areas and allow them to regenerate a forest vegetation.

An ecosystem consisting of secondary species mixed with species from drier environments will become established. Since these species are generally lower in stature at maturity, the vegetation will also be lower in stature than the one the climate can allow. According to this view, some of the vegetation types that have these characteristics—for example, some low semi-evergreen selvas, savanna woodlands, and savannas in Mexico (29), as well as some in Asia (28), Africa (30), and South America (27, 31)—may be the product of an extensive and intensive shifting agriculture of old cultures. Some of the old arguments on the effects of fire for the explanation of many of the anthropogenous savannas and savanna woodlands can be explained better with the idea that we propose of the mass extinction of many tall tree species.

Thus there may be a possibility of bringing back some of these areas to a tall forest condition by introduction of the proper trees. There is, however, a great lack of information about population differentiation in tropical tree species, and research in this field is urgently needed for making basic recommendations in tropical rain forest management (16, 18).

An example of this problem is the failure of a project of one of us (A. G.-P.) for study of population differentiation in a tropical rain forest species, Terminalia amazonia. Seed populations were collected from Central America and Mexico, and, after germination, the young plants were transplanted to an introduction station at the Mexican site of the collections. The study had to be discontinued, for all the seedlings from the populations from Central America were exterminated by predators, especially ants, and comparisons could not be made. It is interesting that in this case probably there has been evolution in connection with the chemical protection against local predators which is not reflected in the morphology because morphological differences can hardly be distinguished. This study needs to be repeated with other species, but it shows that there is a potential problem in induced regeneration by the introduction of seeds from distant populations. This idea has also been developed from the problem of biological control of tropical pests (24).

Other Implications

All the facts and ideas mentioned lead us to the conclusion that, with the present rate of destruction of the tropical rain forests throughout the world, there is great danger of mass extinction of thousands of species. This is due to the simple fact that primary tree species from the tropical rain forests are incapable of recolonizing large areas opened to intensive and extensive agriculture. There has been a long controversy among persons responsible for the intensive use of land in the tropics, and it seems that the most important argument has been that countries like those of Europe, the United States, and some temperate Asiatic ones (Japan) have used the land intensively and extensively and there is not much evidence of mass disappearance of species. In view of the successional processes already discussed and with respect to the understanding of some biological properties of the species of northern temperate and cold areas, the explanation seems evident. In temperate areas the primary tree species are in many cases represented by a great number of individuals, and the distribution of many of the temperate species is large, and, in addition, many of them possess seeds adapted to long periods of inactivity, thereby conserving their vitality (dormancy and long life-span) (*32*) for periods of time while buried in the soil. Even though there are no reliable records of the life-span of seeds of trees buried in the soil of temperate regions, the available data known to us (*33*) suggest strongly the possibility that seeds stored in soils long retain their potential for growth. All these aspects yield a very different general behavior of the land cleared for agriculture and its possible future regeneration. It is also important to note that an isolated tree from a primary temperate forest has greater probability of survival than an isolated tree from a tropical rain forest (*18*); this is due to the complex and delicate net of relationships of each individual with the environment. This means that a gene pool of primary trees can be maintained along roads, near houses, and the like, for temperate areas but not for the tropical rain forest. If we add to these ideas the great difference in number of national parks, arboreta, botanic gardens, and storage facilities in many temperate areas in contrast with the virtual absence of such resources in the tropics, the problem grows to an even larger and more critical dimension.

All that we have said is applicable to tropical evergreen rain forests in the warm and humid areas of the world. In drier tropical areas with a definite long dry season the problem is very different, and the plants behave in connection with the problems of regeneration under intensive exploitation, in a manner more similar to those of temperate areas. The reason for this is that these plants are in some ways preadapted to great disturbances since they possess better characteristics for survival during periods of adverse conditions (drought, fire).

Conclusions

All the evidence available supports the idea that, under present intensive use of the land in tropical rain forest regions, the ecosystems are in danger of a mass extinction of most of their species. This has already happened in several areas of the tropical world, and in the near future it may be of even greater intensity. The consequences are nonpredictable, but the sole fact that thousands of species will disappear before any aspect of their biology has been investigated is frightening. This would mean the loss of millions and millions of years of evolution, not only of plant and animal species, but also of the most complex biotic communities in the world.

We urgently suggest that, internationally, massive action be taken to preserve this gigantic pool of germplasm by the establishment of biological gene pool reserves from the different tropical rain forest environments of the world.

References and Notes

1. C. G. G. J. van Steenis, *Micronesia* **2**, 65 (1965); E. J. H. Corner, *New Phytol.* **45**, 192 (1946).
2. P. W. Richards, *Atas Simp. Biota Amazonica* **7**, 49 (1967).
3. A. Gómez-Pompa, *Biotropica* **3**, 125 (1971).
4. C. Vázquez-Yanes and A. Gómez-Pompa, in *Simp. Latinoam. Fisiol. Vegetal. Qua. Lima, Perú, 20 to 26 Sept.* (1971), pp. 79–80; C. G. G. J. van Steenis, in *Study of Tropical Vegetation. Proc. Kandy Symp., Kandy, Ceylon, 19 to 21 March 1956* (1958), pp. 212–218.
5. G. Budowski, *Turrialba* **15**, 40 (1965).
6. C. F. Symington, *Malayan Forest.* **2**, 107 (1933); A. Gómez-Pompa, J. Vázquez-Soto, J. Sarukhán, *Publ. Especial Inst. Nacl. Invest. Forest. (México)* **3**, 20 (1964).
7. J. Sarukhán, *Publ. Especial. Inst. Nacl. Invest. Forest. (México)* **3**, 107 (1964).
8. P. W. Richards, *The Tropical Rain Forest* (Cambridge Univ. Press, London, 1952).
9. S. Guevara and A. Gómez-Pompa, *J. Arnold Arboretum Harvard Univ.* **53**, 312 (1972).
10. W. Croker, *Botan. Rev.* **4**, 235 (1938); R. C. Barnard, *Malayan Forest Res. Inst. Res. Pamphlet*, No. 14 (1954); T. B. McClelland, *Proc. Fla. State Hort. Soc.* **57**, 161 (1944); J. Marrero, *Caribbean Forest.* **4**, 99 (1942).
11. J. P. Schulz, *Ecological Studies on Rain Forest in Northern Surinam* (Noord-Hollandsche Uitgevers Maatschappij, Amsterdam, 1960), p. 226.
12. L. van der Pijl, *Proc. Kon. Ned. Akad. Wetensch.* **69**, 597 (1966); *Principles of Dispersal in Higher Plants* (Springer-Verlag, Berlin, 1969), p. 87.
13. ——, *Biol. J. Linnean Soc.* **1**, 85 (1969).
14. D. H. Janzen, *Evolution* **23**, 1 (1969).
15. N. Smythe, *Amer. Natur.* **104**, 23 (1970).
16. A. Gómez-Pompa, *J. Arnold Arboretum Harvard Univ.* **48**, 106 (1967).
17. ——, *Bol. Divulgación Soc. Mex. Hist. Nat.* **6**, 5 (1971).
18. G. N. Baur, *The Ecological Basis of Rainforest Management* (Blight Government Printer, New South Wales, 1968).
19. H. C. Conklin, *FAO Forest. Develop. Pap.* **12**, 1 (1957); E. Hernández X., *Chapingo* **2**, 1 (1962); M. A. Martínez, *Bol. Especial Inst. Nacl. Invest. Forest. (México)* **7**, 1 (1970); P. H. Nye and D. J. Greenland, *The Soil under Shifting Cultivation* (Commonwealth Bureau of Soils, Harpenden, 1960); M. Sousa, *Publ. Especial Inst. Nacl. Invest. Forest. (México)* **3**, 91 (1964).
20. J. M. Blaut, in *Symposium on the Impact of Man on Humid Tropics Vegetation, Goroka* (1960), pp. 185–198.
21. J. Kadlec, Coordinator, *Man and the Living Environment, Workshop on Global Ecological Problems* (Univ. of Wisconsin Press, Madison, 1972), p. 176.
22. W. R. Geddes, in *Symposium on the Impact of Man on Humid Tropics Vegetation, Goroka* (1960), pp. 42–56; A. Dilmy, *ibid.*, pp. 119–122; H. Sioli and H. Klinge, in *Int. Symp. Stalzenau*, The Hague (1961), pp. 357–363.
23. F. Tamesis, in *Proc. Forest. Congr. 5th Seattle* (1960), pp. 2025–2032; H. C. Conklin, *Trans. N.Y. Acad. Sci.* **17**, 133 (1954).
24. D. H. Janzen, *Bull. Ecol. Soc. Amer.* **54**, 4 (1970).
25. G. A. Black, Th. Dobzhansky, C. Pavan, *Botan. Gaz.* **111**, 413 (1950); M. E. D. Poore, *J. Ecol. (Jubilee Symposium)* **52**, 213 (1964); F. R. Fosberg, *Trop. Ecol.* **11**, 162 (1970); V. M. Toledo and M. Sousa, *Bol. Soc. Botán. Méx.*, in press; other papers on the same subject appear in *Speciation in Tropical Environments*, R. H. Lowe-McConnell, Ed. (Academic Press, New York, 1969).
26. A. Gómez-Pompa, *Estudios Botánicos en la Región de Misantla, Veracruz* (Ediciones del Instituto Mexicano de Recursos Naturales Renovables, A. C., Mexico, D.F., 1966), p. 104.
27. G. Budowski, *Turrialba* **6**, 22 (1956); F. W. Went and N. Stark, *BioSci.* **18**, 1035 (1968).
28. M. Schmid, in "Study of tropical vegetation," *Proc. Kandy Symp., Kandy, Ceylon, 19 to 21 March 1956* (1958), pp. 183–192.
29. F. Miranda and E. Hernández X., *Bol. Soc. Botán. Méx.* **28**, 29 (1963); A. Gómez-Pompa, *ibid.* **29**, 76 (1964).
30. R. Sillans, thesis, Faculté des Sciences de l'Université de Montpellier (1958), p. 176.
31. M. G. Ferri, *Simposio o Cerrado* (Edit. Univ. de São Paulo, São Paulo, 1963).
32. L. V. Barton, *Seed Preservation and Longevity* (Hill, London, 1961); Anon., *Misc. Publ. U.S. Dept. Agri.* **654**, 290, 361 (1948).
33. N. W. Olmstead and J. D. Curtis, *Ecology* **28**, 49 (1946); H. J. Oosting and M. E. Humphreys, *Bull. Torrey Bot. Club* **67**, 253 (1940).
34. Flora of Veracruz, contribution number 12. A joint project of the Institute of Biology of the National University of Mexico and the Arnold Arboretum and Gray Herbarium of Harvard University, to prepare an ecological floristic study of the state of Veracruz, Mexico. Information about this project was published in *Anal. Inst. Biol. Univ. Nacl. Mex. Ser. Bot.* **41**, 1–2. Partially supported by NSF grant GB-20267X.

11

Reprinted from *Biol Conserv.* 7:129–146 (1975)

THE ISLAND DILEMMA: LESSONS OF MODERN BIOGEOGRAPHIC STUDIES FOR THE DESIGN OF NATURAL RESERVES

JARED M. DIAMOND

Physiology Department, University of California Medical Center, Los Angeles, California 90024, USA

ABSTRACT

A system of natural reserves, each surrounded by altered habitat, resembles a system of islands from the point of view of species restricted to natural habitats. Recent advances in island biogeography may provide a detailed basis for understanding what to expect of such a system of reserves. The main conclusions are as follows:

The number of species that a reserve can hold at equilibrium is a function of its area and its isolation. Larger reserves, and reserves located close to other reserves, can hold more species.

If most of the area of a habitat is destroyed, and a fraction of the area is saved as a reserve,[2] the reserve will initially contain more species than it can hold at equilibrium. The excess will gradually go extinct. The smaller the reserve, the higher will be the extinction rates. Estimates of these extinction rates for bird and mammal species have recently become available in a few cases.

Different species require different minimum areas to have a reasonable chance of survival.

Some geometric design principles are suggested in order to optimise the function of reserves in saving species.

INTRODUCTION

For terrestrial and freshwater plant and animal species, oceanic islands represent areas where the species can exist, surrounded by an area in which the species can survive poorly or not at all and which consequently represents a distributional barrier. Many situations that do not actually involve oceanic islands nevertheless possess the same distributional significance for many species. Thus, for alpine species a mountain top is a distributional 'island' surrounded by a 'sea' of lowlands;

for an aquatic species a lake or river is a distributional island surrounded by a sea of land; for a forest species a wooded tract is a distributional island surrounded by a sea of non-forest habitat; and for a species of the intertidal or shallow-water zones, these zones represent distributional islands compressed between seas of land and of deep water.

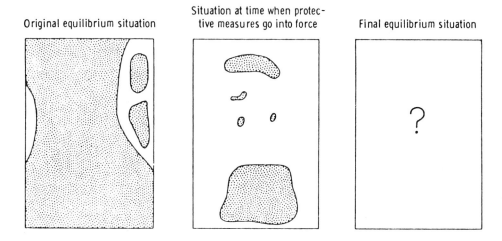

Original equilibrium situation   Situation at time when protective measures go into force   Final equilibrium situation

Fig. 1.   Illustration of why the problems posed by designing a system of natural reserves are similar to the problems of island biogeography. In the situation before the onset of accelerating habitat destruction by modern man, many natural habitats were present as continuous expanses covering large areas (indicated by shaded areas of sketch on left). Species characteristic of such habitats were similarly distributed over large, relatively continuous expanses. By the time that extensive habitat destruction has occurred and some of the remaining fragments are declared natural reserves, the total area occupied by the habitat and its characteristic species is much reduced (centre sketch). The area is also fragmented into isolated pieces. For many species, such distributions are unstable. Applying the lessons of modern island biogeography to these islands of natural habitat surrounded by a sea of disturbed habitat may help predict their future prospects.

Throughout the world today the areas occupied by many natural habitats, and the distributional areas of many species, are undergoing two types of change (Fig. 1). First, the total area occupied by natural habitats and by species adversely affected by man is shrinking, at the expense of area occupied by man-made habitats and by species benefited by man. Second, formerly continuous natural habitats and distributional ranges of man-intolerant species are being fragmented into disjunctive pieces. If one applies the island metaphor to natural habitats and to man-intolerable species, island areas are shrinking, and large islands are being broken into archipelagos of small islands. These processes have important practical consequences for the future of natural habitats and man-intolerant species (Preston, 1962; Willis, 1974; Diamond, 1972, 1973; Terborgh, in press, *a, b*; Wilson & Willis, in press). Ecologists and biogeographers are gaining increasing understanding of these processes as a result of the recent scientific revolution stemming from the work of MacArthur & Wilson (1963, 1967) and MacArthur (1972). In this paper I shall explore four implications of recent biogeographic work for conservation policies: (1) The ultimate *number* of species that a natural reserve will save is likely to be an increasing function of the reserve's area. (2) The *rate* at which species go extinct in a reserve is likely to be a decreasing function of the reserve's area. (3) The relation between reserved area and probability of a species' survival is characteristically different for different species. (4) Explicit suggestions can be made for the optimal geometric design of reserves.

HOW MANY SPECIES WILL SURVIVE?

Let us first examine the relation between reserve area and the number of species that the reserve can hold at equilibrium. As a practical illustration of this problem, consider the fact that we surely cannot save all the rain forest of the Amazon Basin. What fraction of Amazonia must be left as rain forest to guarantee the survival of half of Amazonia's plant and animal species, and how many species will actually survive if only 1% of Amazonia can be preserved as rain forest? Numerous model

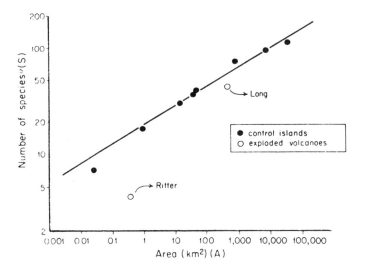

Fig. 2. Example of the relation between species number and island area in an archipelago. The ordinate is the number of resident, non-marine, lowland bird species (S) on the islands of Vitiaz and Dampier Straits near New Guinea in the south-west Pacific Ocean, plotted as a function of island area (A, in km²) on a double logarithmic scale. The points • represent relatively undisturbed islands. The straight line $S = 18.9 A^{0.18}$ was fitted by least mean squares through the points for these islands. Note that species number increases regularly with island area. The two points O refer to Long and Ritter Islands, whose faunas were recently destroyed by volcanic explosions and which have not yet regained their equilibrium species number.

systems to suggest answers to these questions are provided by distributional studies of various plant or animal groups on various archipelagos throughout the world. If one compares islands of different size but with similar habitat and in the same archipelago, the number of species S on an island is usually found to increase with island area A in a double logarithmic relation:

$$S = S_0 A^z \qquad (1)$$

where S_0 is a constant for a given species group in a given archipelago, and z usually assumes a value in the range 0.18-0.35 (Preston, 1962; MacArthur & Wilson, 1963, 1967; May, in press). A rough rule of thumb, corresponding to a z value of 0.30, is that a tenfold increase in island area means a twofold increase in the number of species. Figure 2 illustrates the species/area relation for the breeding land and freshwater bird species on the islands of the Bismarck Archipelago near New Guinea and shows that the number of bird species increases regularly with island area. If one compares islands of similar area but at different distances from the continent or large island that serves as the main source of colonisation, then one finds that the number of species on an island decreases with increasing distance. This feature is illustrated by Fig. 3, which shows that the number of bird species on

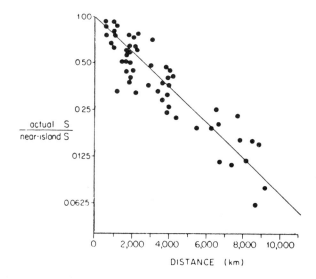

Fig. 3. Example of the relation between species number and island distance from the colonisation source in an island archipelago. The ordinate is the number of resident, non-marine, lowland bird species S on tropical south-west Pacific islands more than 500 km from New Guinea, divided by the number of species expected on an island of equivalent area less than 500 km from New Guinea. The expected near-island S was read off the species/area relation for such islands (Fig. 5). The abscissa is the island distance from New Guinea. Note that S decreases by a factor of 2 per 2600 km distance from New Guinea. (After Diamond, 1972.)

islands of the south-west Pacific decreases by a factor of 2 for each 2600 km of distance from New Guinea. For plants or animals with weaker powers of dispersal than birds, the fall-off in species number with distance is even more rapid.

Similar findings are obtained if, instead of oceanic islands, one compares habitat 'islands' within a continent or large island. For example, isolated as enclaves within the rain forest that covers most of New Guinea are two separate areas of savanna, which received most of their plant and animal species from Australia (Schodde & Calaby, 1972; Schodde & Hitchcock, 1972). The savanna which is larger and also closer to Australia supports twice as many savanna bird species as the smaller and more remote savanna (Fig. 4). Other examples are provided by mountains rising out of the 'sea' of lowlands, such as the isolated mountain ranges of Africa, South America, New Guinea and California. Thus, the number of bird species on each 'island' of alpine vegetation at high elevations in the northern Andes increases with area of alpine habitat and decreases with distance from the large alpine source area in the Andes of Ecuador (Vuilleumier, 1970).

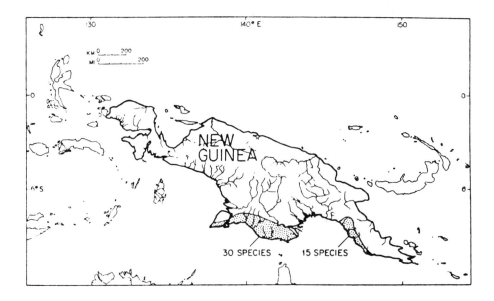

Fig. 4. Example of the relation between area of 'habitat islands' and the number of characteristic species they support. Most of New Guinea is covered by rain forest, but two separate areas on the south coast (shaded in the figure) support savanna woodland. The characteristic bird species of these savannas are mostly derived from Australia (the northern tips of Australia are just visible at the lower border of the figure). The so-called Trans-Fly savanna (left) not only has a larger area than the so-called Port Moresby savanna (right), but is also closer to the colonisation source of Australia. As a result, the Trans-Fly savanna supports twice as many bird species characteristic of savanna woodland (c. 30 compared with 15 species) as does the Port Moresby savanna.

Why is it that species number increases with increasing area of habitat but decreases with increasing isolation? In explanation of these findings, Preston (1962) and MacArthur & Wilson (1963, 1967) suggested that species number S on an island is set by (or approaches) an equilibrium between immigration rates and extinction rates. Species immigrate into an island as a result of dispersal of colonists from continents or other islands; the more remote the island, the lower is the immigration rate. Species established on an island run the risk of extinction due to fluctuation in population numbers; the smaller the island, the smaller is the population and the higher the extinction rate. Area also affects immigration and extinction rates in several other ways: through its relation to the regional magnitude of spatial and temporal variation in resources; by being correlated with the variety of available habitats as stressed by Lack (1973); and by being correlated with the number of 'hot spots', or sites of locally high utilisable resource production for a particular species (Diamond, in press). On a given island, extinction rates increase, and immigration rates decrease, with increasing S. The S value on an island in the steady state is the number at which immigration and extinction rates become equal. The larger and less isolated the island, the higher is the species number at which it should equilibrate.

The correctness of this interpretation has been established by several types of study. One has involved observing the increase in species number on an island whose fauna and/or flora have been destroyed. The most famous such study was provided by a 'natural experiment', the colonisation by birds of the vocanic island of Krakatoa after its fauna had been destroyed by an eruption in 1883 (Dammerman, 1948; see MacArthur & Wilson, 1967, pp. 43-51). Similar 'natural experiments' are provided by the birds of Long Island near New Guinea, whose fauna was destroyed by a volcanic eruption two centuries ago (see Fig. 2), and by the birds of seven coral islets in the Vitiaz-Dampier group near New Guinea, when a tidal wave destroyed the fauna in 1888 (Diamond, 1974). Simberloff & Wilson (1969) created an analogous 'artificial experiment' by fumigating several mangrove trees standing in the ocean off the coast of Florida and observing the recolonisation of these trees by arthropods. In all these studies, the number of species on the island returned within a relatively short time to the value appropriate to the island's area and isolation, confirming that this value really was an equilibrium value. Naturally, the rate of approach to equilibrium depends on the plant or animal group studied and the island's location: for example, successive surveys have shown the number of plant species on Krakatoa still to be rising and not yet to have reached equilibrium (Docters van Leeuwen, 1936; MacArthur & Wilson, 1967, p. 49).

Another type of test of the MacArthur-Wilson interpretation is provided by turnover studies at equilibrium. According to the MacArthur-Wilson interpretation, although the *number* of species on an island may remain near an equilibrium value, the *identities* of the species need not remain constant, because new species are continually immigrating and other species are going extinct. Estimates of immigration and extinction rates at equilibrium have been obtained by comparing surveys of an island in separate years. Such studies have been carried out for the birds of the Channel Islands off California (Diamond, 1969; Hunt & Hunt, 1974; Jones & Diamond, in press), Karkar Island off New Guinea (Diamond, 1971), Vuatom Island off New Britain (Diamond, in press), and Mona Island off Puerto Rico (Terborgh & Faaborg, 1973). All these studies found that a certain number of species present in the earlier survey had disappeared by the time of the later survey, but that a similar number of other species immigrated in the intervening years, so that the total number of species remained approximately constant unless there was a major habitat disturbance. As expected from considering the risk of extinction in relation to population size, most of the populations that disappeared had initially consisted of few individuals. The turnover rates per year (immigration or extinction rates) observed in these studies have been in the order of 0.2-6% of the island's bird species for islands of 300-400 km^2 area.

Thus, the number of species that a reserve can 'hold' at equilibrium is likely to be set by a balance between immigration rates and extinction rates. The set-point will be at a larger number of species, the larger the reserve or the closer it is to a source of colonists:

1. If 90% of the area occupied by a habitat is converted by man into another habitat and the remaining 10% is saved as an undivided reserve, one might expect to save roughly about half of the species restricted to the preserved habitat type, while the populations of the remaining half of the species will eventually disappear from the reserve. It should be stressed explicitly that increased habitat diversity is part of the reason, but not the only one (cf. p. 134 for others), why larger areas hold more species. Thus, even if a reserve does include some of the type of habitat preferred by a threatened species, the species may still disappear because of population fluctuations, spatial or temporal variation in resources, and too few or too small 'hot spots'.

2. If one saves two reserves, the smaller reserve will retain fewer species if it is remote from the larger reserve than it would if it were near the larger reserve.

3. As the contrast increases between the preserved habitat types and the surrounding habitat types, or between the ecological requirements of a threatened species and the resources actually available in areas lying between reserves, the results of island biogeographic studies become increasingly relevant. The greater this contrast, the lower will be the population density of the threatened species in the area between reserves, and the lower will be the species' dispersal rate between the reserves. To some species the intervening area may be no barrier at all, while to other species it may be as much of a barrier as the ocean is to a flightless mammal.

HOW RAPIDLY WILL SPECIES GO EXTINCT ?

Suppose that 90% of a habitat is destroyed and the remaining 10% is saved as a faunal reserve. The reserve will initially support most, though not all, species restricted to the original expanse of habitat. (The actual proportion of the species present in such a portion of a larger habitat is discussed on pp. 9-10 and 16 of MacArthur & Wilson (1967).) However, we have just seen that at equilibrium the reserve will support only about half the species of the original expanse of habitat. Thus, at the time that the reserve is set aside, it will contain more species than its area can support at equilibrium as an island. Species will go extinct until the new equilibrium number is reached. Such a reserve will constitute the exact converse of an island which has had its fauna destroyed: equilibrium of species number will be approached from above, by an excess of extinction over immigration, rather than from below, by an excess of immigration over extinction. The important practical question thus arises: how rapidly will species number 'relax' to the new equilibrium value? If equilibrium times were of the order of millions of years, these extinctions would not be a matter of practical concern, whereas a reserve that lost half of its species in a decade would be unacceptable.

A natural experiment that permits one to assess 'relaxation rates' as a function of the reserve's area is provided by so-called land-bridge islands (Diamond, 1972, 1973). During the late Pleistocene, when much sea-water was locked up in glaciers, the ocean level was about 100-200 m lower than at present. Consequently, islands separated from continents or from larger islands by water less than 100 m deep formed part of the continents or larger islands, and shared the continental faunas and floras. Examples of such 'land-bridge islands' are Britain off Europe, Aru and other islands off New Guinea, Tasmania off Australia, Trinidad off South America, Borneo and Java off south-east Asia, and Fernando Po off Africa. When rising sea levels severed the land-bridges about 10,000 years ago, these land-bridge islands must have found themselves supersaturated; they initially supported a species-rich continental fauna rather than the smaller number of species appropriate to their area at equilibrium. Gradually, species must have been lost by an excess of extinctions over immigrations. Figure 5 illustrates how far the avifaunas of the satellite land-bridge islands of New Guinea have returned towards equilibrium in 10,000 years. The larger land-bridge islands, with areas of several hundred to several thousand km^2, still have more bird species than predicted for their area from the species/area relation based on islands at equilibrium, though they do have considerably fewer bird species than New Guinea itself. That is, the larger land-bridge islands have lost many but not all of their excess species in 10,000 years. However, land-bridge islands smaller than about 250 km^2 at present have the same number of bird species as similar-sized oceanic islands that never had a land-bridge. Thus, the smaller land-bridge islands have lost their entire excess of bird species in 10,000 years.

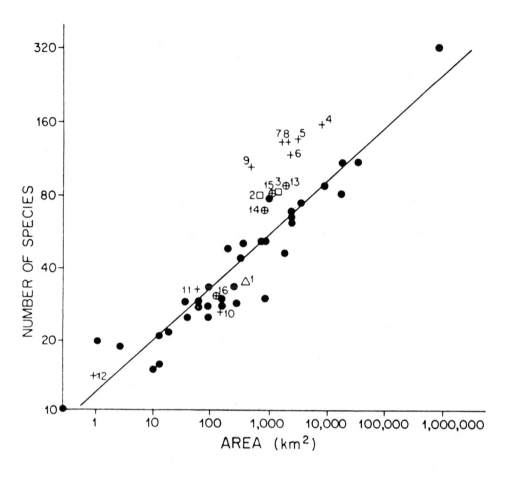

Fig. 5. Example of how one can use land-bridge islands to estimate extinction rates in the faunas of natural reserves. The ordinate is the number of resident, non-marine, lowland bird species on New Guinea satellite islands, plotted as a function of island area on a double logarithmic scale. The points ● are islands which have not had a recent land-connection to New Guinea and whose avifaunas are presumed to be at equilibrium. The numbered point △ (1) refers to a recently exploded volcano whose avifauna has not yet returned to equilibrium; points + (4-12), to islands connected to New Guinea by land-bridges at times of lower sea-level 10,000 years ago; points ⊕ (13-16), to islands formerly connected by land-bridges to some other large island but not to New Guinea itself; and points ⊓ (2-3), to islands that lie on a shallow shelf and had a much larger area at times of lower sea-level. Up to the time that the land-bridges were severed by rising sea-level. the New Guinea land-bridge islands (+, 4-12) must have supported nearly the full New Guinea quota of 325 lowland species (point in the upper right-hand corner). At present none of these land-bridge islands supports anything close to 325 species; the larger ones (+, 4-9) do, however, still have more species than expected at equilibrium (as given by points ● and the straight line); and the smaller ones (+, 10-12) already have about the number of species expected at equilibrium. The conclusion is that no land-bridge island has been able to hold more than half its initial number of species, but that the larger islands have been able to hold an excess of species for longer. The same conclusion follows from points ⊕ and ⊓. (From Diamond, 1972.)

The re-equilibration of land-bridge islands is the resultant of the extinction rate E (in species/year) exceeding the immigration rate I (in species/year) until an equilibrium species number S_{eq} is attained. Both I and E depend on the instantaneous species number $S(t)$, where t represents time (in years). As a highly simplified model, let us assume constant coefficients K_i and K_e (in year^{-1}) of immigration and extinction, respectively:

$$E = K_e S(t) \tag{2}$$

$$I = K_i[S^* - S(t)] \tag{3}$$

where S^* is the mainland species pool, and $[S^* - S(t)]$ is the number of species in the pool not present on the island at time t, hence available as potential immigrants. At equilibrium, when $dS/dt = I - E = 0$, S_{eq} is given by

$$S_{eq} = K_i S^*/(K_i + K_e) \tag{4}$$

If a land-bridge island initially (at $t = 0$) supports a species number $S(0)$ that exceeds S_{eq}, the rate at which $S(t)$ declines from $S(0)$ towards S_{eq} is obtained by integrating the differential equation

$$dS/dt = I - E = (K_i + K_e)[K_i S^*/(K_i + K_e) - S(t)]$$

with the boundary condition $S(t) = S(0)$ at $t = 0$, to obtain:

$$[S(t) - S_{eq}]/[S(0) - S_{eq}] = \exp(-t/t_r) \tag{5}$$

The relaxation time t_r is the length of time required for the species excess $[S(t) - S_{eq}]$ to relax to $1/e$ or 36.8% of the initial excess $[S(0) - S_{eq}]$, where e is the base of natural logarithms. Relaxation is 90% complete after 2.303 relaxation times.

As an example of the use of this formula, consider the land-bridge island of Misol near New Guinea. At the time 10,000 years ago when it formed part of New Guinea, Misol must have supported nearly the full New Guinea lowlands fauna of 325 bird species. With an area of 2040 km^2, Misol should support only 65 species at equilibrium, by comparison with the species/area relation for islands that lacked land-bridges and are at equilibrium. The present species number on Misol is 135, much less than the initial value of 325 but still in excess of the final equilibrium value. Substituting $S(0) = 325$, $S(t) = 135$, $S_{eq} = 65$, $t = 10,000$ years into eqn. (5) yields a relaxation time of 7600 years for the avifauna of Misol.

Similar calculations have been carried out for other land-bridge islands formerly connected to New Guinea, for islands formerly connected to some other large satellite island but not to New Guinea itself, and for islands that lie on a shallow-water shelf and that formerly must have been much larger in area although without

149

connection to a larger island. A similar analysis in a continental situation was made
by Brown (1971), who studied distributions of small non-volant mammals in
forests which are now isolated on the tops of mountains rising out of western North
American desert basins but which were formerly connected by a continuous forest
belt during times of cooler Pleistocene climates. Terborgh (in press, *a, b*) has made
a similar analysis of the avifaunas of Caribbean islands and has dramatically
confirmed the accuracy of his calculations by showing that they correctly predict
the extinction rates observed within the present century on Barro Colorado Island
(Willis, 1974). Both Terborgh's analyses of Caribbean birds and mine of New
Guinea birds show that relaxation times increase with increasing island area. Both
analyses also show that eqns. (2) and (3) are oversimplified: K_i actually increases
with $S(t)$, and K_i decreases with $S(t)$.

Thus, the gradual decline of species number from a high initial value to a lower
equilibrium value on land-bridge islands may furnish a model for what could
happen when a fraction of an expanse of habitat is set aside as a reserve and the
remaining habitat is destroyed. A small reserve not only will eventually contain few
species but will also initially lose species at a high rate. For reserves of a few km^2,
extinction rates of sedentary bird and mammal species unable to colonise from one
reserve to another are so high as to be easily measurable in a few decades. Within a
few thousand years even a reserve of 1000 km^2 will have lost most such species
confined to the reserve habitat. These estimates assume that man's land-use
practices do not grossly alter the preserved habitat. More rapid changes in species
composition are likely to occur if sylviculture or other human use changes the
habitat structure.

WHAT SPECIES WILL SURVIVE?

In the preceding pages we have considered the problem of survival from a statistical
point of view: what fraction of its initial fauna will a reserve eventually save, and
how rapidly will the remainder go extinct? We have not yet considered the survival
probabilities of individual species. If each species had equal probabilities of
survival, then it would be a viable conservation strategy to be satisfied with large
numbers of small reserves. Each such vest-pocket reserve would lose most of its
species before reaching equilibrium, but with enough reserves any given species
would be likely to be among the survivors in at least one reserve. In this section we
shall examine the flaw in this strategy: different species have very different area
requirements for survival.

The survival problem needs to be considered from two points of view: the chance
that a reserve where a species has gone extinct will be recolonised from another
reserve, and the chance that a species will go extinct in an isolated reserve. Consider
the former question first. Suppose that there are many small reserves. Suppose next

that a given species is incapable of dispersing from one reserve to another across the intervening sea of unsuitable habitat. The isolated populations in each reserve run a finite risk of extinction. If there is no possibility of recolonisation, each extinction is irrevocable, and it is only a question of time before the last population of the species disappears. Suppose on the other hand that dispersal from one reserve to another is possible. Then, although a species temporarily goes extinct in one reserve, the species may have recolonised that reserve by the time it goes extinct in another reserve. If there are enough reserves or high enough recolonisation rates or low enough extinction rates, the chances of the species disappearing simultaneously from all reserves are low, and the long-term survival prospects are bright. Dispersal ability obviously differs enormously among plant and animal species. Flying animals tend to disperse better than non-flying ones; plants with wind-borne seeds tend to disperse better than plants with heavy nuts. The more sedentary the species, the more irrevocable is any local extinction, and the more difficult will it be to devise a successful conservation strategy. Thus, conservation problems will be most acute for slowly dispersing species in normally stable habitats, such as tropical rain forest. Even power of flight cannot be assumed to guarantee high dispersal ability. For instance, 134 of the 325 lowland bird species of New Guinea are absent from all oceanic islands more than a few km from New Guinea, and are confined to New Guinea plus islands with recent land-bridge connections to New Guinea. Similarly, many neotropical bird families with dozens of species have not even a single representative on a single New World island lacking a recent land-bridge to South or Central America; and not a single member of many large Asian bird families has been able to cross Wallace's Line separating the Sunda Shelf land-bridge islands from the oceanic islands of Indonesia. Such bird species have insuperable psychological barriers to crossing water gaps, and are generally characteristic of stable forest habitats. Thus, low recolonisation rates may mean either that a species *cannot* cross unsuitable habitats (a mountain forest rodent faced by a desert barrier), or that it *will not* cross unsuitable habitats (some tropical forest birds faced by a water gap).

Having seen that species vary in their ability to recolonise, let us now consider how species vary in extinction rates of local populations. The New Guinea land-bridge islands again offer a convenient test situation (Diamond, 1972, in press). Recall that these islands initially supported most of the New Guinea lowlands fauna, that the land-bridges were severed about 10,000 years ago, that 134 New Guinea lowlands bird species do not cross water gaps, and that any extinctions of populations of these species on the land-bridge islands cannot therefore have been reversed by recolonisation. Virtually all these species are now absent from all land-bridge islands smaller than 50 km², because extinction rates on small islands are so high that virtually no isolated population survives 10,000 years. However, these 134 species vary greatly today in their distribution on the seven larger (450-8000 km²) land-bridge islands. At the one extreme, some species, such as the frilled monarch flycatcher *(Monarcha telescophthalmus)*, have survived on all seven islands. At the other extreme, 32 species have disappeared from all seven islands, and must be especially prone to extinction in isolated populations. Most of these 32 species fit into one or more of three categories: birds whose initial populations must have numbered few individuals because of very large territory requirements (*e.g.* the New Guinea harpy eagle *(Harpyopsis novaeguineae)*); birds whose initial populations must have numbered few individuals because of specialised habitat requirements (*e.g.* the swamp rail *(Megacrex inepta)*); and birds which are dependent on seasonal or patchy food sources and normally go through drastic population fluctuations (*e.g.* fruit-eaters and flower-feeders).

Another natural experiment in differential extinction is provided by New Hanover, an island of 1200 km² in the Bismarck Archipelago near New Guinea. In the late Pleistocene, New Hanover was connected by a land-bridge to the larger island of New Ireland and must then have shared most of New Ireland's species. Today New Hanover has lost about 22% of New Ireland's species, a fractional loss that does not sound serious. However, among these lost species are 19 of the 26 New Ireland species confined to the larger Bismarck islands, including every endemic Bismarck species in this category. That is, New Hanover differentially lost those species most in need of protection. As a faunal reserve, New Hanover would rate as a disaster. Yet its area of 1200 km² is not small by the standards of many of the tropical rain forest parks that one can realistically hope for today.

As a further example of a natural experiment in differential extinction, consider the mammals isolated on mountain tops rising from North American desert basins, mentioned in the previous section. Like the bird species restricted to the New Guinea land-bridge islands, the isolated populations of these mammal species have been exposed to the risk of extinction for the past 10,000 years, without opportunity for recolonisation. Today, some of these mammal species are still present on most of the mountains, while other species have disappeared from all but a few mountains. The species with the highest extinction rates are those whose initial populations must have numbered few individuals: either because the species is a carnivore rather than a herbivore, or because it has specialised habitat requirements, or because it is a large animal (Brown, 1971).

A method of quantifying the survival prospects of a species is to determine its so-called incidence function (Diamond, in press). On islands of the New Guinea region one notes that some bird species occur only on the largest and most species-rich islands; other species also occur on medium-sized islands; and others also occur on small islands. To display these patterns graphically, one groups islands into classes containing similar numbers of bird species (*e.g.* 1-4, 5-9, 10-20, 21-35, 36-50, etc.); calculates the *incidence J* or fraction of the islands in a given class on which a particular species occurs; and plots incidence against the total species number S on the island (Fig. 6). Since S is closely correlated with area, in effect these graphs represent the probability that a species will occur on an island of a particular size.

For most species, *J* goes to zero for *S* values below some value characteristic of the particular species, meaning that there is no chance of survival on islands below a certain size. These incidence functions can be interpreted in terms of the biology of the particular species (*e.g.* its population density, reproductive strategy, and dispersal ability). From these incidence functions one can estimate what chance a certain species has of surviving on a reserve of a certain size.

Thus, different species have different probabilities of persisting on a reserve of a given size. These probabilities depend on the abundance of the species and 'the

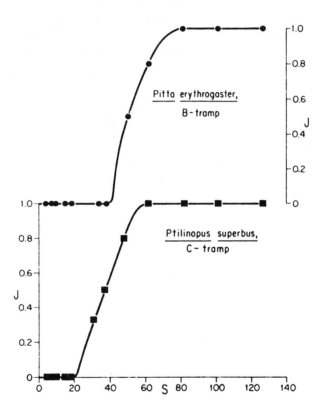

Fig. 6. So-called incidence functions for two bird species of the Bismarck Archipelago near New Guinea. The incidence *J(S)* is defined as the fraction of the islands with a given total number of bird species *S* that a given species occurs on. For example, the so-called B-tramp *Pitta erythrogaster* (•) is on all islands (*i.e. J* = 1.0) with *S* > 80, on about half of the islands (*J* = 0.5) with *S* around 55, and on no island (*J* = 0) with *S* < 40. Other bird species of the Bismarck Archipelago have different incidence functions: for example, the so-called C-tramp *Ptilinopus superbus* (■) is on all islands with *S* > 60 and on many islands (*J* = 0.3-0.8) with *S* = 30-50. Since *S* is mainly a function of island area, the message is that each species requires some characteristic minimum area of island for it to have a reasonable chance of surviving.

magnitude of its population fluctuations, and also on its ability to recolonise a reserve on which it has once gone extinct. Even on reserves as large as 10,000 km², some species have negligible prospects of long-term survival. Such species would be doomed by a system of many small reserves, even if the aggregate area of the system were large.

WHAT DESIGN PRINCIPLES WILL MINIMISE EXTINCTION RATES IN NATURAL RESERVES?

In the preceding sections we have examined how the eventual number of species that a reserve can hold is related to area, how extinction rates are related to area, and how area-dependent survival prospects vary among species. Given this background information, let us finally consider what the designer of natural

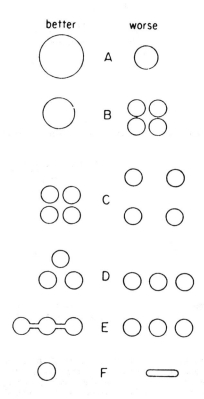

Fig. 7. Suggested geometric principles, derived from island biogeographic studies, for the design of natural reserves. In each of the six cases labelled A to F, species extinction rates will be lower for the reserve design on the left than for the reserve design on the right. See text for discussion.

reserves can do to minimise extinction rates (Diamond, 1972, 1973; Terborgh, in press, a, b, Wilson & Willis, in press). Figure 7 (modified from Wilson & Willis, in press) summarises a series of design principles, identified as A, B, C, D, E and F.

A large reserve is better than a small reserve (principle A), for two reasons: the large reserve can hold more species at equilibrium, and it will have lower extinction rates.

In practice, the area available for reserves must represent a compromise between competing social and political interests. Given a certain total area available for reserves in a homogeneous habitat, the reserve should generally be divided into as few disjunctive pieces as possible (principle B), for essentially the reasons underlying principle A. Many species that would have a good chance of surviving in a single large reserve would have their survival chances reduced if the same area were apportioned among several smaller reserves. Many species, especially those of tropical forests, are stopped by narrow dispersal barriers. For such species even a highway swath through a reserve could have the effect of converting one large island into two half-sized islands. Principle B needs to be qualified by the statement that separate reserves in an inhomogeneous region may each favour the survival of a different group of species; and that even in a homogeneous region, separate reserves may save more species of a set of vicariant similar species, one of which would ultimately exclude the others from a single reserve.

If the available area must be broken into several disjunctive reserves, then these reserves should be as close to each other as possible, if the habitat is homogeneous (principle C). Proximity will increase immigration rates between reserves, hence the probability that colonists from one reserve will reach another reserve where the population of the colonist species has gone extinct.

If there are several disjunctive reserves, these should ideally be grouped equidistant from each other rather than grouped linearly (principle D). An equidistant grouping means that populations from each reserve can readily recolonise, or be recolonised from, another reserve. In a linear arrangement, the terminal reserves are relatively remote from each other, reducing exchange of colonists.

If there are several disjunctive reserves, connecting them by strips of the protected habitat (Preston, 1962; Willis, 1974) may significantly improve their conservation function at little further cost in land withdrawn from development (principle E). This is because species of the protected habitat can then disperse between reserves without having to cross a sea of unsuitable habitat. Especially in the case of sedentary species with restricted habitat preferences, such as understorey rain forest species or some bird species of California oak woodland and chaparral, corridors between reserves may dramatically increase dispersal rates over what would otherwise be negligible values.

Any given reserve should be as nearly circular in shape as other considerations permit, to minimise dispersal distances within the reserve (principle F). If the reserve is too elongate or has dead-end peninsulas, dispersal rates to outlying parts of the reserve from more central parts may be sufficiently low to perpetuate local extinctions by island-like effects.

ACKNOWLEDGEMENTS

Field work in the south-west Pacific was supported by the National Geographic Society, Explorers Club, American Philosophical Society, Chapman Fund and Sanford Trust of the American Museum of Natural History, and Alpha Helix New Guinea Program of the National Science Foundation.

REFERENCES

BROWN, J. H. (1971). Mammals on mountaintops: nonequilibrium insular biogeography. *Am. Nat.*, 105, 467-78.

DAMMERMAN, K. W. (1948). The fauna of Krakatau 1883-1933. *Verh. Koninkl. Ned. Akad. Wetenschap. Afdel. Natuurk.*, 44(2), 1-594.

DIAMOND, J. M. (1969). Avifaunal equilibria and species turnover rates on the Channel Islands of California. *Proc. natn. Acad. Sci. USA*, 64, 57-63.

DIAMOND, J. M. (1971). Comparison of faunal equilibrium turnover rates on a tropical island and a temperate island. *Proc. natn. Acad. Sci. USA*, 68, 2742-5.

DIAMOND, J. M. (1972). Biogeographic kinetics: estimation of relaxation times for avifaunas of southwest Pacific islands. *Proc. natn. Acad. Sci. USA*, 69, 3199-203.

DIAMOND, J. M. (1973). Distributional ecology of New Guinea birds. *Science, N.Y.*, 179, 759-69.

DIAMOND, J. M. (1974). Colonization of exploded volcanic islands by birds: the supertramp strategy. *Science, N.Y.*, 184, 802-6.

DIAMOND, J. M. (in press). Incidence functions, assembly rules, and resource coupling of New Guinea bird communities. In *Ecological structure of species communities*, ed. by M. L. Cody & J. M. Diamond. Cambridge, Mass., Harvard University Press.

DOCTERS VAN LEEUWEN, W. M. (1936). Krakatau, 1833 to 1933. *Ann. Jard. Bot. Buitenzorg*, 56-57, 1-506.

HUNT, G. J. Jr & HUNT, M. W. (1974). Trophic levels and turnover rates: the avifauna of Santa Barbara Island, California. *Condor*.

JONES, H. L. & DIAMOND, J. M. (in press). *Species equilibrium and turnover in the avifauna of the California Channel Islands*. Princeton, N.J., Princeton University Press.

LACK, D. (1973). The numbers and species of hummingbirds in the West Indies. *Evolution*, 27, 326-7.

MACARTHUR, R. H. & WILSON, E. O. (1963). An equilibrium theory of insular zoogeography. *Evolution*, 17, 373-87.

MACARTHUR, R. H. & WILSON, E. O. (1967). *The theory of island biogeography*. Princeton, N.J., Princeton University Press.

MACARTHUR, R. H. (1972). *Geographical ecology*. New York, Harper & Row.

MAY, R. M. (in press). Patterns of species abundance and diversity. In *Ecological structure of species communities*, ed. by M. L. Cody & J. M. Diamond. Cambridge, Mass., Harvard University Press.

PRESTON, F. W. (1962). The canonical distribution of commonness and rarity. *Ecology*, 43, 185-215, 410-32.

SCHODDE, R. & CALABY, J. H. (1972). The biogeography of the Australo-Papuan bird and mammal faunas in relation to Torres Strait. In *Bridge and barrier: the natural and cultural history of Torres Strait*, ed. by D. Walker, 257-300. Canberra, Australian National University.

SCHODDE, R. & HITCHCOCK, W. B. (1972). Birds. In *Encyclopedia of Papua and New Guinea*, 1, ed. by P. A. Ryan, 67-86. Melbourne, Melbourne University Press.

SIMBERLOFF, D. S. & WILSON, E. O. (1969). Experimental zoogeography of islands: the colonization of empty islands. *Ecology*, **50**, 278-96.

TERBORGH, J. W. (in press, *a*). Faunal equilibria and the design of wildlife preserves. In *Trends in tropical ecology*. New York, Academic Press.

TERBORGH, J. W. (in press, *b*). Preservation of natural diversity: the problem of extinction prone species. *BioScience*.

TERBORGH, J. W. & FAABORG, J. (1973). Turnover and ecological release in the avifauna of Mona Island, Puerto Rico. *Auk*, **90**, 759-79.

VUILLEUMIER, F. (1970). Insular biogeography in continental regions. I. The northern Andes of South America. *Am. Nat.*, **104**, 373-88.

WILLIS, E. O. (1974). Populations and local extinctions of birds on Barro Colorado Island, Panama. *Ecol. Monogr.*, **44**, 153-69.

WILSON, E. O. & WILLIS, E. O. (in press). Applied biogeography. In *Ecological structure of species communities*, ed. by M. L. Cody & J. M. Diamond. Cambridge, Mass., Harvard University Press.

Part II

FUNCTIONING OF TROPICAL ECOSYSTEMS

An energy strategy based on biomass is a natural for Brazil. We have lots of land, lots of water, and an ideal climate for growth.

J. Goldemberg. Quoted in
"Alcohol: A Brazilian answer
to the energy crisis" by J. Walsh
Science 195:564–565.

The persistence of the myth of boundless productivity in spite of the ignominious failure of every large-scale effort to develop the (Amazon) region constitutes one of the most remarkable paradoxes of our time.

*Amazonia: Man and Culture
in a Counterfeit Paradise,*
B. J. Meggers

Editor's Comments
on Papers 12 Through 16

THE IDEA OF HIGH TROPICAL PROCESS RATES

"The primeval forests of the equatorial zone are grand and overwhelming by their vastness, and by the display of a force of development and vigour of growth rarely or never witnessed in temperate climates." This statement by Wallace (Paper 12) reflects a belief in the high productive potential of the tropics that has remained almost unquestioned for a century.

Comparisons of actual measurements of forest productivity in temperate and tropical regions validated the impression that plant production was relatively high in tropical forests. Murphy (Paper 13) has carried out a comprehensive review of these productivity studies.

High rates of productivity suggest that the tropics should also have high rates of decomposition compared to temperature zones. This indeed was found in the pioneering work by Jenny et al. (Paper 14). Reviews of decomposition rates in Africa (Bernhard-Reversat. 1972) and of worldwide rates of decomposition (Olson, 1963) also showed relatively high decomposition rates in the tropics.

The relatively high rates of productivity and decomposition characteristc of the tropics are reflected in rates of nutrient cycling, which are also relatively high. A classic paper on nutrient cycling in the tropics by Nye (Paper 15) demonstrated this for Ghanian tropical moist forests. Nye's work is especially notable since it was the first major ecosystem-scale nutrient cycling study, and because it was carried out in an underdeveloped country where, as experienced tropical ecologists well know, logistical problems are much more difficult than in developed countries.

The high rates of ecosystem processes in the tropics have quite understandably led to the idea that the tropics have great potential for agriculture and forestry, if only large scale "efficient" temperate-zone methods would be used to exploit the potential. Thurston (Paper 16) presents this point of view. Note particularly the statement in the summary of the paper that "The most realistic short-range method of solving the world's food crisis is to increase food production through conventional agriculture in the tropics."

REFERENCES

Bernhard-Reversat, F., 1972, Decomposition de la litière de fevilles en forêt ombrophile de basse Cote-D'Ivoire, *Oecol. Plant.* 7:279–300.
Olson, J. S., 1963, Energy Storage and the Balance of Producers and Decomposers in Ecological Systems, *Ecology* **44**:322–331.

12

Reprinted from pages 29–34, 37–40, and 65–68 of *Tropical Nature, and Other Essays*, Macmillan and Co., London, 1878, 372 p.

EQUATORIAL VEGETATION

A. R. Wallace

[*Editor's Note:* In the original, material precedes this excerpt.]

General Features of the Equatorial Forests.—It is not easy to fix upon the most distinctive features of these virgin forests, which nevertheless impress themselves upon the beholder as something quite unlike those of temperate lands, and as possessing a grandeur and sublimity altogether their own. Amid the countless modifications in detail which these forests present, we shall endeavour to point out the chief peculiarities as well as the more interesting phenomena which generally characterise them.

The observer new to the scene would perhaps be first struck by the varied yet symmetrical trunks, which rise up with perfect straightness to a great height without a branch, and which, being placed at a considerable average distance apart, give an impression similar to that produced by the columns of some enormous building. Overhead, at a height, perhaps, of a hundred feet, is an almost unbroken canopy of foliage formed by the meeting together of these great trees and their interlacing branches; and this canopy is usually so dense that but an indistinct glimmer of the sky is to be seen, and even the intense tropical sunlight only penetrates to the ground subdued and broken up into scattered fragments. There is a weird gloom and a solemn silence, which combine to produce a sense of the vast—the primeval—almost of the infinite. It is a world in which man seems an intruder, and where he feels overwhelmed by the contemplation of the ever-acting forces, which, from the simple elements of the atmosphere, build up the great mass of vegetation which overshadows, and almost seems to oppress the earth.

Characteristics of the Larger Forest-trees.—Passing from the general impression to the elements of which the scene is composed, the observer is struck by the great diversity of the details amid the general uniformity. Instead of endless repetitions of the same forms of trunk such as are to be seen in our pine, or oak, or beech woods, the eye wanders from one tree to another and rarely detects two of the same species. All are tall and upright columns, but they differ from each other more than do the columns of Gothic, Greek, and Egyptian temples. Some are almost cylindrical, rising up out of the ground as if their bases were concealed by accumulations of the soil ; others get much thicker near the ground like our spreading oaks ; others again, and these are very characteristic, send out towards the base flat and wing-like projections. These projections are thin slabs radiating from the main trunk, from which they stand out like the buttresses of a Gothic cathedral. They rise to various heights on the tree, from five or six, to twenty or thirty feet ; they often divide as they approach the ground, and sometimes twist and curve along the surface for a considerable distance, forming elevated and greatly compressed roots. These buttresses are sometimes so large that the spaces between them if roofed over would form huts capable of containing several persons. Their use is evidently to give the tree an extended base, and so assist the subterranean roots in maintaining in an erect position so lofty a column crowned by a broad and massive head of branches and foliage. The buttressed trees belong to a variety of distinct groups. Thus, many of the Bombaceæ or silk-cotton trees, several of the Leguminosæ, and perhaps many trees belonging to other natural orders, possess these appendages.

There is another form of tree, hardly less curious, in which the trunk, though generally straight and cylindrical, is deeply furrowed and indented, appearing as if made up of a number of small trees grown together at the centre. Sometimes the junction of what seem to be the component parts, is so imperfect, that gaps or holes are left by which you can see through the trunk in various places. At first one is disposed to think this is

caused by accident or decay, but repeated examination
shows it be due to the natural growth of the tree.　The
accompanying outline sections of one of these trees that
was cut down, exhibits its character.　It was a noble

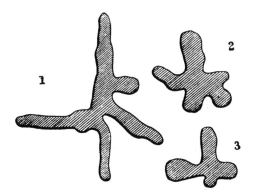

Sections of trunk of a Bornean Forest-tree.
1. Section at seven feet from the ground.
2. 3. Sections much higher up.

forest-tree, more than 200 feet high, but rather slender
in proportion, and it was by no means an extreme
example of its class.　This peculiar form is probably
produced by the downward growth of aerial roots, like
some New Zealand trees whose growth has been traced,
and of whose different stages drawings may be seen at
the Library of the Linnean Society.　These commence
their existence as parasitical climbers which take root in
the fork of some forest-tree and send down aerial roots
which clasp round the stem that upholds them.　As
these roots increase in size and grow together laterally
they cause the death of their foster-parent.　The climber
then grows rapidly, sending out large branches above
and spreading roots below, and as the supporting tree
decays away the aerial roots grow together and form a

new trunk, more or less furrowed and buttressed, but exhibiting no other marks of its exceptional origin. Aerial-rooted forest-trees—like that figured in my *Malay Archipelago* (vol. i. p. 131)—and the equally remarkable fig-trees of various species, whose trunks are formed by a miniature forest of aerial roots, sometimes separate, sometimes matted together, are characteristic of the Eastern tropics, but appear to be rare or altogether unknown in America, and can therefore hardly be included among the general characteristics of the equatorial zone.

Besides the varieties of form, however, the tree-trunks of these forests present many peculiarities of colour and texture. The majority are rather smooth-barked, and many are of peculiar whitish, green, yellowish, or brown colours, or occasionally nearly black. Some are perfectly smooth, others deeply cracked and furrowed, while in a considerable number the bark splits off in flakes or hangs down in long fibrous ribands. Spined or prickly trunks (except of palms) are rare in the damp equatorial forests. Turning our gaze upwards from the stems to the foliage, we find two types of leaf not common in the temperate zone, although the great mass of the trees offer nothing very remarkable in this respect. First, we have many trees with large, thick, and glossy leaves, like those of the cherry-laurel or the magnolia, but even larger, smoother, and more symmetrical. The leaves of the Asiatic caoutchouc-tree (*Ficus elastica*), so often cultivated in houses, is a type of this class, which has a very fine effect among the more ordinary-looking foliage. Contrasted with this is the fine pinnate foliage of some of the largest forest-trees which, seen far aloft against the sky, looks as delicate as that of the sensitive mimosa.

Forest-trees of Low Growth.—The great trees we have hitherto been describing form, however, but a portion of the forest. Beneath their lofty canopy there often exists a second forest of moderate-sized trees, whose crowns, perhaps forty or fifty feet high, do not touch the lowermost branches of those above them. These are of course shade-loving trees, and their presence effectually prevents the growth of any young trees of the larger kinds, until, overcome by age and storms, some monarch of the forest falls down, and, carrying destruction in its fall, opens up a considerable space, into which sun and air can penetrate. Then comes a race for existence among the seedlings of the surrounding trees, in which a few ultimately prevail and fill up the space vacated by their predecessor. Yet beneath this second set of medium-sized forest-trees there is often a third undergrowth of small trees, from six to ten feet high, of dwarf palms, of tree-ferns, and of gigantic herbaceous ferns. Coming to the surface of the ground itself we find much variety. Sometimes it is completely bare, a mass of decaying leaves and twigs and fallen fruits. More frequently it is covered with a dense carpet of selaginella or other lycopodiaceæ, and these sometimes give place to a variety of herbaceous plants, sometimes with pretty, but rarely with very conspicuous flowers.

[*Editor's Note*: Material has been omitted at this point.]

The Climbing Plants of the Equatorial Forests.—Next to the trees themselves the most conspicuous and remarkable feature of the tropical forests is the profusion of woody creepers and climbers that everywhere meet the eye. They twist around the slenderer stems, they drop down pendent from the branches, they stretch tightly from tree to tree, they hang looped in huge festoons from bough to bough, they twist in great serpentine coils or lie in entangled masses on the ground. Some are slender, smooth, and root-like; others are rugged or knotted; often they are twined together into veritable cables; some are flat like ribands, others are curiously waved and indented. Where they spring from or how they grow is at first a complete puzzle. They pass overhead from tree to tree, they stretch in tight cordage like the rigging of a ship from the top of one tree to the base of another, and the upper regions of the forest often seem full of them without our being able to detect any earth-growing stem from which they arise. The conclusion is at length forced upon us that these woody climbers must possess the two qualities of very long life and almost indefinite longitudinal growth, for by these suppositions alone can we explain their characteristic features. The growth of climbers, even more than all other plants, is upward towards the light. In the shade of the forest they rarely or never flower, and seldom even produce foliage ; but when they have reached the summit of the tree that supports them, they expand under the genial influence of light and air, and often cover their foster-parent with blossoms not its own. Here, as a rule, the climber's growth would cease ; but the time comes when the supporting tree rots and falls, and the creeper comes with it in torn and tangled masses to the ground. But though its foster-parent is dead it has itself received no permanent injury, but shoots out again till it finds a fresh support, mounts another tree, and again puts forth its leaves and flowers. In time the old tree rots entirely away and the creeper remains tangled on the ground. Sometimes branches only fall and carry a portion of the creeper tightly stretched to an adjoining tree ; at other times the whole tree is arrested by a neighbour to which the creeper soon transfers itself in order to reach the upper light. When by the fall of a branch the creepers are left hanging in the air, they may be blown about by the wind and catch hold of trees growing up beneath them, and thus become festooned from one tree to another. When these accidents and changes have been again and again repeated the climber may have travelled very far from its parent stem, and may have mounted to the tree tops and descended again to the earth several times over. Only in this way does it seem possible to explain the wonderfully complex manner in which these climbing plants wander up and down the forest as if guided by the strangest caprices, or how they become so crossed and tangled together in the wildest confusion.

The variety in the length, thickness, strength and toughness of these climbers, enables the natives of tropical countries to put them to various uses. Almost every kind of cordage is supplied by them. Some will stand in water without rotting, and are used for cables, for lines to which are attached fish-traps, and to bind and strengthen the wooden anchors used generally in the East. Boats and even large sailing vessels are built, whose planks are entirely fastened together by this kind of cordage skilfully applied to internal ribs. For the better kinds of houses, smooth and uniform varieties are chosen, so that the beams and rafters can be bound together with neatness, strength and uniformity, as is especially observable among the indigenes of the Amazonian forests. When baskets of great strength are required special kinds of creepers are used; and to serve almost every purpose for which we should need a rope or a chain, the tropical savage adopts some one of the numerous forest-ropes which long experience has shown to have qualities best adapted for it. Some are smooth and supple; some are tough and will bear twisting or tying; some will last longest in salt water, others in fresh; one is uninjured by the heat and smoke of fires, while another is bitter or otherwise prejudicial to insect enemies.

Besides these various kinds of trees and climbers which form the great mass of the equatorial forests and determine their general aspect, there are a number of forms of plants which are always more or less present, though in some parts scarce and in others in great profusion, and which largely aid in giving a special character to tropical as distinguished from temperate vegetation: Such are the various groups of palms, ferns, ginger-worts, and wild plantains, arums, orchids, and bamboos ; and under these heads we shall give a short account of the part they take in giving a distinctive aspect to the equatorial forests.

[*Editor's Note*: Material has been omitted at this point.]

Concluding Remarks on Tropical Vegetation.—In concluding this general sketch of the aspect of tropical vegetation we will attempt briefly to summarize its main features. The primeval forests of the equatorial zone are grand and overwhelming by their vastness, and by the display of a force of development and vigour of growth rarely or never witnessed in temperate climates. Among their best distinguishing features are the variety of forms and species which everywhere meet and grow side by side, and the extent to which parasites, epiphytes, and creepers fill up every available station with peculiar modes of life. If the traveller notices a particular species and wishes to find more like it, he may often turn his eyes in vain in every direction. Trees of varied forms, dimensions, and colours are around him, but he rarely sees any one of them repeated. Time after time he goes towards a tree which looks like the one he seeks, but a closer examination proves it to be distinct. He may at length, perhaps, meet with a second specimen half a mile off, or may fail altogether, till on another occasion he stumbles on one by accident.

The absence of the gregarious or social habit, so

general in the forests of extra-tropical countries, is
probably dependent on the extreme equability and per-
manence of the climate. Atmospheric conditions are
much more important to the growth of plants than any
others. Their severest struggle for existence is against
climate. As we approach towards regions of polar cold
or desert aridity the variety of groups and species regularly diminishes ; more and more are unable to sustain
the extreme climatal conditions, till at last we find only
a few specially organized forms which are able to
maintain their existence. In the extreme north, pine or
birch trees ; in the desert, a few palms and prickly shrubs
or aromatic herbs alone survive. In the equable equatorial zone there is no such struggle against climate.
Every form of vegetation has become alike adapted to
its genial heat and ample moisture, which has probably
changed little even throughout geological periods ; and
the never-ceasing struggle for existence between the
various species in the same area has resulted in a nice
balance of organic forces, which gives the advantage,
now to one, now to another, species, and prevents any
one type of vegetation from monopolising territory to
the exclusion of the rest. The same general causes have
led to the filling up of every place in nature with some
specially adapted form. Thus we find a forest of smaller
trees adapted to grow in the shade of greater trees.
Thus we find every tree supporting numerous other forms
of vegetation, and some so crowded with epiphytes of
various kinds that their forks and horizontal branches
are veritable gardens. Creeping ferns and arums run
up the smoothest trunks ; an immense variety of climbers
hang in tangled masses from the branches and mount over

the highest tree-tops. Orchids, bromelias, arums, and ferns grow from every boss and crevice, and cover the fallen and decaying trunks with a graceful drapery. Even these parasites have their own parasitical growth, their leaves often supporting an abundance of minute creeping mosses and hepaticæ. But the uniformity of climate which has led to this rich luxuriance and endless variety of vegetation is also the cause of a monotony that in time becomes oppressive. To quote the words of Mr. Belt : " Unknown are the autumn tints, the bright browns and yellows of English woods ; much less the crimsons, purples, and yellows of Canada, where the dying foliage rivals, nay, excels, the expiring dolphin in splendour. Unknown the cold sleep of winter ; unknown the lovely awakening of vegetation at the first gentle touch of spring. A ceaseless round of ever-active life weaves the fairest scenery of the tropics into one monotonous whole, of which the component parts exhibit in detail untold variety and beauty." [1]

To the student of nature the vegetation of the tropics will ever be of surpassing interest, whether for the variety of forms and structures which it presents, for the boundless energy with which the life of plants is therein manifested, or for the help which it gives us in our search after the laws which have determined the production of such infinitely varied organisms. When, for the first time, the traveller wanders in these primeval forests, he can scarcely fail to experience sensations of awe, akin to those excited by the trackless ocean or the alpine snowfields. There is a vastness, a solemnity, a gloom, a sense of solitude and of human insignificance

[1] *The Naturalist in Nicaragua*, p. 58.

which for a time overwhelm him ; and it is only when the novelty of these feelings have passed away that he is able to turn his attention to the separate constituents that combine to produce these emotions, and examine the varied and beautiful forms of life which, in inexhaustible profusion, are spread around him.

13

Net Primary Productivity
in Tropical Terrestrial
Ecosystems

Peter G. Murphy

Ranging from lowland evergreen rain forest to alpine tundra, the variety of
terrestrial ecosystems lying within tropical latitudes exceeds that of any other
region on earth. Our knowledge of net primary productivity (NPP) rates in
tropical ecosystems must be described as fragmentary. The relatively few available data pertain to a diverse assortment of samples subject to different levels
of precipitation and disturbance.

The published data on organic productivity are dispersed widely in the literature; recent efforts to review and summarize this information have, therefore,
been welcome. Notable among the treatments of productivity on a worldwide
basis are reviews by Odum and Odum (1959), Pearsall (1959), Lieth (1962),
Westlake (1963), Rodin and Bazilevich (1967), Art and Marks (1971), Jordan
(1971a), and Lieth (1972, 1973). Productivity in tropical ecosystems has been
reviewed by Golley (1972), Golley and Lieth (1972), and Golley and Misra
(1972). Other papers concern specific areas within the tropics: India (Misra,
1972), Nigeria (Hopkins, 1962), and the western Pacific region (Kira and
Shidei, 1967). A paper by Bourlière and Hadley (1970) on the ecology of
savannas reviews the productivity data for that important category of tropical
ecosystem.

The objective of this chapter is to present and summarize the available data
relating to annual NPP in tropical terrestrial ecosystems, including data too
recently collected to have been included in earlier reviews.

Available Data

Table 11–1 contains data relative to NPP in a variety of tropical ecosystems. It should be emphasized that each category of tropical ecosystem included in the table is composed of a large variety of subtypes. Tropical grassland, for example, varies from short, sparse herbaceous communities in which bare soil is clearly visible, to tall and dense communities, depending upon local conditions. Because of variation within the categories of ecosystems, and in order to allow a more accurate interpretation of the data, each value of NPP is accompanied by information on the site from which it was obtained. The table includes geographic location in addition to annual rainfall and approximate length of growing season as defined by rainfall pattern when the data were available or could be estimated.

The estimates of total NPP in Table 11–1 are based upon a variety of methods of measurement. In many instances total NPP (aboveground + belowground) had to be estimated from information on some component of the total, such as aboveground NPP in grasslands and leaf-litter production in forests. The factors used in adjusting the original data to obtain total NPP are specified in the footnotes to Table 11–1.

Grassland

The NPP of grasslands varies widely depending on the total annual rainfall and its distribution by seasons. Walter (1954) demonstrated a direct relationship between water availability and aboveground productivity for arid and semiarid desert and grassland in southwest Africa where annual rainfall ranges from 100 to 600 mm. In certain geographic areas, India for example, a prolonged dry season of up to 9 months duration greatly restricts the growing season and consequently the total annual NPP.

Figure 11–1 shows the relationship between total annual rainfall and total annual NPP for tropical grasslands in India, Australia, and Africa. Most of the published reports of productivity in tropical grasslands are based upon the periodic harvesting of aboveground replicated samples and do not include data on belowground parts. Varshney (1972), however, reported that belowground parts accounted for ~ 40% of total NPP in grassland near Varanasi, India. For lack of more extensive information, Varshney's value is assumed to be representative for tropical grassland; data on aboveground NPP were adjusted accordingly for inclusion in Table 11–1 and Figure 11–1. It is apparent from Fig. 11–1 that grassland productivity on sites that receive less than 700 mm of annual rainfall is low. The lowest value reported is 40 g/m²/year for grassland at Jodhpur, India, for a dry year in which rainfall totaled only 92.7 mm (Gupta *et al.*, 1972). On sites that receive between 700 and 1000 mm of rain annually, total annual NPP ranged from 650 to 3810 g/m²/year. The large variations in NPP within this relatively small range of rainfall may be related to any one or a combination of factors including periodicity of rainfall, rate of evapotranspiration, soil permeability and fertility, species characteristics, and grazing pressure. Of the published data for unirrigated grassland, the maximum site value is

Table 11–1 Total annual net primary productivity (oven-dry basis) in tropical terrestrial ecosystems[a,b]

Ecosystem	Location	Approximate growing season (days)	Annual rainfall (mm)	Total Annual NPP (g/m²/year)	Reference
Desert	Southwest Africa	—	155	200*	Walter (1954)
Grassland, *Dichanthium annulatum*-dominated	Varanasi, India 25°18'N, 83°1'E	120	725	1420	Ambasht et al. (1972)
Grassland, *Dichanthium annulatum*-dominated (protected 1 year from grazing)	Varanasi, India 25°18'N, 83°1'E	120	725	1060	Choudhary (1972)
Grassland, *Dichanthium annulatum*-dominated (protected 3 years from grazing)	Varanasi, India 25°18'N, 83°1'E	120	725	650**	Choudhary (1972)
Grassland	Varanasi, India 25°18'N, 83°1'E	120	725	790*	Singh (1968)
Grassland, *Heteropogon contortus*-dominated	Chakia, Varanasi, India 25°18'N, 83°1'E	150	1000	3810	Ambasht et al. (1972)
Grassland, *Panicum miliare*-dominant, forbs important	Kurukshetra, India 29°58'N, 76°50'E	120	770	2980	Singh and Yadava (1972)
Grassland	Jodhpur, India 26°15'N, 73°3'E	90	289	180*	Gupta et al. (1972)

[a] Adjustments have been made in the original data as indicated by the following symbols:

* Includes an estimated 40% for unmeasured belowground parts (based on the data of Varshney, 1972).

** Includes an estimated 52% for unmeasured belowground parts (based on data of Choudhary, 1972).

† Includes an estimated 30% for unmeasured belowground parts.

‡ Based all or in part on rates of CO_2 exchange.

§ Estimated by multiplying annual leaf litter production or annual leaf litter respiration by three (based on data presented by Bray and Gorham, 1964). Some of these estimates may be high because in some cases litter samples included small branches.

[b] OD, Oven-dry weight estimated as 50% of fresh weight.

Table 11-1 *continued*

Ecosystem	Location	Approximate growing season (days)	Annual rainfall (mm)	Total Annual NPP (g/m²/year)	Reference
Grassland, *Heteropogon contortus*-dominated	Delhi, India 28°54′N, 77°13′E	120	800	1330*	Varshney (1972)
Grassland	Udaipur, India 24°32′N, 73°25′E	90	627	300*	Vyas et al. (1972)
Grassland	South West Africa	—	360	520*	Walter (1954)
Grassland, *Andropogon* sp.-dominated	Pretoria, South Africa 25°43′S, 28°16′E	—	607	150*	Bourlière and Hadley (1970)
Grassland	Springbok Flats, South Africa 29°35′S, 17°55′E	—	162	170*	Louw (1968)
Grassland, (native pasture)	Katherine, Australia 14°15′S, 132°20′E	110	660	250*	Norman (1963)
Savanna, *Prosopis* sp.-dominated	Jodhpur, India 26°15′N, 73°3′E	90	361	1450	Bazilevich and Rodin (1966)
Savanna	Rajastan, India 31°20′N, 72°00′	—	610	730	Rodin and Bazilevich (1966)
Savanna, *Dichanthium* sp.-dominated	Gir Forest, India	120	820	680*	Bourlière and Hadley (1970) (data of Hodd)
Savanna, *Themeda* sp.-dominated	Gir Forest, India	120	820	480*	Bourlière and Hadley (1970) (data of Hodd)
Savanna (derived)	Olokemeji, Nigeria 8°57′N, 6°30′E	270	1168	1130*	Hopkins (1965, 1968)
Savanna	Shika, Nigeria 8°57′N, 6°30′E	200	1118	570*	Rains (1963)
Savanna (derived)	Eruja, Ghana 8°N, 2°W (approx)	—	1500	1450*	Nye and Greenland (1960)

Savanna/forest mosaic	Lamto, Ivory Coast 7°43'N, 6°30'W	270	1370	1660*.OD	Roland (1967)
Savanna, *Cenchrus* sp.- and *Chloris* sp.-dominated	Richard-Toll, Senegal 14°53'N, 14°58'W	60	300	70*	Morel and Bourlière (1962)
Savanna, *Aristida papposa*-dominated	Lidney, Chad 17°N, 19°E (approx)	40	320	70*	Gillet (1967)
Savanna, *Cenchrus biflorus*-dominated	Mohi, Chad 17°N, 19°E (approx)	40	320	200*	Gillet (1967)
Savanna, *Brachiaria deflexa*-dominated	Rimé, Chad 17°N, 19°E (approx)	40	320	230*	Gillet (1961)
Savanna, *Cassia tora*-dominated	Tebede, Chad 17°N, 19°E (approx)	40	320	530*	Gillet (1967)
Savanna *Themeda* sp.- and *Heteropogon* sp.-dominated	Matopos, Rhodesia 17°50'S, 29°30'E	—	650	230*	Bourlière and Hadley (1970) (data of West)
Savanna (tall grass)	Serengeti, Tanzania 6°48'S, 33°58'E	—	700	870*	Bourlière and Hadley (1970) (data of Verschuren)
Savanna, *Themeda* sp.- and *Heteropogon* sp.-dominated	Kivu, Albert Park	—	860	800*	Bourlière and Hadley (1970) (data of Verschuren)
Savanna, *Imperata* sp.-dominated	Kivu, Albert Park	—	860	2920*	Bourlière and Hadley (1970) (data of Verschuren)
Savanna, *Trachypogon* sp.-dominated	Llanos of Venezuela	130	1300	530*	Blydenstein (1962)
Savanna (derived)	Costa Rica 10°30'N, 84°30'W	—	2044	2320*	Daubenmire (1972)

* Adjustments have been made in the original data as indicated by the following symbols:
* Includes an estimated 40% for unmeasured belowground parts (based on the data of Varshney, 1972).
** Includes an estimated 52% for unmeasured belowground parts (based on data of Choudhary, 1972).
† Includes an estimated 30% for unmeasured belowground parts.
‡ Based all or in part on rates of CO_2 exchange.
§ Estimated by multiplying annual leaf litter production or annual leaf litter respiration by three (based on data presented by Bray and Gorham, 1964). Some of these estimates may be high because in some cases litter samples included small branches.
OD, Oven-dry weight estimated as 50% of fresh weight.

Table 11–1 *continued*

Ecosystem	Location	Approximate growing season (days)	Annual rainfall (mm)	Total Annual NPP (g/m²/year)	Reference
Savanna (irrigated), *Heteropogon contortus*-dominated	Jhansi, India 25°29′N, 78°32′E	—	—	3400	Ann. Report Indian Forest and Grassland Res. Inst. (1969)
Savanna (irrigated) *Schima* sp.-dominated	Jhansi, India 25°29′N, 78°32′E	—	—	4900	Ann. Report Indian Forest and Grassland Res. Inst. (1969)
Dry deciduous forest	Varanasi, India 25°18′N, 83°1′E	120	1040	1550	Bandhu (1971)
Seasonal forest	Ivory Coast 5°N, 5°W (approx)	—	1500	1340	Müller and Nielsen (1965)
Mixed dry forest	Ibadan, Nigeria 7°26′N, 3°48′E	—	1230	1140§	Madge (1965)
Deciduous forest in Savanna	Calabozo Plains, Venezuela 8°48′N, 67°27′W	—	1200	2460§	Medina and Zelwer (1972)
Rain forest	Thailand 7°35′N, 99°00′E	365	> 2000	2860	Kira *et al.* (1967)
Montane (1460 m) rain forest	Tjibodas, Java 7°S, 107°E (approx)	365	> 2000	2430§	Wanner (1970)
Lowland rain forest	Sarawak 3°N, 112°E (approx)	365	3800	3210§	Wanner (1970)
Rain forest	Ivory Coast 5°N, 5°W (approx)	365	—	2460§	Bernhard-Reversat *et al.* (1972)
Rain forest	Ghana 8°N, 2°W (approx)	365	—	2430	Nye (1961)
Rain forest	Congo 3°S, 13°48′E	365	—	3150	Bartholomew *et al.* (1953)

Rain forest	Manaus, Brazil 3°01'S, 60°00'W	365	1800	1680§	Klinge (1968)
Evergreen Cloud Forest (1000 m)	Rancho Grande, Venezuela 5°N, 65°W (approx)	365	1750	2340§	Medina and Zelwer (1972)
Montane (500 m) rain forest, *Dacryodes excelsa*-dominated	El Verde, Puerto Rico 18°N, 66°W (approx)	365	3800	1030	Jordan (1971b)
Montane (500 m) rain forest, *Dacryodes excelsa*-dominated	El Verde, Puerto Rico 18°N, 66°W (approx)	365	3800	1230‡	H. T. Odum and Jordan (1970) and Odum (1970)
Montane Successional Rain Forest (500 m)	El Verde, Puerto Rico 18°N, 66°W (approx)	365	3800	540	Jordan (1971b)
Montane rain forest, *Goethalsia meiantha*-dominated	Costa Rica 10°30'N, 84°30'W	365	—	350‡	(This is net *ecosystem* production.) Lemon *et al.* (1970)
Bamboo Brake in monsoon forest, *Oxytenanthera albociliata*-dominated	Burma 23°N, 95°E (approx)	—	—	2780†	Rozanov and Rozanov (1964)
Bamboo Brake in rain forest, *Dendrocalamus brandisii*-dominated	Burma 23°N, 95°E (approx)	—	—	2300†	Rozanov and Rozanov (1964)
Bamboo Brake (arid area) *Dendrocalamus strictus*-dominated	Burma 23°N, 95°E (approx)	—	—	1530†	Rozanov and Rozanov (1964)
Mangrove, *Rhizophora mangle*-dominated	Puerto Rico 18°N, 67°W (approx)	365	—	930‡	Golley *et al.* (1962)

* Adjustments have been made in the original data as indicated by the following symbols:

* Includes an estimated 40% for unmeasured belowground parts (based on the data of Varshney, 1972).

** Includes an estimated 52% for unmeasured belowground parts (based on the data of Choudhary, 1972).

† Includes an estimated 30% for unmeasured belowground parts.

‡ Based all or in part on rates of CO_2 exchange.

§ Estimated by multiplying annual leaf litter production or annual leaf litter respiration by three (based on data presented by Bray and Gorham, 1964).
Some of these estimates may be high because in some cases litter samples included small branches.

ᵇ OD, Oven-dry weight estimated as 50% of fresh weight.

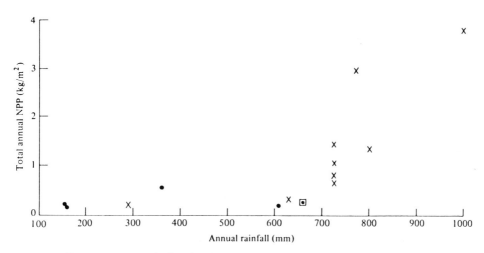

FIGURE 11–1. Relationship between annual rainfall and total annual NPP in dry matter for tropical grasslands. ●, Africa; ×, India; and □, Australia.

3810 g/m^2/year measured in a successional *Heteropogon contortus*-dominated grassland near Varanasi, India; it is estimated to receive in excess of 1000 mm of rainfall annually, most of which is distributed over a 3-month period (Ambaṣht *et al.*, 1972). This exceptionally productive grassland becomes very dense and tall (> 1.5 m) when protected from grazing (Ambasht, *personal communication*). In many areas of India where annual rainfall is high, woodland or forest is the ultimate end point of succession, but grassland is maintained by the pressures of grazing and other disturbances. Such successional grasslands appear to be the most productive. Based on 11 representative samples, the average total annual NPP for tropical grassland is estimated to be 1080 g/m^2/year.

Savanna

Bourlière and Hadley (1970) define savanna as ". . . a tropical formation where the grass stratum is continuous and important but is interrupted by trees and shrubs; the [grass] stratum is burnt from time to time, and the main growth patterns are closely associated with alternating wet and dry seasons." Figure 11–2 plots total annual NPP as a function of annual rainfall for savannas in India, Venezuela, Costa Rica, and Africa. As indicated in Fig. 11–2, savanna may exist in areas that receive an annual amount of rainfall as low as that received in some grassland areas. The dramatic effects of irrigation on two savanna areas in Jhansi, India, is apparent in Figure 11–2. The irrigated savannas in India are three to four times as productive as unirrigated savannas in that country.

As in the case of grassland, total NPP was estimated from data on above-ground productivity. Because most of the productivity estimates for savanna were based on measurements of peak standing crop (trees excluded), the esti-

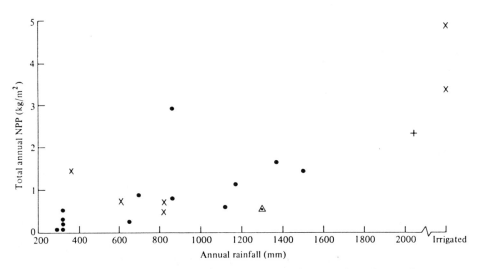

FIGURE 11–2. Relationship between annual rainfall and total annual NPP in Savanna. ●, Africa; ×, India; +, Costa Rica; and △, Venezuela.

mates of total NPP are assumed to be low. Mathews and Westlake (1969) demonstrated that actual NPP may exceed peak standing crop by a factor of 1.5–3.5 in communities with high rates of turnover. The data in Table 11–1 and Fig. 11–2 were not adjusted for this error.

The maximum site value for tropical savanna, excluding irrigated areas, is 2920 g/m²/year for a sample in the Congo where annual rainfall averages 860 mm (Bourlière and Hadley, 1970). The minimum site value is 70 g/m²/year for savannas in Chad (Gillet, 1967) and Senegal (Morel and Bourlière, 1962) that receive only ~ 300 mm of rainfall annually. On the basis of 19 representative samples, the average annual NPP for unirrigated savanna is estimated to be 890 g/m²/year.

Seasonal forest (Raingreen forest)

Forests displaying conspicuous seasonal properties, such as leaf fall and temporary cessation of growth, are found throughout the tropics wherever seasonal drought alternates with relatively wet periods. All or just a few of the tree species of a given stand may show seasonal properties, depending upon the duration of the dry period.

Forest productivity has been estimated allometrically and from measurements of annual litter fall, annual leaf fall, and rates of litter respiration. Based upon data presented by Bray and Gorham (1964) annual leaf fall is considered to represent one-third of total annual NPP. The minimum value of NPP for seasonal forest is 1140 g/m²/year, estimated from data on leaf-litter production for a forest at Ibadan, Nigeria (Madge, 1965). Based on litter-production measurements by Medina and Zelwer (1972) in deciduous forest patches in the savanna of the Calabozo Plains of Venezuela, the maximum site value is 2460 g/m²/year. Average NPP, based on only four samples, is 1620 g/m²/year.

Evergreen rain forest

Forests receiving abundant year-round rainfall and lacking distinct seasonality in leaf fall are, on the average, the most productive of any of the tropical terrestrial ecosystems measured to date. The forests grouped in this category range from lowland types to montane types. Total annual NPP, based on nine representative samples, averages 2400 g/m²/year. The maximum site value is 3210 g/m²/year for lowland forest in Sarawak, estimated from rates of litter respiration measured by Wanner (1970). From this value productivities range far downward to the value of 540 g/m²/year for a successional montane rain forest.

Productivity of tropical rain forest has been of special interest, and some very high estimates have been published. It should be kept in mind that extensive areas on tropical podzol soils are apparently of low productivity (Janzen, 1974) and that in some montane forests productivity may be limited by high precipitation and humidity, with intense soil leaching and restricted transpiration. The mean of 2400 g/m²/year seems a reasonable value, but even this could be revised downward. If the 3:1 ratio of total to litter productivity used for some values in Table 11–1 is too high (as may well be the case in these forests), a ratio of 2.5:1 reduces the mean for the same nine samples to 2170 g/m²/year. Using a 2.0:1 ratio reduces the mean to 1960 g/m²/year; adding two more montane samples to the set gives an 11-sample mean of 2120 g/m²/year, with the high ratio of 3:1, or of 1930 g/m²/year with the ratio of 2.5:1. Brünig (1974) estimates 2100 g/m²/year, aboveground. A more accurate mean will depend not only on more reliable measurements, but on a weighting of different kinds of rain forests with different productivities by their relative areas.

Two studies have attempted to measure the integrated metabolism of tropical rain forest. H. T. Odum and Jordan (1970) measured rates of CO_2 exchange in a lower montane rain forest in Puerto Rico. They estimated total daily respiration to be 16.4 g C/m². The ecosystem was considered to be near steady state, and total gross photosynthesis was therefore assumed to be equal to total respiration. NPP was estimated as 1230 g/m²/year by subtracting autotrophic respiration from gross photosynthesis. This value agrees reasonably with the allometrically derived value of 1030 g/m²/year of Jordan (1971b) for the same site. A 3-year-old successional rain forest in the same area of Puerto Rico was found to have a total annual NPP of 540 g/m²/year (Jordan, 1971b). Lemon *et al.* (1970) measured rates of CO_2 exchange in a 50-year-old rain forest in Costa Rica and found that net *ecosystem* production equaled 350 g/m²/year, indicating that the ratio of gross photosynthesis to total respiration in that particular ecosystem was greater than unity.

Bamboo brake

Total annual NPP in three bamboo brakes occurring in forest openings in Burma was high, ranging from 1530 to 2780 g/m²/year and averaging 2200 g/m²/year (based on the data of Rozanov and Rozanov, 1964). In arriving at these estimates it was assumed that the aboveground productivity, which was the only portion measured, accounted for 70% of total NPP.

Mangrove

Total NPP in a *Rhizophora mangle*-dominated ecosystem in southeastern Puerto Rico was estimated to be 930 g/m²/year. The estimate is based on rates of CO_2 exchange reported by Golley *et al.* (1962).

Summary

Figure 11–3 summarizes rates of NPP in tropical terrestrial ecosystems. Mean levels of productivity range widely, from 200 g/m²/year in desert to 2000–2400 g/m²/year in evergreen rain forest. The wide range in values of NPP exhibited within several categories of ecosystems, particularly savanna, grassland, and evergreen rain forest, represent, in part, a probable result of differences in methods used in measuring NPP. But the variation also reflects the pooling of widely varying communities into very general categories. As more data become available these general categories should be partitioned into more meaningful subtypes. Nevertheless, the available data are sufficient to give an idea of the order of magnitude of NPP in a variety of tropical terrestrial ecosystems.

Comparisons of NPP between ecosystems of different climatic areas are difficult, but it does appear that tropical ecosystems are more productive on an annual basis than their temperate counterparts. Tropical forests as a whole, with a mean annual NPP of 2160 g/m²/year, exceed temperate forests, averaging 1300 g/m²/year, by a factor of 1.7; boreal forests, averaging only 800 g/m²/year, by a factor of 2.7. The annual NPP of tropical grassland (much of it in

FIGURE 11–3. Average annual net primary productivity (dry matter) in various tropical ecosystems. ●, Mean (or individual value); —, range; and ☐, standard error of the mean.

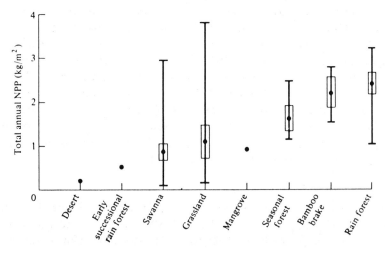

forest climates), averaging 1080 g/m²/year, exceeds that of temperate grass-land, averaging 500 g/m²/year, by a factor of 2.2. The mean NPP of tropical savannas (890 g/m²/year), which may be more directly comparable with temperate grasslands, is 1.8 times that of the latter. The estimates of temperate rates of NPP are those of Whittaker and Likens in Whittaker and Woodwell (1971). Further discussions of correlations between NPP and environmental parameters follow in Chapter 12.

References

Ambasht, R. S., A. N. Maurya, and U. N. Singh. 1972. Primary production and turn-over in certain protected grasslands of Varanasi, India, In *Tropical Ecology with an Emphasis on Organic Production*, P. M. Golley and F. B. Golley, eds., pp. 43–50. Athens, Georgia: Univ. of Georgia.

Annual report of the Indian Forest and Grassland Research Institute at Jhansi. 1969. (*Unpublished data.*)

Art, H. W., and P. L. Marks. 1971. A summary of biomass and net annual primary production in forest ecosystems of the world. In *Forest Biomass Studies*, H. E. Young, ed., pp. 1–34. (*Internat. Union Forest Research Organizations Conf., Sect. 25. Gainesville, Florida.*) Orono, Maine: Univ. of Maine.

Bandhu, D. 1971. A study of the productive structure of tropical dry deciduous forest at Varanasi. Ph.D. thesis. Banaras: Banaras Hindu Univ.

Bartholomew, M. V., J. Meyer, and H. Laudelot. 1953. Mineral nutrient immobiliza-tion, under forest and grass fallow in the Yangambi (Belgian Congo) region. *Publ. INEAC Ser. Sci.* 57:27 p.

Bazilevich, N. I., and L. E. Rodin. 1966. The biological cycle of nitrogen and ash elements in plant communities of the tropical and sub-tropical zones. *Forestry Abstr.* 27:357–368.

Bernhard-Reversat, F., C. Huttel, and G. Lemee. 1972. Some aspects of the seasonal ecologic periodicity and plant activity in an evergreen rain forest of the Ivory Coast. (In French with English summ.), In *Tropical Ecology with an Emphasis on Organic Production*, P. M. Golley and F. B. Golley, eds., pp. 217–218; 219–234. Athens, Georgia: Univ. of Georgia.

Blydenstein, J. 1962. The *Trachypogon* savanna in the high plains. Ecological study of the area surrounding Calatgo, Guárico State, Venezuela. *Bol. Soc. Venez. Cienc. Nat.* 23:139–206.

Bourlière, F., and M. Hadley. 1970. The ecology of tropical savannas. In *Ann. Rev. Ecol. Syst.* Vol. 1, pp. 125–152.

Bray, R., and E. Gorham. 1964. Litter production in forests of the world. In *Advan. Ecol. Res.*, J. B. Cragg, ed., Vol. 2, pp. 101–157. New York: Academic Press.

Brünig, E. F. 1974. Ökosysteme in den Tropen. *Umschau* 74(13):405–410.

Choudhary, V. B. 1972. Seasonal variation in standing crop and net above-ground production in *Dichanthium annulatum* grassland at Varanasi. In *Tropical Ecol-ogy with an Emphasis on Organic Production*, P. M. Golley and F. B. Golley, eds., pp. 51–57. Athens, Georgia: Univ. of Georgia.

Daubenmire, R. 1972. Standing crops and primary production in savanna derived from semideciduous forest in Costa Rica. *Bot. Gaz.* 133:395–401.

Gillet, H. 1961. Pâturages sahéliens de Ranch de l'ouadi rimé. *J. Agr. Trop. Bot. Appl.* 8:465–536.

————. 1967. Essai d'évaluation de la biomasse végétale en zone sahélienne (végétation annuelle). *J. Agr. Trop. Bot. Appl.* 14:123–258.

Golley, F. B. 1972. Summary. In *Tropical Ecology with an Emphasis on Organic Production*, P. M. Golley and F. B. Golley, eds., pp. 407–413. Athens, Georgia: Univ. of Georgia.

————, and H. Lieth. 1972. Bases of organic production in the tropics. In *Tropical Ecology with an Emphasis on Organic Production*, P. M. Golley and F. B. Golley, eds., pp. 1–26. Athens, Georgia: Univ. of Georgia.

————, and R. Misra. 1972. Organic production in tropical ecosystems. *BioScience* 22:735–736.

Golley, F., H. T. Odum, and K. Wilson. 1962. The structure and metabolism of a Puerto Rican red mangrove forest in May. *Ecology* 43:9–19.

Gupta, R. K., S. K. Saxena, and S. K. Sharm. 1972. Above-ground productivity of grasslands at Jodhpur, India. In *Tropical Ecology with an Emphasis on Organic Production*, P. M. Golley and F. B. Golley, eds., pp. 75–93. Athens, Georgia: Univ. of Georgia.

Hopkins, B. 1962. Biological productivity in Nigeria. *Sci. Assoc. Nigeria Proc.* 1/3: 20–28.

————. 1965. Observations on savanna burning in the Olokemeji Forest Reserve, Nigeria. *J. Appl. Ecol.* 2:367–381.

————. 1968. Vegetation of the Olokemeji Forest Reserve, Nigeria. V. The vegetation of the savanna site with special reference to its seasonal changes. *J. Ecol.* 56:97–115.

Janzen, D. H. 1974. Tropical blackwater rivers, animals, and mast fruiting by the Dipterocarpaceae. *Biotropica* 6:69–103.

Jordan, C.?F. 1971a. A world pattern in plant energetics. *Amer. Sci.* 59:425–433.

————. 1971b. Productivity of a tropical rain forest and its relation to a world pattern of energy storage. *J. Ecol.* 59:127–142.

Kira, T., and T. Shidei. 1967. Primary production and turnover of organic matter in different forest ecosystems of western Pacific. *Jap. J. Ecol.* 17:70–87.

————, H. Ogawa, K. Yoda, and K. Ogino. 1967. Comparative ecological studies in three main types of forest vegetion in Thailand. IV. Dry-matter production, with special reference to the Khao Chong rain forest. *Nature Life SE Asia* 5:149–174.

Klinge, H. 1968. Litter production in an area of Amazonian terra firma forest. I. Litter-fall, organic carbon and total nitrogen contents of litter. *Amazonia* 1(4): 287–301.

Lemon, E. R., L. H. Allen, and L. Muller. 1970. Carbon dioxide exchange of a tropical rain forest. II. *BioScience* 20:1054–1059.

Lieth, H. 1962. *Die Stoffproduktion der Pflanzendecke.* 156 p. Stuttgart: Fischer.

————. 1972. Über die Primärproduktion der Pflanzendecke der Erde. *Zeit. Angew. Bot.* 46:1–37.

————. 1973. Primary production: Terrestrial ecosystems. *Human Ecol.* 1:303–332.

Louw, A. J. 1968. Fertilizing natural veld on red loam soil of the Springbok flats. 2. Effect of sulphate of ammonia and superphosphate on air-dry yield and mineral content. *S. Afr. J. Agr. Sci.* 11:629–636.

Madge, D. S. 1965. Leaf fall and litter disappearance in a tropical forest. *Pedobiologia* 5:273–288.

Mathews, C. P., and D. F. Westlake. 1969. Estimation of production by populations of higher plants subject to high mortality. *Oikos* 20:156–160.

Medina, E., and M. Zelwer. 1972. Soil respiration in tropical plant communities. In *Tropical Ecology with an Emphasis on Organic Production*, P. M. Golley and F. B. Golley, eds., pp. 245–267. Athens, Georgia: Univ. of Georgia.

Misra, R. 1972. A comparative study of net primary productivity of dry deciduous forest and grassland of Varanasi, India. In *Tropical Ecology with an Emphasis on Organic Production*, P. M. Golley and F. B. Golley, eds., pp. 279–293. Athens, Georgia: Univ. of Georgia.

Morel, G., and F. Bourlière. 1962. Relations écologiques des avifaunes sédentaires et migratices dans une savane sahélienne du bas Sénégal. *Terre et Vie* 16:371–393.

Müller, D., and J. Nielsen. 1965. Production brute, pertes par respiration et production nette dans la forêt ombrophile tropicale. *Forstl. Forsøgsv. Danm.* 29:69–160.

Norman, M. J. T. 1963. The pattern of dry matter and nutrient content changes in native pastures at Katherine, N.T. *Austral. J. Exp. Agr. Animal Husbandry* 3:119–124.

Nye, P. H. 1961. Organic matter and nutrient cycles under moist tropical forest. *Plant Soil* 8:333–346.

———, and D. J. Greenland. 1960. The soil under shifting cultivation. *Tech. Commonw. Bur. Soils* 51:1–156.

Odum, E. P., and H. T. Odum. 1959. *Fundamentals of Ecology*, Chapter 3, 546 pp. Philadelphia, Pennsylvania: Saunders.

Odum, H. T. 1970. Summary: An emerging view of the ecological system at El Verde. In *A Tropical Rain Forest*, H. T. Odum and R. F. Pigeon, eds., Vol. I, pp. 191–289. Washington, D.C.: U.S. Atomic Energy Commission, Div. of Technical Information.

———, and C. F. Jordan. 1970. Metabolism and evapotranspiration of the lower forest in a giant plastic cylinder. In *A Tropical Rain Forest*, H. T. Odum and R. F. Pigeon, eds., Vol. I, pp. 165–189. Washington, D.C.: U.S. Atomic Energy Commission, Div. of Technical Information.

Pearsall, W. H. 1959. Production ecology. *Sci. Progr.* 47:106–111.

Rains, A. B. 1963. Grassland research in northern Nigeria 1952–1962. *Misc. Pap. Inst. Agr. Res. Samaru* 1:1–67.

Rodin, L. E., and N. I. Bazilevich. 1967. *Production and Mineral Cycling in Terrestrial Vegetation*, 288 pp. London: Oliver and Boyd.

Roland, J.-C. 1967. Recherches écologiques dans la savane de Lamto (Côte d'Ivoire): Données préliminaires sur le cycle annuel de la végétation herbacée. *Terre et Vie* 21:228–248.

Rozanov, B. G., and I. M. Rozanova. 1964. The biological cycle of nutrient elements of bamboo in the tropical forests of Burma. *Bot. Zh.* 49(3):348–357.

Singh, J. S. 1968. Net above-ground community productivity in the grasslands at Varanasi. In *Proc. Symp. Rec. Advan. Trop. Ecol. 2*, R. Misra and B. Gopal, eds., pp. 631–653. Varanasi: International Society of Tropical Ecology.

———, and P. S. Yadava. 1972. Biomass structure and net primary productivity in the grassland ecosystem at Kurukshetra. In *Tropical Ecology with an Emphasis on Organic Production*, P. M. Golley and F. B. Golley, eds., pp. 59–74. Athens: Georgia: Univ. of Georgia.

Varshney, C. K. 1972. Productivity of Delhi grasslands. In *Tropical Ecology with an Emphasis on Organic Production*, P. M. Golley and F. B. Golley, eds., pp. 27–42. Athens: Georgia: Univ. of Georgia.

Vyas, L. N., R. K. Garg, and S. K. Agarwal. 1972. Net above-ground production in the monsoon vegetation at Udaipur. In *Tropical Ecology with an Emphasis on Organic Production*, P. M. Golley and F. B. Golley, eds., pp. 95–99. Athens, Georgia: Univ. of Georgia.

Walter, H. 1954. Le facteur eau dans les régions arides et sa signification pour l'organisation de la végétation dans les contrées sous-tropicales. In *Colloques Internationaux des Centre National de la Recherche Scientifique*. Vol. 59, pp. 27–39, Paris: Centre National dela Recherche Scientifique. Les Divisions Ecologiques du Monde.

Wanner, H. 1970. Soil respiration, litter fall and productivity of tropical rain forest. *J. Ecol.* 58:543–547.

Westlake, D. F. 1963. Comparisons of plant productivity. *Biol. Rev.* 38:385–425.

Whittaker, R. H., and G. M. Woodwell. 1971. Measurement of net primary production of forests. (French summ.) In *Productivity of Forest Ecosystems: Proc. Brussels Symp. 1969*, P. Duvigneaud, ed. *Ecology and Conservation*, Vol. 4:159–175. Paris: UNESCO.

14

COMPARATIVE STUDY OF DECOMPOSITION RATES OF ORGANIC MATTER IN TEMPERATE AND TROPICAL REGIONS

HANS JENNY, S. P. GESSEL, AND F. T. BINGHAM

University of California

Received for publication May 14, 1949

In comparison to the soils of the United States, the well-drained upland soils of the equatorial regions of Colombia, South America, are very rich in total nitrogen and organic matter. In particular, from localities with identical mean annual temperatures and precipitation the nitrogen and organic matter levels of Colombian soils are severalfold higher than those of United States soils (7).

The present paper shows that Central American as well as Colombian soils contain large amounts of organic matter. To facilitate quantitative comparisons with midlatitudes, organic matter levels of Californian soils also are reported. Furthermore, measurements of production and decomposition of organic matter in various climates are discussed.

NITROGEN AND ORGANIC MATTER LEVELS OF CENTRAL AMERICAN SOILS

In the fall of 1946 the senior author collected soil samples at various elevations in eastern Costa Rica.[1] The soils were analyzed at Berkeley, California, according to methods previously reported (7). All analytical results are recorded on an oven-dry basis.

In figure 1 are plotted the nitrogen-depth functions of seven well-drained Costa Rican upland soil profiles. They group themselves into the climatically warm Turrialba set and the relatively cool Juan Viñas set. Though all profiles indicate high levels of soil nitrogen, those of the Juan Viñas group display especially high values. The organic matter contents also are high, as may be judged from the high carbon-nitrogen ratios shown. The profile features of the Juan Viñas soils are akin to the humic yellow-brown soils of Colombia (6).

In table 1 are presented nitrogen values, carbon-nitrogen ratios, pH values, and textures of 23 well-drained soils collected to a depth of 0 to 8 inches. The sites range in elevation from 300 to 4,200 feet. In all localities the mean annual rainfall is high, 70 to 150 inches. Mean annual temperature varies from about 76°F. in the hot lowlands to 65°F. on the cooler Central Plateau.

The nitrogen contents of these Costa Rican surface soils are very high. On the basis of mean annual temperature and precipitation, their nitrogen and organic matter contents are comparable to those of Colombian soils. Wheeting (13) has reported high nitrogen contents of Guatemalan soils also.

[1] The senior author is indebted to R. H. Allee, Director of the International Institute of Agricultural Sciences at Turrialba, Costa Rica, for arranging the author's sojourn at the Institute, and to N. C. Ives and S. Bonilla for assistance in collecting soil samples and conducting experiments.

SOIL NITROGEN LEVELS IN THE SIERRA NEVADA MOUNTAINS OF CALIFORNIA

In the region of Friant Dam, Shaver Lake, Huntington Lake, and Mt. Hilgard in California, a granitic batholith extends over a range of elevation from 500 feet to more than 13,000 feet. From an altitude of 2,000 feet to the timber-

TABLE 1

Analytical data of 23 well-drained Costa Rican soils

Depth 0–8 inches

NUMBER	N	C/N	pH*	TEXTURE	REMARKS
	per cent				
Cairo, Elevation 300 feet					
21	0.26	10.2	5.1	Sandy loam	Rubber plantation.
22	0.35	11.0	5.5	Sandy loam	Virgin forest, mineral soil only.
23	0.32	9.6	5.1	Sandy loam	Pasture.
Av.	0.31	10.3	5.2		
Vicinity of Turrialba, Elevation 1,837 feet					
1	0.33	12.7	6.3	Sandy loam	Young rubber plantation.
2	0.38	10.3	5.2	Loam	Coffee grove.
3	0.28	10.0	5.0	Silty clay loam	Sugar cane field.
4	0.61	11.3	6.6	Sandy loam	Coffee grove.
7	0.50	11.1	5.7	Sandy loam	Sugar cane field, 10 per cent slope.
8	0.46	12.3	5.8	Fine sandy loam	Best sugar cane field of Exp. Sta.
12	0.17	12.2	6.0	Loamy fine sand	Virgin forest, mineral soil only.
13	0.14	11.3	6.0	Loamy fine sand	Virgin forest, mineral soil only.
17	0.42	10.7	5.6	Loam	Abandoned coffee grove
18	0.38	13.9	6.1	Sandy loam	Plowed field
24	0.25	12.8	4.9	Loam	Field crops, 30 per cent slope.
25	0.37	14.3	5.0	Loam	Field crops, 15 per cent slope.
Av.	0.36± 0.04	11.9± 0.39	5.7		
Vicinity of Juan Viñas, Elevation about 4,200 feet					
9	1.23	13.4	5.6	Organic loam	Sugar cane, 30 per cent slope.
10	0.81	17.4	5.5	Organic sandy loam	Pasture, 30 per cent slope.
11	0.91	14.5	5.3	Organic loamy sand	Pasture, 50 per cent slope.
14	0.99	14.8	5.3	Organic loamy sand	Pasture, 10 per cent slope.
15	0.92	13.2	6.0	Organic sandy loam	Sugar cane, 20 per cent slope.
16	1.05	18.1	5.4	Organic sandy loam	Sugar cane, level spot on knoll.
Av.	0.99± 0.06	15.2± 0.84	5.5		

* 1:2 soil-water ratio.

line at about 10,000 feet, the soils belong to the "podzolized yellow-red soils," having well-defined light gray A_2 horizons and yellow or reddish B horizons.

In this mountain range were duplicated the profile studies and the measurements of production and decomposition of organic matter carried out simultaneously in Costa Rica and Colombia. In figure 2 is portrayed the variation of the nitrogen content of the soil (0–8 inches, exclusive of forest floor) with elevation and type of vegetation. All analyses pertain to the fine earth fraction (<2 mm.) of the soil. No corrections for stone contents were made. For comparison, the corresponding relationships between soil nitrogen and elevation in Colombia and Costa Rica also are included in the graph.

The great contrasts in the nitrogen contents of the selected equatorial (Colombia and Costa Rica) and midlatitude soils are clearly brought out. The authors believe that the soil samples collected are representative of large areas.

MEASUREMENTS OF PRODUCTION OF ORGANIC MATTER BY VEGETATION

There is in the Americas a deplorable lack of data on the production of organic matter by forest vegetation. This deficiency is especially noticeable in tropical regions. Whereas it is relatively easy, because of the shedding of leaves in October and November, to collect freshly fallen leaf litter of deciduous trees in

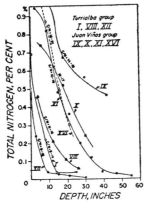

FIG. 1. NITROGEN-DEPTH FUNCTIONS OF SOILS FROM COSTA RICA

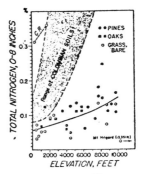

FIG. 2. RELATIONSHIP BETWEEN NITRO-GEN CONTENT OF SOIL (0–8 INCHES) AND ELEVATION IN CALIFORNIA

Range of Colombian soils and Costa Rican (C.R.) averages also are indicated.

North America, collection of litter in equatorial regions must be extended over an entire year.[2]

Leaf collectors were placed in virgin forests of Colombia (2) and California. The collectors consisted of square wooden frames 4 to 6 inches high. To the bottom of the frame was fastened a metal mosquito screen which rested directly on the soil. The Colombian collectors had an inside area of 1 square meter; the California ones measured half a square meter. The frames were placed 6 feet from a tree trunk. At frequent intervals the leaves, needles, fruits, twigs, and all other plant materials that fell into the collectors were removed and weighed. This material is here designated as *litter*. The forest floor below the collectors was also sampled.

The seasonal variation of litter production in tropical forests is depicted in figure 3. The two Chinchiná samplers were several hundred feet apart, one being

[2] The senior author was fortunate in obtaining the cooperation of L. O. Souffront in Chinchiná and of V. M. Patiño in Cali in measuring leaf fall in tropical rain forests during 1947.

below a large mestizo tree (*Guarca Gigantea*), the other under medium-sized broad-leaved trees and shrubs.

In table 4 are listed the annual values of litter production (*A*) on an over-dry basis (105°C.). The California data were collected during 1947–48. It is important to note that the annual productions of litter in the tropical locations exceed those in California. On an acre basis the tropical forests dropped in a year 8,000 to 11,000 pounds of leaves and twigs, whereas the Sierra trees produced only 800 to 3,000 pounds.

DECOMPOSITION VELOCITIES OF ALFALFA IN VARIOUS CLIMATES

Quantitative information on decomposition velocities of organic matter in various climates was obtained (table 2) by putting into natural soils metal cans containing dried alfalfa leaves (3.5 per cent nitrogen).

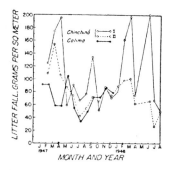

FIG. 3. VARIATION OF MONTHLY LITTER FALL IN TROPICAL RAINFORESTS

Multiplying grams per square meter by 8.92 gives pounds per acre. Summations are given in table 4.

The metal containers were constructed from two types of cans obtained from the Continental Can Co. They were type L. Plate, No. 211–301 (7 cm. height, 6.7 cm. diameter) and No. 211–400 (10 cm. height, 6.7 cm. diameter). The cans were stripped of their top and bottom parts. The short cylinder was then soldered on the long one, the two being separated by a metal screen. The bottom cylinder was left empty; the top cylinder was filled with the following layers of material (see figure 4): glass wool, alfalfa leaves (10 or 15 gm.), filter paper, glass wool. The top cylinder was capped with a galvanized wire net consisting of material from commercial pot cleaners.

These receptacles were placed vertically in the soil, so that only the top half inch emerged above the surface. Alfalfa was inoculated with soil suspension.

In every locality eight cans were placed within an area of 1 square meter. At 3-month intervals two cans were removed, and their contents were dried and sent by air mail to Berkeley where they were weighed and analyzed. One hundred and twelve cans were placed in Costa Rica (in 1946) and Colombia (in 1946–47) and more than 600 in various parts of California (1946–1948). Because of human and animal interferences the mortality rate of the cans was fairly high, especially in the tropics, where even sites were lost as a result of rapidly growing vegetation.

TABLE 2

Localities, climate, and names of cooperators of alfalfa decomposition studies

Figures in parentheses denote estimated values

NUMBER	LOCALITY	ELEVATION		ANNUAL CLIMATE OR CLIMATE DURING EXPERIMENT				DATE OF SETTING OUT CANS	COOPERATORS
				Temperature		Precipitation			
		m.	feet	°C.	°F.	mm.	inches		
	Costa Rica (N. C. Ives, Turrialba, in charge). Ten grams of alfalfa exposed.								
1*	Diamantes, lawn	250	820	24.4	75.9	3,007	118.4	Oct. 2, 1946	H. Echeverri
2	Diamantes, under tree	250	820	24.4	75.9	3,007	118.4	Oct. 2, 1946	H. Echeverri
3	Cairo, bare soil	91	300	24.2	75.5	3,747	147.5	Oct. 3, 1946	H. Echeverri
4	Turrialba, jungle	500	1,640	(22.8)	(73)	(1,870)	(74)	Oct. 4, 1946	G. Bonilla
5	Turrialba, rubber plantation	500	1,640	(22.8)	(73)	(1,870)	(74)	Oct. 4, 1946	G. Bonilla
6	Turrialba, Hulera (lawn)	560	1,837	22.5	72.5	1,872	73.7	Oct. 6, 1946	A. Lizano
7	Juan Viñas, bare soil	1,280	4,200	18.4	65.1	3,581	141.0	Oct. 5, 1946	G. Bonilla, and
8	Juan Viñas, under tree	1,311	4,300	(18.3)	(65)	(3,581)	(141)	Oct. 5, 1946	E. A. Strauman
	Colombia (L. O. Souffront, Chinchiná, in charge). Fifteen grams of alfalfa exposed.								
9	Calima, lawn	30	98	26.6	79.9	9,123	359	Dec. 15, 1946	V. M. Patiño, C. Pontón Rangel
10	Palmira, lawn	1,066	3,498	24.0	75.2	956	37.6	Dec. 18, 1946	G. Ramirez
11	Villavicencio, lawn	498	1,633	26.3	79.3	4,033	158.8	Jan. 25, 1947	Marston Bates
12	Chinchiná, lawn	1,433	4,702	22.0	71.6	2,769	109.0	Dec. 9, 1946	L. O. Souffront
13	Chinchiná, jungle	1,630	5,350	(21)	(70)	(2,800)	(110)	Dec. 9, 1946	L. O. Souffront
14	Bogotá, lawn	2,640	8,659	(14.5)	(58.1)	(976)	(38.4)	Feb. 4, 1947	J. Ancízar-Sordo

* Samples 1, 2, 3, 6, 7, 9, 10, 11, 12 were placed within a few feet of rain gauges and thermographs.

Results from Colombia

Figure 5 shows the accumulative losses of alfalfa (percentage of oven-dry weight) in relation to time of exposure. A high inital rate of loss is followed by lower rates. In the perhumid tropical area of Calima nearly 100 per cent of the alfalfa disappeared in 1 year. Chinchiná, having a subtropical humid climate, also provided large losses of organic matter. Palmira and Bogatá have lower rates of loss. Presumably the relatively low annual rainfall, 38 inches for both localities, mainly controlled the rates of decomposition.

For the sake of comparison, the decay curve of alfalfa at 25°C. obtained in a laboratory thermostat at optimum moisture conditions also is plotted in figure 5. It lies below the Calima (26°C.) and Chinchiná (24°C.) curves. Since only

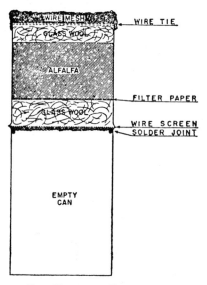

FIG. 4. TYPE OF CAN USED FOR STUDIES ON DECOMPOSITION OF ALFALFA IN SOILS

gaseous losses are measured in the laboratory experiments, the differences among the curves must be attributed to losses caused by leaching. In both field and laboratory the nitrogen losses were nearly proportional to the weight changes.

Results from Costa Rica

To portray more accurately the diversified pattern of the eight Costa Rican decay curves, the scale on the vertical axis of figure 6 is enlarged. All alfalfa samples exhibit heavy losses of material, especially those at the Goodyear Rubber Experiment Station in hot and humid Cairo. Even at "cool" Juan Viñas the losses are high, possibly as the result of heavy leaching. In all instances the initial losses exceed the initial rates of decay of the 25°C. laboratory set.

Results from Sierra Nevada Mountains, California

In contrast to the humid tropical and subtropical climates, the rates of decomposition of alfalfa in the Sierra Nevada Mountains are much lower (fig. 7). The results of many of field experiments suggest the following explanations:

a) At low elevations (Madera) decomposition occurs mainly during the rainy season, normally from September to March. The high summer temperatures are ineffective because of lack of moisture.

b) At high elevations decomposition is restricted to early fall and late spring. The winter months are too cold and the summer months are too dry to influence the rate of decay.

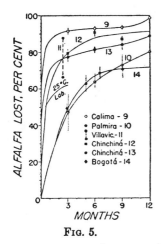

FIG. 5.

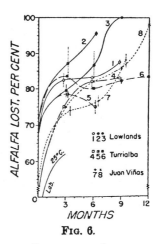

FIG. 6.

FIG. 5. ACCUMULATIVE SEASONAL DECOMPOSITION AND LOSSES OF ALFALFA IN THE FIELD IN COLOMBIA

The dotted vertical lines show the variations between duplicate cans. From Villavicencio only one collection was available.

FIG. 6. ACCUMULATIVE SEASONAL DECOMPOSITION AND LOSSES OF ALFALFA IN THE FIELD IN COSTA RICA

Note the change in ordinates in comparison with figure 5. The dotted vertical lines indicate the variation between duplicate cans. Absence of these lines denotes variability too small to be plotted.

The dependency of the annual losses of alfalfa on elevation, annual precipitation, and annual temperature is shown in figure 8. The seemingly erratic decomposition curve faithfully records the variation in annual precipitation and temperature with elevation. From 300 to 3,000 feet elevation, decomposition rises because rainfall increases. Annual temperatures vary but little. Between 3,000 and 7,000 feet, precipitation remains nearly constant, whereas temperature falls notably. Accordingly, the rate of decomposition is reduced from 74 per cent to 59 per cent. In the range of 7,000 to 9,000 feet, precipitation rises sharply, causing an acceleration in decay, in spite of a lowering of temperature. Between 9,000 and 10,000 feet, precipitation remains constant, temperature decreases, and the rate of decomposition declines.

Summary of decomposition rates of alfalfa in situ

To summarize conveniently the decomposition studies, the residues, that is, the weight percentages of alfalfa left in the cans at the end of one year's exposure, are plotted in a temperature–precipitation diagram (fig. 9). The height of each

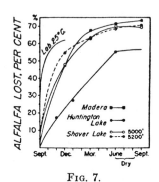

FIG. 7. 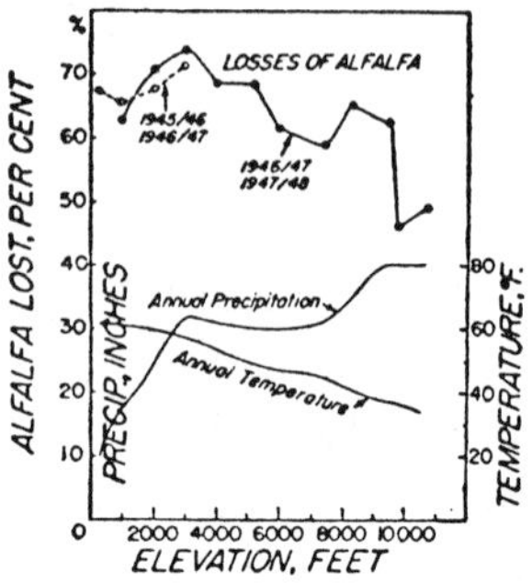   FIG. 8.

FIG. 7. ACCUMULATIVE SEASONAL DECOMPOSITION AND LOSSES OF ALFALFA IN THE FIELD IN THE SIERRA NEVADA MOUNTAINS, CALIFORNIA

FIG. 8. ANNUAL LOSSES OF ALFALFA IN THE SIERRA NEVADA MOUNTAINS IN RELATION TO ELEVATION, ANNUAL TEMPERATURE, AND PRECIPITATION

At higher elevations the precipitation values were obtained from rain gauges and snow survey data of the California Edison Company.

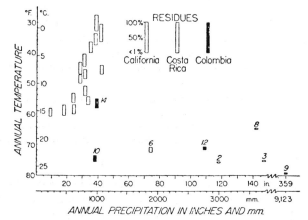

FIG. 9. FRACTION OF ALFALFA LEFT IN CANS (RESIDUES) AFTER 1 YEAR'S EXPOSURE

Length of bar indicates residue in percentage of original amount. Coordinates correspond to the climate of the sites. Numbers denote site number (table 2).

bar denotes the residue of alfalfa. The locus of the center of the base of each bar indicates the annual temperature and precipitation of the site. Whereas the tropical sets represent 1-year studies (1946–47), the California values are averages of two experiments (1946–47 and 1947–48).

In its broader aspects the diagram reflects the climatically controlled organic matter content of soils. In accordance with North American and Colombian climate functions, the residues are small at high temperatures and high moistures and large at low temperatures. Clearly, in the hot, humid tropics decomposition of alfalfa is very marked. Abundant rainfall seems to be especially effective in reducing the amount of alfalfa left in the cans.

The alfalfa studies permit an examination of Corbet's assertion (1) that insola-

TABLE 3

Role of type of plant cover (microclimate) on the losses of alfalfa

LOCALITY	TYPE OF PLANT COVER	AMOUNT OF ALFALFA LOST	
		Period	Amount
		months	*per cent*
Chinchiná, Colombia	Jungle	9	85
	Lawn (cut)	9	90
Diamantes, Costa Rica	Large tree and ferns	6	95
	Adjacent lawn (cut)	6	83
Cairo, Costa Rica	Bare soil	6	87
Turrialba, Costa Rica	Jungle	6	82
	Young rubber plantation	6	75
	Lawn (cut)	6	80
Berkeley, California	Canary pine (dense stand)	12	71
	Adjacent grass	12	70
Whittacker Forest, California	White fir, very shady	13	74
	Adjacent, sunny opening	13	72
Mt. Givens, California	Western white pine (dense stand)	12	43
	Adjacent bare ground, open	12	43
O'Neils, California	Under chaparral	12	73
	Adjacent grass, sunny	12	69

tion is the prime factor which determines the rate of loss of organic matter in tropical regions. To the contrary, according to the comparisons listed in table 3, there is neither a marked nor a consistent difference between losses of alfalfa in jungle, under isolated trees, on lawns frequently cut, or in bare soil. If such differences do exist, they are of secondary importance in comparison to the effects produced by temperature and precipitation of the macroclimate.

DECOMPOSITION VELOCITIES OF FOREST FLOORS

It appears possible to measure *in situ* the rate of decomposition of forest floors, provided certain conditions are fulfilled.[3]

[3] The ideas are not new in principle (3, 5, 9, 10, 11) but they are in their specific applications. Rode (12) mentions Kostychev's work in 1885.

In a deciduous climax forest which has reached a quasi-equilibrium stage, the forest floor, that is, all organic material lying on the mineral soil, exhibits a rhythmical variation. The amount of forest floor is lowest (F_E) in autumn, just prior to the annual fall of leaves. It is highest ($\mathfrak{F}_E$) immediately after the drop of leaves. In an equilibrium forest the annual rate of addition of organic material is equal to the annual rate of loss.

The *rate of addition* is given by the average annual fall of leaves and twigs and other parts of the tree. For a large area within a climax forest it is presumably a constant, say A pounds of organic matter per acre per year.

The *rate of loss* is given by the average amount of forest floor that disappears during 1 year. The losses consist of evolution of gas and of migration of soluble and dispersed humus substances into the mineral soil. In a stabilized forest the average rate of loss also is a constant, say, L pounds of organic matter per acre per year.

A schematic illustration of the rhythmic variation is given in figure 10. To simplify quantitative treatment, it is postulated that the annual fall of leaves takes place very rapidly, in the relatively short time interval dt. Accordingly, in figure 10, the maximum ($\mathfrak{F}_E$) and the minimum (F_E) values of the equilibrium forest floor have practically the same x-coordinates, t, $t+1$, $t+2$, etc.

During the 1-year time interval from t to $(t+1)$ a certain fraction (k') of the forest floor decomposes. In an annual equilibrium system this loss in balanced by the addition A. We may write

$$A = k'\mathfrak{F}_E, \text{ or}$$
$$A = k'(F_E + A) \tag{1}$$

We may also write, more precisely,

$$A_n = k'_\mathfrak{F} F_E; \text{ since } A_n = A - k'_a A,$$

we obtain
$$A = k'_\mathfrak{F} F_E + k'_a A$$

The loss constants k', $k'_\mathfrak{F}$, k'_a, are functions of the soil-forming factors;[4] in other words:

$$k' = \frac{A}{\mathfrak{F}_E} = \frac{A}{F_E + A} = f(cl, o, r, p, t) \tag{2}$$

This constant k' is an average for the various layers of the forest floor.

In a tropical rainforest, leaves drop continuously (fig. 3). Equilibrium may be visualized as being maintained at every instant. During the short time interval dt the following equality exists:

$$A \, dt = k(F_E + A) \, dt, \text{ or}$$
$$A = k(F_E + A)$$

which is identical with equation (1).

The magnitude of k' may be calculated for any given locality, provided near-equilibrium conditions of litter fall and forest floor obtain. Experimental values

---

[4] Unfortunately the authors failed to determine k'_a in the field. All calculations in this paper are based on k' only.

of k' for volatile matter (loss of ignition) are given in table 4. They were calculated from 1-year collections of litter fall and from the forest floor underlying the collectors. Near-equilibrium conditions were assumed to exist. In the perhumid tropical rainforest of Calima the annual loss constant reaches a high value of $\dfrac{730}{730 + 432} = 63$ per cent. In the moist subtropical region of Chinchiná the value of k' is 39 per cent. In contrast, the California values are much lower, varying from 6 to 12 per cent for oaks and 1 to 3 per cent for pines. The variations

TABLE 4

Litter fall, forest floors, and their rates of loss (k')

To convert gm./sq.m. to lb./A. multiply by 8.92

Sequence of localities arranged according to decreasing k' values.

NUMBER	LOCALITY	TYPE OF VEGETATION	LITTER FALL (A)		FOREST FLOOR (F_B)		k' VOLATILE MATTER*	TIME
			Total	Volatile matter*	Total	Volatile matter*		
			gm./sq.m.	gm./sq.m.	gm./sq.m.	gm./sq.m.	per cent	years
	Colombia							
76	Calima	Broad-leaved rain forest	852	730	504	432	62.8	3
—	Chinchiná I	Large mestizo tree	1,205	1,115	—	—	—	—
73	Chinchiná II	Broad-leaved forest	1,011	935	1,648	1,455	39.1	6
	California							
19	Elevation 4,000 feet	Black oak† (50–100 years)	128	122	—	924	11.7	24
41	" 6,000 feet	" " (75–100 years)	92	88	—	724	10.8	26
23	" 5,000 feet	" " "	155	149	4,381	2,517	5.59	52
18	" 4,000 feet	Ponderosa pine‡ (150 years)	160	143	—	4,741	2.93	101
22	" 5,000 feet	" " "	314	305	18,837	12,635	2.36	126
24	" 7,300 feet	" " "	249	246	—	12,351	1.95	152
43	" 6,000 feet	" " "	133	129	—	8,705	1.46	203
36,37	" 9,800 feet	Lodgepole pine§ (200 years)	116	101	—	11,081	0.90	332
—	" 5,000 feet	Mixed conifers (50 years)	454	435	—	14,453	2.92	102

* Loss on ignition.

† *Quercus kellogii.*

‡ *Pinus ponderosa.*

§ *Pinus contorta.*

of k' within the oak and pine groups are conditioned by elevation, local climate, density of stand, and other variables. By selection of trees free of fire scars, it was hoped to minimize possible disturbances of forest floors by fires.

TIME FUNCTIONS OF ACCUMULATION OF FOREST FLOORS

Foresters (4, 8) have calculated the time necessary to reach equilibrium of forest floor accumulation by dividing the amount of annual leaf fall into the amount of forest floor. An additional mode of computation is proposed and its limitations are stressed here.

Suppose we enter a climax forest that has reached a quasi-equilibrium state. Let us select a spot or area that represents average conditions, and let us there remove the forest floor until the mineral soil is exposed. Now, we wish to investigate, over a long period, the accumulation of litter and the formation of the forest floor.

We may consider the special case of an oak tree (table 4) which sheds annually 1,000 pounds of leaves per acre. A decomposition constant of 10 per cent is assumed to obtain throughout the life of the forest floor. The floor oscillates as suggested in figure 10. On the basis of these special conditions and simplifications, the build-up of the forest floor is depicted in figure 11. By a zigzag path the floor eventually approaches the steady state shown in figure 10. The inset in figure 11 displays the annual rate of increase.

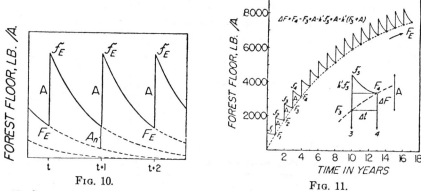

FIG. 10.   FIG. 11.

FIG. 10. SCHEMATIC ILLUSTRATION OF RHYTHMIC VARIATION OF FOREST FLOOR IN A
DECIDUOUS FOREST AT EQUILIBRIUM

FIG. 11. BUILDING UP OF A FOREST FLOOR UNDER CONDITIONS OF INVARIANT k

We may develop the following simple geometrical series:

$$\mathfrak{F}_1 = A$$
$$F_1 = \mathfrak{F}_1(1-k') = A(1-k')$$
$$\mathfrak{F}_2 = F_1 + A = A + A(1-k')$$
$$F_2 = \mathfrak{F}_2(1-k') = A(1-k') + A(1-k')^2$$
$$F_{(t)} = \frac{A(1-k')[1-(1-k')^t]}{k'} = F_E[1-(1-k')^t] \tag{3}$$

This equation describes the sequence of the $F_1, F_2, F_3, \ldots$ values indicated in figure 11 by small circles.

In regions such as the humid tropics where the litter fall occurs through the year (fig. 3), the F values form a nearly continuous curve, as illustrated by the broken line in figure 11. For such conditions the increase in the forest floor during the time dt is given by

$$dF = A\, dt - k(F + A)\, dt = [A(1-k) - k\, F]dt.$$

Upon integration

$$F = \frac{A(1 - k)}{k}(1 - e^{-kt}) = F_E(1 - e^{-kt}), \tag{4}$$
$$k = -ln(1 - k')$$

Figure 12 illustrates equation (4) for tropical and Sierran profiles.

Another way of expressing the trend of the broken line in figure 11 is as follows: The rate with which the forest floor reaches equilibrium is proportional to the difference between the amount of forest floor at any time and its equilibrium value, or, in the form of a differential equation

$$\frac{dF}{dt} = k(F_E - F).$$

Integration of this equation also yields equation (4). Again it is postulated that k remains constant throughout the life of the forest floor.

According to equation (4) final equilibrium will be reached only after a very

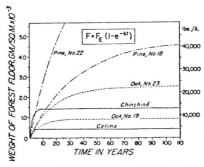

FIG. 12. TIME FUNCTIONS OF FORMATION OF FOREST FLOORS (VOLATILE MATTER)

long time. For practical purposes we may calculate the time necessary to reach, say, 95 per cent of the equilibrium value, according to the equation

$$t = -\frac{ln\left(\frac{F_E - F}{F_E}\right)}{k} = -\frac{ln\left(\frac{100 - 95}{100}\right)}{k} = \frac{3.0}{k} \qquad (5)$$

These time periods are given in table 4. Assuming the validity of equation (5), one would conclude that in the two tropical soils near-equilibrium of the forest floor is reached in a few years. In the Sierra Nevada Mountains oak forest floors would be established in a few decades. Pine forest floors would require one or two centuries to approach a steady state.

DISCUSSION

The loss constants of forest floors determined *in situ* corroborate the annual loss rates of alfalfa. Since alfalfa decomposes more readily than palm leaves, oak leaves, or pine needles, the absolute values of the loss rates for alfalfa are high, but for both sets of observations the values are much higher in the tropics than in California.

Although these studies yield conclusive information on the effect of climate on the rate of decomposition of organic residues, they do not provide a direct clue as to the cause of the high organic matter levels in tropical soils. In fact, the alfalfa and the forest floor studies would lead one to expect low contents of carbon and nitrogen in tropical soils.

An attempt to bring into harmony the two conflicting observations—high content of soil organic matter and high rates of decomposition of leaf materials—will be made in a forthcoming paper.

SUMMARY

Many Colombian as well as Costa Rican soils are very rich in nitrogen and organic matter, especially in comparison with Californian soils found in the Sierra Nevada Mountains.

Annual production of organic matter in the form of leaves and twigs is much greater in tropical forests (7,600–10,700 pounds per acre) than in Sierran forests (820–2,800 pounds per acre).

Alfalfa leaves placed in natural soils decompose rapidly in tropical soils and slowly in temperate soils.

Calculations of decomposition rates of forest floors in Colombia and California corroborate the results of the alfalfa studies.

The time required to reach near-equilibrium accumulation of forest floors is calculated to be less than a decade in tropical forests, 30 to 60 years under California oak, and 100 to 200 years under Ponderosa pine.

REFERENCES

(1) CORBET, A. S. 1935 Biological Processes in Tropical Soils. W. Heffer and Sons, Cambridge.

(2) CUATRECASAS, J. 1946 Vistazo a la vegetación natural del Bajo Calima. Presentación de Calima; Departamento del Valle del Cauca, Secretaria de Agricultura y Fomento, Cali, Colombia.

(3) HENIN, S., AND DUPUIS, M. 1945 Essai de bilan de la matière organique du sol. Ann. Agron. 15: 17–29.

(4) HEYWARD, F., AND BARNETTE, R. M. 1936 Field characteristics and partial chemical analysis of the humus layers of long leaf pine forest soils. Fla. Agr. Exp. Sta. Bul. 302.

(5) JENNY, H. 1941 Factors of Soil Formation. New York.

(6) JENNY, H. 1948 Great soil groups in the equatorial regions of Colombia, S. A. Soil Sci. 66: 5–28.

(7) JENNY, H., BINGHAM, F., AND PADILLA-SARAVIA, B. 1948 Nitrogen and organic matter contents of equatorial soils of Colombia, S. A. Soil Sci. 66: 173–186.

(8) KITTREDGE, J. 1948 Forest Influences. New York.

(9) LEEPER, G. W. 1938 Organic matter of soil as determined by climate. Jour. Aust. Inst. Agr. Sci. 4: 145–147.

(10) NIKIFOROFF, C. C. 1937 Some general aspects of the chernozem formation. Soil Sci. Soc. Amer. Proc. (1936) 1: 333–342.

(11) NIKIFOROFF, C. C. 1942 Fundamental formula of soil formation. Amer. Jour. Sci. 240: 847–866.

(12) RODE, A. 1945 Review of Nikiforoff's fundamental formula of soil formation. (In Russian.) Pochvoved. 1945: 222–225.

(13) WHEETING, L. C. 1939 Some observations on the soils of the Pacific slope of Guatemala. Proc. Sixth Pacific Sci. Conf. 4: 885–889.

15

Reprinted from *Plant and Soil* 13:333–346 (1961)

ORGANIC MATTER AND NUTRIENT CYCLES UNDER MOIST TROPICAL FOREST

by P. H. NYE

University College of Ghana

INTRODUCTION

In a recent paper in this journal, Greenland and Kowal[3] described the measurement of the total dry matter and the nutrient capital accumulated in the aerial parts and roots of a mature secondary forest, about 40 years old, at Kade, Ghana. The present paper describes work designed primarily to estimate the rate of turn-over of organic matter and nutrients in the same forest.

The return of organic matter and nutrients from the vegetation to the soil occurs in the following forms:

    I. litter fall

   II. timber fall

 III. root decomposition and nutrient excretion from roots, and, for the nutrients only,

 IV. rain wash from the standing vegetation.

In investigations of the nutrient cycle under forest and woodland it has often been assumed that the nutrients in the litter fall represented all or nearly all the loss. The amount and composition of litter fall have been measured in a number of temperate regions, and the results have been summarized by Lutz and Chandler[10]. In the tropics the only data giving both amount and composition are provided by the measurements of Laudelout and Meyer[9] in the Belgian Congo forest.

In addition to the litter fall, in the present work some allowance is made for the timber fall; that is, the dead branches and stems that are not collected in the litter trays. An allowance is also made for addition of organic matter by roots.

The importance of leaching of nutrients from the forest canopy by rain has only recently been recognized. It has for long been known that salts could be washed out of crop plants by rain, but most of the quantitative data on such leaching has been obtained under artificial conditions with detached leaves, and not on intact plants. Stenlid [14] has recently reviewed the literature on this subject. Data for trees are particularly scarce. Tamm [15], in a study of leaching from six deciduous species and two conifers in Sweden, found considerable amounts of K were leached, but there was little loss of Na or Ca. Will [17] has recently reported the results of a comprehensive study of litter fall and rain leaching from exotic conifers in New Zealand. At least two thirds of the K reaching the soil under radiata pine and Douglas fir, and half the P under Douglas fir do so in the rain water. Leaching of Ca and Na from the needles was slight, except for Na on one site high in exchangeable Na.

Though the forest at Kade is certainly not primary (such examples are rare and inaccessible these days) there is good evidence that little or no further accumulation of organic matter or nutrients is taking place in it. As Greenland and Kowal have indicated, the basal area at breast height of the trees in the example studied is 145 sq. ft. per acre. According to Dawkins [1] this is the value attained by a primary tropical high forest under favourable conditions. Thus if there is no further net growth or accumulation of nutrients in this forest the rate of litter and timber fall represents the total production of organic matter. The total rate of loss of nutrients from the vegetation will also be a measure of their rate of uptake from the soil.

THE SITE

A description of the forest and soil, and some details of location and climate have already been given by Greenland and Kowal [3]. It is sufficient to note here that the forest is in the ecotone between the moist semi-deciduous and the moist evergreen forest. It contains a wide range of species assorted among a fairly open shrub layer, a dense lower storey extending from some 15 to 50 ft., and a more scattered upper storey extending to 130 ft. with occasional massive-crowned emergents rising through it. The present work was carried out in the undisturbed forest on either side of the 1½ acre block that was felled for estimating the nutrients stored, and was to all appearances identical.

The rainfall, which averages 65 inches p.a., is concentrated in a major rainy season from March to July and a minor season from September to December. During the main dry season centred on January there is often a lot of dust in the air carried by the 'Harmattan' wind from the desert regions far to the North, and this is likely to affect the mineral content of the rain-water. During the rest of the year the prevailing wind blows off the sea, which lies 60 miles to the South.

EXPERIMENTAL METHODS

Litter was collected in six nylon mesh trays measuring 3 ft. × 6 ft., which were raised one foot from the ground. The litter falling in these trays was collected every two weeks over a period of a year. After each collection the tray was moved to a fresh site, in order, over the year, to increase the effective sampling of litter from the extremely heterogeneous vegetation. The litter was air dried, weighed, and bulked. At the end of the year it was thoroughly mixed, sampled in triplicate, and the samples ground for analysis. Litter was collected from May 1958 to April 1959.

Rainfall and leaf drips were collected in 8-in. diameter copper rain gauges fitted with a nylon mesh screen and a plug of cotton-wool at the base of the funnel. There were six gauges under the forest and three in the open clearing. The gauges were emptied every week and the bulked samples from each gauge separately analysed every month. During the first six months of the experiment the nylon screens in the gauges were cleared of any fallen litter every week. In the second six months the number of gauges under forest was increased to eight. Four of them were cleared every day, and the other four every two weeks — the same interval as the litter trays were cleared. Over this six month period the difference in the composition of each set of four gauges was insignificant. This shows that the substantial amounts of nutrients collected in the gauges under the forest were little affected by leachings from the small amounts of litter held up by the nylon screens of the gauges. It also indicates that the litter lost only small amounts of nutrients while on the trays.

ANALYTICAL METHODS

N, P, and K in the litter were analysed by the method of Hutton and Nye [6], which uses a sulphuric acid and selenium digestion. Ca and Mg were determined following a digestion with perchloric and nitric acids, since the highly siliceous residue was found to retain small amounts of these elements in a sulphuric acid digest. Ca in the digest solution was determined by titration with versene using Patton and Reader's indicator. Ca and Mg together were also determined by titration with versene using Eriochrome Black T as indicator. The Mg value, determined by difference, was checked by a Titan Yellow method.

Rain-water samples were analysed as follows: NH_4-N by micro-Kjeldahl distillation and titration with 0.001 N HCl; NO_3-N by the phenol disulphonic acid method; P by the ammonium molybdate — stannous chloride method;

K and Na by EEL flame photometer; Ca and Mg as described for the litter samples; HCO_3 by titration with 0.001 N HCl using bromo-phenol blue as indicator; Cl by titration with silver nitrate using potassium chromate as indicator; SO_4 by the turbidimetric method.

EXPERIMENTAL RESULTS

1. *The rate of production and decomposition of litter*

Annual production. The total amount of litter falling during the 12 month period was 9,400 lb. per acre (oven dry). Leaves formed two thirds of the total, the remainder being mainly twigs and small-wood. This result agrees with the amount of 11,000 lb. per acre (dry matter) recorded by Laudelout and Meyer [9] under mixed forest at Yangambi in the Belgian Congo, and the amounts of 7,000 and 9,100 lb. per acre (oven dry) recorded by Jenny et al. [7] under rain forest in Columbia.

Seasonal distribution. The cumulative total of the bi-weekly collections of litter through the year is shown in Fig. 1. It bears out the description of this forest as an integrade between the moist evergreen and the moist semi-deciduous forest. Though the fall of

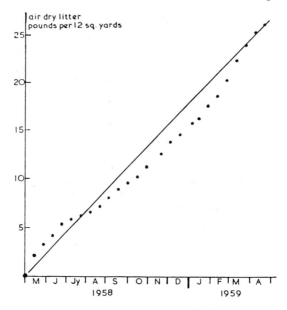

Fig. 1. Litter fall. Cumulative total of 2-weekly collections.

litter is continuous, the influence of the short dry season in January and February may be observed in the relatively high rate of accumulation during February and March. There is a relatively low rate of accumulation from mid-June to mid-August.

The rate of decomposition. Reference is often made to the high rate of decomposition of organic matter in the humid tropics. In the present instance the rate of disappearance of litter from the forest floor can be deduced if the litter fall and the amount of litter on the ground are known. When the litter has built up to an equilibrium value the rate of addition in a small time element dT, will equal the rate of loss over the same period. Hence

$$A \times dT = kL \times dT \qquad (1)$$

where A is the litter fall per annum

L is the amount of litter on the floor

k is the annual decomposition constant.

Greenland and Kowal[3] found the amount of litter on the floor of this forest to be 2,020 lb. per acre (oven dry), a result based on 12 sample plots of 20 sq. yd. each, sampled in late January. The rate of decomposition is therefore $\dfrac{9,400 \times 100}{2,020} = 465$ per cent per annum, or 1.3 per cent per day — a very high figure.

Jenny et al. [7], who first developed this approach, calculated that the annual decomposition constant of litter in moist evergreen forest in Columbia was 40 to 60 per cent and Laudelout and Meyer[9] obtained a value of 76 per cent for the moist evergreen forest at Yangambi. Both workers used the equation

$$A = k(L + A) \qquad (2)$$

to calculate the decomposition constant. This equation was derived by Jenny for conditions with a marked seasonal litter fall. Its use when A is large compared with L, and when litter fall is continuous, leads to a large error. For instance, using Equation (1), the annual decomposition constant for the Belgian Congo is raised from 76 to 316 per cent, a value in fairly good agreement with that obtained for the forest at Kade.

Jenny et al. [7] calculate that the annual decomposition constant for litter under a Californian oak forest is 6 to 12 per cent per annum. The vastly accelerated rate of decomposition of litter in the tropical forest is most striking. At Kade, the difference is probably due in large measure to the intense activity of myriads of termites and ants, which rapidly comminute the litter so that it is readily attacked by smaller organisms.

Though the rate of destruction of litter under tropical forest is so much greater than under temperate forest, it is worth emphasizing that the humus in this tropical forest has been estimated to decompose at a rate of only about 3 per cent per annum (Greenland and Nye [4]).

Nutrient composition of the litter. The nutrient composition of the whole litter fall, and of the leaf fraction only, is shown

TABLE 1

Nutrient composition of litter (%)							
Place	Forest	Condition	N	P	K	Ca	Mg
Fresh litter							
Kade, Ghana	Moist tropical mixed spp.	Oven dry	1.90	0.069	0.65	1.99	0.43
Yangambi, Belgian Congo	Moist tropical mixed spp.	Dry matter	1.8	0.06	0.39	0.85	0.43
Fresh litter-leaf fraction only							
Kade, Ghana		Oven dry	2.10	0.087	1.00	2.02	0.54
Eastern United States	Average value for for 14 broad-leaved forest spp.	Dry weight	0.68	0.14	0.65	2.07	0.35
Litter on the forest floor							
Kade, Ghana		Oven dry	1.54	0.057	0.45	1.98	0.24
Dead wood							
Kade, Ghana		Oven dry	0.32	0.026	0.05	0.73	0.07

in Table 1. It will be seen that the freshly fallen litter is somewhat richer in N, P, K, and Mg than, and similar in Ca to, the litter of mixed age collected from the forest floor by Greenland and Kowal. The litter collected by Laudelout and Meyer [9] from the moist evergreen forest at Yangambi, Belgian Congo, where the soils are very acid, is considerably poorer in Ca and K.

The leaf fraction of the litter, which forms two thirds of the total, is richer than the remainder in all nutrients except Ca. Its composition may be compared with that of the leaf litter from temperate broad-leaved forests. For this purpose the mean of average values for 14 broad-leaved species from the eastern United States (Lutz and Chandler [10], Table 18) has been calculated. The most notable difference is the high nitrogen content of the Kade leaf litter. This appears to be a general effect, for it is confirmed by the high nitrogen content of the litter in the example from the Belgian Congo, and of freshly fallen leaves in two other areas of the Ghanaian forest (Nye [12]). The nitrogen content of the nature leaves in the tropical forest appears to be no greater than in temperate forest (Nye [12]), but the nitrogen in the temperate species must migrate to be branches to a much greater extent before leaf fall.

The contribution of the litter to the complete nutrient cycle is shown in Table 2.

TABLE 2

The nutrient cycle under mature forest (lb. per acre per annum)						
Addition to soil surface from forest	Wt. of material (oven dry)	Nutrient elements				
		N	P	K	Ca	Mg
Litter-fall	9,400	178	6.5	61	184	40
Timber-fall	ca. 10,000	32	2.6	5	73	7
Rain-wash		11	3.3	196	26	16
Total		221	12.4	262	283	63

2. Organic matter and nutrients in timber fall

The timber fall can only be roughly assessed, but even an approximate figure is useful for estimating its importance to the nutrient cycle. In a fully grown forest the annual timber fall will, on the average, equal the annual production of timber. This may be roughly estimated at 10,000 lb. per acre per annum. This is the average arrived at by Weck [16] for production of wood in tropical forest. At Kade, the standing timber represents an average production of 6,500 lb. per acre per annum over a 40-year period. If some allowance is made for timber that has fallen during this period, the estimate of 10,000 lb. is reasonable. The fall over a small area is, however, erratic, since it is greatly influenced by the fall of very large trees.

The composition of the falling timber will be similar to that of the standing dead wood analysed by Greenland and Kowal [3] and shown in Table 1. This data is used to derive the estimate of annual nutrient addition to the soil as timber fall which appears in Table 2.

It will be seen that the falling dead wood may increase the rate of the calcium cycle by around one third, the phosphorus cycle by one quarter, the nitrogen cycle by one sixth, and the magnesium cycle by one eighth.

3. *Root decomposition and nutrient excretion from roots*

The contribution of root decomposition to the addition of organic matter cannot be measured directly, though a rough estimate is possible if the assumption is made that in a mature forest the productions of organic matter below and above the ground are in the same ratio as the weights of the roots and the aerial parts. In this forest Greenland and Kowal [3] found the weight of the roots to be 22,100 lb. per acre, and the weight of vegetation above ground 190,100 lb. per acre. These figures exclude material classified as stumps and dead wood. Since the annual production of litter and wood totals about 20,000 lb. per acre, the annual production of roots will be of the order of 2,300 lb. per acre. Since in this mature forest the net increase in the weight of roots present will be slight, this figure is also an estimate of the annual addition of dead root material to the soil.

A complete statement of the total return of nutrients to the soil has to take into account both the loss of nutrients from dying roots, and the excretion of nutrients from living roots into the soil. Examples of the latter are well known in crops approaching maturity, and may well occur seasonally among forest trees. It also seems likely that nutrients will be absorbed by the roots from spots at high potential in the soil and excreted to spots of low potential. The magnitude of these effects is unknown, though they are probably small compared with the total uptake and return of nutrients to the soil.

4. *The effects of rain wash*

The complete nutrient cycle. The monthly rainfall during 1959, its composition, and the nutrients added to the soil, both under forest and in the open, are shown in Table 3. The total amounts of nutrients leached from the leaves during the year are

TABLE 3

Composition of rainfall and amounts of nutrients falling in the open and under mature forest at Kade, Ghana — 1959

Month	Inches	Ppm						Lb. per acre					
		NO₃-N	NH₄-N	P	K	Ca	Mg	NO₃-N	NH₄-N	P	K	Ca	Mg
In the open													
Jan.	1.24	0.5		0.04	2.3	3.8	1.1	0.1		0.01	0.6	1.1	0.3
Feb.	3.37	0.2		0.03	1.3	1.4	0.8	0.1		0.02	1.0	1.1	0.6
Mar.	5.31	0.2	1.9	0.02	1.5	0.7	0.7	0.2	2.3	0.02	1.8	0.8	0.9
Apr.	4.18	0.4	0.8	0.02	1.4	1.2	0.5	0.4	0.8	0.02	1.2	1.1	0.5
May	12.97	0.0	0.3	0.02	0.8	0.6	0.4	0.1	0.9	0.06	2.4	1.8	1.1
June	11.24	0.1	0.3	0.02	0.5	0.6	0.5	0.2	0.8	0.05	1.3	1.5	1.2
July	10.93	0.1	0.5	0.01	1.0	0.2	0.7	0.2	1.2	0.02	2.5	0.5	1.8
Aug.	2.05	0.1	1.2	0.02	0.6	0.6	0.6	0.0	0.6	0.01	0.3	0.3	0.3
Sep.	4.14	0.1	1.0	0.01	1.1	0.4	0.5	0.1	0.9	0.01	1.0	0.4	0.4
Oct.	5.81	0.2	0.6	0.03	0.9	0.4	1.2	0.3	0.8	0.04	1.2	0.5	1.6
Nov.	8.73	0.1	0.8	0.04	0.6	0.6	0.6	0.2	1.6	0.08	1.2	1.2	1.2
Dec.	2.77	0.4	0.6	0.04	1.7	1.6	0.2	0.3	0.4	0.03	1.1	1.0	0.2
Total	72.7							2.2	(10.3)	0.37	15.6	11.3	10.1
Under forest canopy													
Jan.	1.02	1.5		0.32	24.7	6.0	1.8	0.3		0.07	5.7	1.4	0.4
Feb.	3.48	1.9		0.75	21.4	3.6	1.6	1.5		0.59	16.8	2.8	1.2
Mar.	5.01	0.8	1.6	0.22	15.0	1.6	2.0	0.9	1.8	0.25	17.0	1.8	2.3
Apr.	3.28	1.1	1.9	0.23	15.7	3.6	1.2	0.8	1.4	0.16	11.6	2.7	0.9
May	11.38	0.4	0.8	0.27	15.8	1.8	1.7	1.0	2.1	0.69	40.5	4.6	4.4
June	9.76	0.4	0.6	0.17	10.0	2.6	1.4	0.8	1.4	0.37	22.0	5.8	3.2
July	8.45	0.6	1.1	0.23	16.1	2.6	2.3	1.1	2.1	0.44	30.7	5.0	4.4
Aug.	1.73	1.1	1.2	0.20	27.7	4.0	3.2	0.4	0.5	0.08	10.9	1.6	1.3
Sep.	3.33	0.7	1.0	0.30	16.8	3.2	1.7	0.6	0.7	0.23	12.6	2.4	1.3
Oct.	4.82	0.9	0.9	0.24	12.5	2.2	2.9	1.0	1.0	0.26	13.6	2.4	3.2
Nov.	7.06	0.7	1.0	0.26	15.2	2.4	1.8	1.1	1.6	0.42	24.1	3.8	2.9
Dec.	2.20	1.6	1.1	0.19	12.9	5.6	1.1	0.8	0.6	0.10	6.4	2.8	0.5
Total	61.5							10.4	(13.2)	3.66	211.9	37.1	26.0
Nutrients leached from canopy							8	3	3.3	196	26	16

calculated from this data, and the results are included in Table 2 to complete the data on the nutrient cycle. It will be seen that in comparison with amounts falling in the litter three times as much K, half as much P and Mg, and small amounts only of Ca, and N are leached from the leaves. The amounts of K leached are extraordinarily high. The extreme mobility of K is, however, in accord with Tamm's measurements [15]. In addition, the heavy canopy of leaves, which is maintained throughout the year, ensures that every drip collected has passed over a number of leaf surfaces.

The annual sum of the nutrients in the litter, timber fall, and

rain wash gives the rate of the nutrient cycle. If these totals are
expressed as a percentage of the amounts of nutrients stored in the
vegetation, as reported by Greenland and Kowal, the annual
rate of turn-over of the nutrient capital is seen to be N–12, P–10,
K–32, Ca–12, Mg–18 per cent.

Interception of rainfall by the canopy. At Kade during
the year of collection 72.7 in. fell in the open, and 61.5 in. was
collected under the forest. There was a total of 167 rain days. On
the average therefore, the canopy intercepted 0.07 in. for each rain
day.

The very inadequate data from the tropics on this subject has
been reviewed by Mohr and Van Baren [11] (*l.c.* p. 48 and 96). They
estimated that a dense virginal forest will hold up approximately
0.16 in. for each rain day.

In Fig. 2 the monthly rainfall under forest is plotted against the
corresponding rainfall in the open. It will be seen that in spite of
considerable variation in monthly rainfall, the canopy retained and
evaporated 16 per cent of the rainfall in each month.

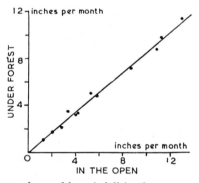

Fig. 2. Comparison of monthly rainfall in the open and under forest.

The possibility that significant amounts of rain flowed down the
trunks of the trees was investigated. Each of the trunks of 12 trees
was encircled with a strip of aluminium sheeting to form an up-
turned frill, whose base was sealed to the trunk with pitch. Water
intercepted by the frill was led into a large container. During April
and May less than 1 per cent of the rain falling (17 in.) flowed down

the trunks. Since its composition was similar to the leaf drips collection was discontinued.

Monthly additions of nutrients from rainfall and leaf wash. Records of the composition of rainfall in other regions, for example in Illinois (Larson and Hettick [8]), show that considerable variation from month to month is to be expected. Though the composition of the rainfall at Kade is likewise rather erratic, there is a general tendency for NO_3-N, P, K, and Ca, to occur in greater concentration during the dry months of December and January when the 'Harmattan' wind blows from the Sahara carrying dust in the air. It may be noted that during the short dry season in August when the wind is from the south-west and the air is clear, nutrient concentrations are the same as for the rainy months.

Analysis of the rainfall at these very high dilutions was subject to a number of positive errors. Blank determinations carried out on distilled water stored one month in the containing vessels indicated that the values for K, Ca, and Mg might be too high by about 0.3, 0.2, and 0.4 ppm. This would decrease the amounts falling in the year by 5, 3, and 6 pounds respectively. Since these errors are common to the analyses of rainfall collected under the forest the calculation of the amounts of nutrients leached from the leaves will not be affected.

The data in Table 2 show that the amounts leached from the leaves are closely related to the rainfall. There is only a slight tendency for each inch of rain to be less effective in wet than in dry months.

The distribution of nutrient concentration among the collecting vessels under forest was markedly asymmetrical, because of one or two notably high values. These were associated with more brown-coloured leachates and probably represented water that had been temporarily arrested by pockets of debris in the canopy. The irregularity in composition of the drips was such that the standard error for the average potassium concentration in any month was 28 per cent.

It will be noticed that insignificant amounts of NH_4-N, and only small amounts of NO_3-N are leached out of the leaves. This is of special interest to the problem of N-fixation under tropical forest. Ruinen [13] has suggested that nitrogen fixed by *Beijerinckia* sp. on the leaves of the trees might offer an essential source of bound nitrogen, which might be taken up directly by the leaves or washed off by the rain. Beijerinckia has indeed been found on the leaves of the Kade forest by Meiklejohn (priv. comm.). The present data indicate, however, that little fixed nitrogen is being washed off the leaves by rain.

The relatively high concentrations of P and NO_3–N in February may be due to a flush of young leaf at that time.

Leaching of sodium, chloride, sulphate, and bicarbonate. In view of the large amounts of K leached from the leaves measurements were made to determine whether large amounts of other mobile ions such as Na, Cl, and SO_4 were leached; and also what anions accompanied the cations. The results of a more complete analysis that was carried out on the October and November rainfalls is given in Table 4.

TABLE 4

Composition of rainfall in the open and under forest canopy in October and November (milli-equiv. per litre)											
Month	Cations					Anions					
	Na	K	Ca+Mg	NH$_4$	Total	HCO$_3$	Cl	NO$_3$	SO$_4$	H$_2$PO$_4$	Total
In the open											
Oct.	0.07	0.02	0.12	0.05	0.25	0.16	0.11	0.01	<.01	0.00	0.28
Nov.	0.04	0.02	0.08	0.05	0.19	0.12	0.09	0.02	<.01	0.00	0.23
Under forest canopy											
Oct.	0.09	0.31	0.35	0.06	0.81	0.50	0.16	0.07	<.01	0.01	0.74
Nov.	0.06	0.38	0.27	0.07	0.78	0.49	0.15	0.05	<.01	0.01	0.70

It is clear from this data that much the most important anion accompanying the cations leached from the leaves is bicarbonate.

The amounts of Na and Cl in the rainfall conform with values obtained in inland areas elsewhere (Eriksson [2]). During the two-month period in which 13.5 in. of rain fell it carried down the following amounts; Na – 3.9, Cl – 11.4, lb. per arcre. During the same period the amounts collected beneath the canopy were Na – 4.5, Cl – 14.7, lb. per acre. Thus only small quantities of these elements are leached from the leaves.

With regard to sulphate, Hesse [5], working in Kenya, noted an increased concentration in rain drips under wattle and eucalyptus during the first rain after the dry season, but not during a period of three days when the rainy season was well advanced. During the two-month period at Kade less than 0,5 lb. per acre of sulphur was collected beneath the canopy.

SUMMARY

The amount and composition of litter and of rain, falling beneath a tropical high forest, have been measured. The litter fell continuously through the year. It was considerably richer in nitrogen than the litter of temperate forests. On the forest floor it decomposed extremely rapidly, at a rate of 1.3 per cent per day. Sixteen per cent of the annual rainfall was intercepted by the canopy and evaporated before it reached the ground.

Compared with amounts falling in the litter, very large amounts of K, significant amounts of P and Mg, but only small amounts of N and Ca were washed out of the canopy by the rain. The main anion accompanying the cations in the leaf drips was HCO_3. Little Na, Cl, or SO_4 was leached from the leaves.

The amounts of nutrients in the litter fall and in the rain wash, together with an estimate of the amounts in the timber fall, have been added together to give the rate of the nutrient cycle under this forest.

ACKNOWLEDGEMENTS

This work was carried out at the Agricultural Research Station of the University College Faculty of Agriculture, at Kade The assistance of D. J. Greenland and J. M. L. Kowal in starting the collection of litter, of S. Boateng, Senior Technical Assistant, in collecting litter and rainfall, and of the management of the station in providing facilities, is gratefully acknowledged.

Received April 26, 1960

REFERENCES

1 Dawkins, H. C., The management of natural tropical high forest. Imp. Forestry Inst. Paper. Oxford (1958).

2 Eriksson, E., Composition of atmospheric precipitation. Tellus 4, 215–232 and 280–303 (1952).

3 Greenland, D. J. and Kowal, J. M., Nutrient content of moist tropical forest. Plant and Soil 12, 154–174 (1960).

4 Greenland, D. J. and Nye, P. H., Increases in the carbon and nitrogen contents of tropical soils under natural fallows. J. Soil Sci. 9, 284–299 (1959).

5 Hesse, P. R., Sulphur and nitrogen changes in forest soils of East Africa. Plant and Soil 9, 86–96 (1957).

6 Hutton, R. G. and Nye, P. H., The rapid determination of the major nutrient elements in plants. J. Sci. Food and Agr. 9, 7–14 (1958).

7 Jenny, H., Gessel, S. P. and Bingham, F. T., Comparative study of decomposition

rates of organic matter in temperate and tropical regions. Soil Sci. **68**, 419–432 (1949).

8 L a r s o n, T. E. and H e t t i c k, I., Mineral composition of rain water. Tellus **8**, 191–201 (1956).

9 L a u d e l o u t, H. and M e y e r, J., Les cycles d'éléments minérales et de matière organique en forêt equatoriale Congolaise. Trans. 5th Int. Cong. Soil Sci. **2**, 267–272 (1954).

10 L u t z, H. J. and C h a n d l e r, R. F., Forest Soils. New York (1947).

11 M o h r, E. C. J. and V a n B a r e n, F. A., Tropical Soils. London (1954).

12 N y e, P. H., The mineral composition of some shrubs and trees in Ghana. J. West African Sci. Assoc., **4**, 91–98 (1958).

13 R u i n e n, J., Occurrence of *Beijerinckia* sp. in the 'phyllosphere'. Nature **177**, 220–221 (1956).

14 S t e n l i d, G., Salt losses and redistribution of salts in higher plants. Encyclopedia of Plant Physiology (Ed. W. Ruhland) **4**, 615–637 (1958).

15 T a m m, C. O., Growth, yield and nutrition in carpets of a forest moss (*Hylocomium splendens*). Medd. Stat. Skogsforskn. Inst. **43**, No. 1, 1–140 (1953).

16 W e c k, J., Über die Grossenordnung der Substanzerzeugung in Baumbestanden verschiedener Vegetationsgebiete. Allg. Forst- u. Jagdz. **127**, 76–80 (1956).

17 W i l l, G. M., Nutrient return in litter and rainfall under some exotic conifer stands in New Zealand. New Zeland J. Agr. Research **2**, 719–724 (1959).

16

Reprinted from *BioScience* 19:29–34 (1969)

Tropical Agriculture

A Key to the World Food Crises

H. David Thurston

After reading the papers by Drs. McElroy and Robinson, little additional discussion seems necessary on whether or not the world faces a food crisis. President Lyndon B. Johnson (Hornig, 1967), reflecting our government's judgment, has stated "The world food problem is one of the foremost challenges of mankind today." This challenge is an especially strong one for all biologists. If one is interested in making the world a better place in which to live, or, for more selfish reasons, is interested in lessening the prospects for international conflicts, one concrete way of making a real contribution is to alleviate the gravity of the world food situation.

Where is the food crisis? Martin E. Abel and Anthony S. Rojko (1967) have stated, "Two-thirds of the world's people live in countries with national average diets that are nutritionally inadequate. A country is classified as diet-deficient if the average annual per capita consumption of food results in a deficiency of calories, proteins, or fat below minimum levels recommended by nutritionists. The diet-deficient areas include all of Asia, except Japan, Israel,

The author is affiliated with the Department of Plant Pathology, New York State College of Agriculture, Cornell University, Ithaca, New York.

and the Asian part of USSR, all but the southern tip of Africa, part of South America, and almost all of Central America and the Caribbean." In other words the problem is largely in the tropics, subtropics, and contiguous areas. Failure to recognize the world food crisis as a tropical problem often deludes people into thinking that our temperate zone agricultural technology can be transplanted directly. The principles and methods from temperate zones often work, but there must be adaptive research to make them work under tropical conditions.

The tropics consist of the zone bounded on the north by the tropic of Cancer and on the south by the tropic of Capricorn. The latitude at these points is 23°27′. Some would extend the tropics to include the area between 30° North and 30° South latitude. A few minutes of study with a globe will show that most "developing" countries are in or on the border of the tropics. Uruguay, for example, is the only country of Latin America with its boundaries entirely within the temperate zone. In Africa, only Morocco and Tunisia are entirely within the temperate zone.

Because of the effect of altitude, many crops grown in temperate zones

are also grown in the tropics. Seasonality exists with respect to rainfall, but the extremes and seasonality of temperature we experience in temperate zones does not. Daylength also has a striking effect on crops and animals in the tropics, but changes in length of day and in solar radiation are small in comparison with the corresponding changes in temperate zones. Thus, even when working with crop plants which are common in temperate zones, adaptive research is necessary when working with them in the tropics. The tropics should not be thought of as a single unit, however, since a wide diversity of ecological and climatic regions can be found within them.

George Harrar (1961) has stated "Rice, wheat, maize, sorghum, the millets, rye, and barley are principal foods in cereal-producing areas, but elsewhere cassava, the sweet potato, potatoes, coconuts, and bananas are basic foods. Although over 3000 plant species have been used for food and over 300 are widely grown, only about 12 furnish nearly 90% of the world's food." Many more of these crop plants are grown in the tropics than in the temperate zones, and we of the temperate zone

219

often do not know much about these plants.

Agricultural research in tropical areas in the past has been primarily on cash and plantation crops while food crops — the plants that people eat, especially in the hot, humid tropics — have been largely ignored by research workers. The Food and Agricultural Organization of the United Nations (FAO) (1966) renders a valuable service by compiling statistics on food production throughout the world. If one studies the 1966 figures for world food production, one finds that such crops as sweet potatoes, yams, and cassava are ranked after rice and maize as important food sources in the tropics. Plantains (cooking bananas) are also of great importance as a food crop in tropical areas.

These crops, with the possible exception of sweet potatoes which grow well in temperate zones, have been largely ignored by agricultural researchers. Perhaps I can illustrate this point with research on cassava in my field, plant pathology. *The Review of Applied Mycology (R.A.M.)* is probably the best abstracting journal for information on plant diseases throughout the world. From 1945 to 1965 there were 119 references to cassava in the *R.A.M.* or six references per year. Carnations were on the same page and I noted 350 references to carnations during the same period or 17 references per year. In 1966 the *R.A.M.* had two references to cassava diseases and 17 to carnation diseases. There were 234 references to research on tobacco diseases in 1966 in the *R.A.M.* These figures illustrate how research (on a world-wide basis) is not giving some of the important food crops in the tropics the attention they need.

I do not wish to imply that research on important plants such as carnations and tobacco is not valuable, but I do wish to make clear the point that research on some of man's basic food crops such as cassava and plantains is insignificant in comparison. Lack of research on cassava is not due to lack of disease problems. Chevangeon (1956) has listed 73 species of fungi that have been found associated with cassava. In addition, many destructive diseases caused by bacteria, viruses, nematodes, and various abiotic factors are reported in the literature. In 1956 G. W. Padwick brought together the

available information on losses caused by one virus disease, cassava mosaic, and estimated that the yearly loss in yield due to the disease was equivalent to 11% of the yearly production of cassava in Africa.

The starchy crops such as cassava, sweet potatoes, yams, and plantains hardly enter into world commerce when compared to the cereals. One of the reasons is that they can seldom be stored for any appreciable time, but they are probably far more important as food in the tropics than available figures indicate.

Nutritionists generally dislike these crops because they are primarily carbohydrate and have a low protein content, yet varieties exist with an appreciable protein content and much might be done in a breeding program to increase it. These crops also can be converted to protein when used as feeds for livestock. Yields of cassava in Brazil, expressed as calories per hectare, are about three times those from corn or rice (Jones, 1959). The starchy root and tuber crops and plantains have a real potential for rapidly reducing food shortages in tropical countries.

Numerous books and references can be found on the cash and plantation crops of the tropics such as sugar cane, rubber, coffee, tea, cacao, citrus fruits, and bananas for fruit, but there is a paucity of information on tropical food crops as they grow in the tropics. Before World War II, most of the research done by the Dutch, English, French, Belgians, and Americans in the tropics was done on cash or plantation crops. Since World War II, other entities such as US-AID (United States Agency for International Development), universities, private foundations, the UN-FAO (United Nations Food and Agricultural Organization), and the Inter-American Institute of Agricultural Science of the OAS (Organization of American States), have entered the field of research in agriculture in the tropics, but even so a great enough emphasis has not been focused on tropical food crops. Outstanding contributions have been made in research on rice, corn, wheat, and potatoes, research which has had and will continue to have an impact on alleviating food shortages, but opportunities for research on other tropical food crops are now largely ignored.

Most of our success stories about

agricultural research in developing countries pertain to the subtropics or the semi-temperate border of the tropics. Success stories about agricultural research in the hot, humid tropics are hard to find. The development of IR-8 and IR-5 rice by the International Rice Research Institute in the Philippines is an outstanding exception. Vicente-Chandler (1967) has noted that there are 2 billion acres in the hot-humid area of tropical Latin America, almost as much as the entire United States. In the entire Amazon Basin (half the size of the USA) there are only 10 experiment stations, some with only one researcher with the equivalent of a high school education. From 1960-62 the U.S. Government and international and regional agencies together spent less than 8 million dollars on agricultural research in all of Latin America and only a small part of this in the hot-humid tropics. Two billion dollars were spent on agricultural research during the same period in the United States by the government and agricultural industries. This does not mean that agricultural research in Latin America is the responsibility of the government of the United States, but these observations do bring into focus the relative emphasis on agricultural research in the two areas.

Most of the students from developing countries who come to the United States to study come from tropical countries. Yet few courses are available at most U.S. universities on tropical agriculture. Furthermore, only a superficial introduction to tropical agriculture can be given in one course. In addition, very few students from developing countries study agriculture and few of these take courses related to tropical agriculture. To quote William C. Paddock (1967) "in 1962 Central America had only 187 university students out of 10,546 studying agriculture." He also adds "In nearly all Latin American countries, for example, the percentage of university students studying agriculture has decreased during the past decade. For instance, Mexico has declined from to 1 percent; Panama from 4 to 2 percent; Dominican Republic from 2 to percent. Of 105,000 Latin American students enrolled in the United States during the decade 1956 to 1965, only 5% studied agriculture." It should b pointed out, to put this situation into proper perspective, that the U.S. government and the other agencies which send students from developing countries to the United States to study give high priority to agriculture.

In plant pathology, I know of no course in the United States (including Hawaii and Puerto Rico) in tropical plant pathology. Little or no emphasis is given in existing courses to tropical problems. In other words, students from developing countries trained in plant pathology in the United States return with a training consisting almost entirely of a study of temperate crop plants and their problems. I suspect the situation is similar in other biological disciplines.

Let us return to the world food crisis. What are the possible solutions for the world food crisis?

1) Atomic war is one solution. We as biologists can do much to see that this is not the solution.

2) Birth Control. Dr. McElroy has covered the many ramifications of this solution. The food crisis may be upon us, however, before this solution is able to bring about real change or slow down the population increase.

3) Massive food shipments. At present, only a very small percentage of the world's total food supply consists of food shipments to developing countries. In 1966 the United States shipped over 15 million tons of grain to developing countries (Abel & Rojko 1967). However, according to FAO statistics (1966), the world production of cereals in 1966 was over 1 billion tons. In addition to the unfortunate side effects of this type of solution, which allows developing countries to delay facing their food problems realistically and thus undertaking to solve their agricultural problems, it appears that in a decade the United States, Canada, and Australia may not be able to continue this type of aid. As recently as 1966, A. H. Moseman stated that "the estimated 42 million ton gap ten years hence exceeds the total of one annual U.S. wheat crop, and the 88 million ton shortfall in 1968 will be beyond U.S. cropping capacity, even if we were to put into production the 55 to 60 million acres now in reserve. It is abundantly clear that the U.S. cannot feed the world." Don Paarlberg who was the U.S. government's Food for Peace coordinator,

stated in 1968, "If we were to remove all acreage restrictions, our grain production might increase the world total by some four percent." A recent study by Martin E. Abel and Anthony S. Rojko (1967) of the Economic Research Service of the United States Department of Agriculture gives a less pessimistic prediction: "the results of this study imply that the world probably will continue to have excess production capacity by 1980." Massive food shipments can be of help but they are not the solution to the world food crisis.

4) Increased food production. Some of this may come from fresh and salt water fish, from desalinization of sea water for the deserts, from synthetic foods, and from microorganisms and other nonagricultural food sources. These are extremely important and promising areas of endeavor for biologists. Breakthroughs may occur, but it is doubtful that these sources will produce food fast enough to solve the world food crisis. They are, however, long-range solutions that in several decades may be the key to mankind's food problem.

For the short-range solution, most of the food needed will have to come from more efficient conventional agriculture in tropical countries.

How can food production be increased in developing countries which, as I have pointed out, are primarily tropical? Part of the solution is massive overseas technical programs. Another aspect of the solution is to train the young people of developing nations both in their home countries and abroad so that they can increase food production.

First, let us examine the concept of massive overseas technical programs. By "massive" is meant a commitment by the people of the United States and their government large enough to make reasonable progress in reducing the gravity of the world food crisis. However, we must keep in mind that the United States cannot possibly do the job alone. Other favored countries of the temperate zones should make similar commitments. England, France, Holland, Germany, Russia, Canada, Japan, and many others have much to offer and should join this effort.

Since World War II, many agencies such as U.S.-AID, U.S. universities, the UN-FAO, the Peace Corps, private

foundations such as the Rockefeller Foundation and the Ford Foundation, and many other entities have had overseas technical programs with some commitment in agriculture. An expansion, with modifications, of the activities of these agencies could have an ever-increasing impact on the world food crisis.

Those who have studied the phenomenal success story of U.S. agriculture know that much of the credit is due to the combination of research, education, and extension in our land-grant system of colleges and universities. Unfortunately, this combination and cooperation is seldom found in developing nations. Successful extension is complicated by the sheer numbers of farmers in many tropical countries where as many as 50-80% of the population may be actively engaged in farming versus 4-5% in the United States. The level of education is also a problem in an effective extension of new agricultural technology since largely illiterate populations cannot easily assimilate new knowledge. Lester Brown (1965) has clearly shown the importance of this and other basic factors in obtaining a "yield take-off" in an agricultural country. This does not mean that the job cannot be done, but that experimentation and innovation in extension in developing countries must find methods of accomplishing the task.

Even in the fields of education and research in agriculture in developing countries a poor record of cooperation is more often the rule than the exception. Agencies, both national and international, often duplicate efforts, facilities, and manpower even to the point of interagency animosity. Cooperation and coordination of efforts between agencies should be studied and encouraged. As in taxonomy, splitters usually win in the long run over the lumpers, but the effort should be made to focus these massive efforts on common goals.

During the past two decades, a number of guidelines or principles have been observed which can be very useful for overseas technical programs. There seems to be general agreement on most of these points, but seldom is this agreement translated into action. I would like to give a few examples.

1) When sending people to developing countries it is absolutely essential to

send the very best people available. Second raters cannot do the job. The added difficulty of language problems, cultural adjustments, and working in a new and different climate with new crops and technical problems means that only the very best are able to make significant progress. It is necessary to send men who are not only technically competent but by temperament and interest have the patience to work under commonly frustrating circumstances.

2) Continuity of effort is absolutely essential. One- or two-year stints are unfair to the agency, to the individual, and to the developing country. Far too often, about the time an investigator arrives at the point where he can begin to produce he returns home or goes to another country. Few research programs or universities in the United States would have such an impossible and self-defeating personnel policy. Scientists need to be trained and hired to spend decades, not years, in the tropics. To do this, the agencies, such as universities, need long-term commitments of funds for continuity of personnel on overseas assignments.

3) Far too many scientists go to tropical countries as advisors. If one cannot go as a working partner or colleague it may be better to stay home. Few problems in agricultural science exist on which all the advice possible cannot be given in a few weeks. Frequent short visits over a period of many years would be better than a stay of a year or two if one's only function is to give advice.

4) Human relationships are of primary importance. Customs and traditions in tropical lands are often strange and different to people of temperate zones. It is not only important, but essential, to learn as much as possible about the culture, customs, traditions, history, and sociology of the tropical country where one will work in order to be able to partially understand its reason for existence before one can · hope to bring about change, if indeed change is necessary for progress.

The ability to meet one's hosts and treat them as equals and coworkers often is more important than one's scientific knowledge. Jose Nolla (1962) has pointed out that even the poorest farmer and laborer often has great pride and human dignity. The least suggestion of inferiority will be re-

sented and may ruin all one's future work. On the other hand, when treated as equals, the poor farmers can become the warmest of friends and loyal supporters. We in the United States can learn much from them about how to live and enjoy life without losing our much prized reputation for "getting the job done."

5) Any worth-while research effort to increase food production in a developing country should have as one of its aims a training program for local personnel. This enables a tropical country to develop the competent leadership necessary to assume the direction of its own food production efforts. Too often in overseas technical assistance programs several years of intelligent and productive effort by an outside scientist may be essentially lost because no local personnel were assigned to work and learn at his side in the development of a program which will continue on after he leaves.

The United Nations, the United States, and many other nations have been sending thousands of technical personnel overseas during the past two decades to work in increasing food production. Seldom have these personnel received more than a short orientation program to prepare them for their assignments. They may be well prepared in their field of specialization as regards the temperate regions, but they do not have a sound understanding of how to most effectively work in an overseas tropical environment. Professional State Department personnel usually spend many years preparing for their assignments. It should be no less important for young people interested in international agriculture to receive several years of special education and training in preparation for an overseas career. This education should include not only a sound professional training but also language competence, courses and seminars dealing with the tropics as related to their subject matter field, and an opportunity to work and live in a developing country before graduation. Thesis research (especially at the Ph.D. level) should be done overseas in a tropical environment. This type of training should produce men with a real background of competence, training, and aptitude for working abroad in the tropics.

One of the best and most lasting so-

lutions to the world food crisis is to train the young people of developing nations so that they can help to increase food production. This should be done both in their own countries and abroad.

We have been training students from foreign countries in the United States for probably more than 100 years and in ever-increasing numbers since World War II. The question I wish to ask is, "Have we been doing a good job of training them to return home to increase food production?" I don't think we have for many reasons. If we examine some of the reasons why we have not been doing a good job, perhaps we can improve our future performance.

In agriculture, the great majority of foreign students from tropical countries are "asphalt" farmers. Students from Latin America, Africa, and Asia usually come from a stratum of society in which people do not actually till the soil with their hands. Many departments in U.S. colleges of agriculture require their undergraduates to give proof of farm experience or obtain such experience before or during their time in college as a prerequisite for graduation. We should do the same for foreign graduate students who plan to major in an agricultural science. There is no classroom substitute for practical farm experience.

The foreign students that do manage to come to the United States are usually a highly selected group. When they return home, they often end up as administrators. Poor administration is one of the facts of life in most tropical countries. Yet we make no attempt to give even passing reference to this important aspect in our training programs.

Increasing numbers of tropical countries are setting up graduate and undergraduate schools to train the people their countries need for increasing agricultural production. Foreign students seldom, if ever, are exposed to a study of teaching methods or given practical experience in teaching, yet they may return to a teaching position or may be expected to set up a graduate program. We have given almost no attention to this aspect of training.

Increasing specialization and fragmentation of biological sciences is the rule in the United States at the present time. Students from developing countries need a far broader training than we are now generally able or willing to give.

At present in the United States, the emphasis in most fields of biology is on basic research. Applied or practical research has somehow developed a connotation of being "second class." Considerable basic research is needed if breakthroughs are to be made in tropical agriculture, but there is a far more pressing need for practical and applied research. Students from foreign countries, and U.S. students interested in international agriculture, soon find out that the prestige and the challenging forefront of human knowledge is in basic research, and the best of them frequently seek training and do their thesis in "basic" research. What they need, however, is training in how to apply the methods and principles they learn here in solving the problems of increased food production in a tropical environment.

Frequently, a well-trained scientist may return to his country with glowing letters of recommendation from his department and advisor after a brilliant academic career and after completing a thesis which is a real contribution in his field. He insists on, and gets, a fine laboratory, expensive equipment, and technical assistants. After a few years his superiors begin to ask embarrassing questions on practical applications and expect him to give them figures in dollars, rupees, or pesos on how his research has increased food production. When they are not forthcoming, administrators become gun-shy of future brilliant returnees and it becomes increasingly difficult to support good research.

I have previously mentioned the lack of training with a tropical emphasis. Much could be done here to at least expose students from tropical countries to the pertinent literature in their field of specialization as regards the tropics. In addition, a few universities now encourage students from developing countries to do their thesis in a tropical environment. This type of training should be far more effective than that which most students receive at present.

Every major university department in the agricultural sciences in the United States turns down large numbers of qualified students from developing countries every year. At present, there

is no method of helping these students, no central clearinghouse to help them find the type of training they need, with the result that there is an obvious duplication of effort and seeming chaos in placing students from foreign countries. Selection of students is equally confused. The present process is similar to a lottery since the information available about the students is usually not sufficient for an adequate judgment. A study of these problems by our organization and other similar bodies should result in worthwhile recommendations which could improve this chaotic situation. Only the most highly qualified and motivated students, those who plan to return to the tropics and work in increasing food production there should be given priority. This may be asking the impossible, but we certainly could do a far better job than we are doing at present. Our resources, and those of tropical countries, are not so great that an essentially hit and miss system of student training and selection will meet the challenge of increasing food production in these critical days.

There is a tremendous ferment in our society today. Our youth are confused and unhappy with the world as it is. There is a great backlog of youth in our nation who want to make the world a better place in which to live. Unfortunately, they don't know how, and they have few skills. Love of humanity and the best of intentions will not make a better life for others. You cannot will it. Our society, and biologists in general, have a challenging task — namely, to interest our best young people in biological science and show them a way whereby they can have a real impact on making the world a better place in which to live. To do this job, however, I believe we have to go outside of our profession and enlist the aid of professionals in communication. We can help, but we cannot do it alone. Once we have students in the classroom or laboratory we can do much to show them the opportunities, but we have to get them into the classroom first. If their high school biology is deadly dull and no connection is made between frog eggs and a better world, we will not be doing our job.

International activities, especially those in the biological sciences, are often thought of as an activity which universities should engage in as a phil-

anthropic cause. As Charles Frankel (1968) has pointed out "it is that; but it is also a commitment that can sustain universities in their honesty and their honor. It is through the shock of new perspectives that people can veraciously look at themselves. It is through the exchange of ideas across the frontiers that people come to look at their own traditions objectively. International education involves the simple practical business of keeping universities true to their own vocation.

"It is from this point of view that one can speak of international education not only as a duty but as an opportunity, not only as a service to the world but as a way of learning from the world. Properly conceived and organized, international education can represent precisely the sort of marriage of self-interest and social purpose, of social influence and emancipated intelligence, which should characterize all relationships between the university and the world."

Summary

The world faces a food crisis and much of the problem is in the tropics and subtropics. Tropical agricultural research in the past has concentrated primarily on cash and plantation crops, and the food crops have been largely neglected. Great opportunities exist for biologists interested in tropical food crops to make significant contributions to solving some of the problems of food shortages in developing countries.

The most realistic short-range method of solving the world's food crisis is to increase food production through conventional agriculture in the tropics. This can be done in part by overseas technical programs originating in developed countries, but training the young people of developing nations to do the job of increasing food production themselves is the better and more lasting method.

The efficiency of overseas technical programs can be greatly increased by stimulating closer cooperation among overseas technical agencies, by sending only the most competent personnel overseas, by establishing continuity of effort and personnel for decades and not months or years, and by greater cognizance of the importance of human relationships when working in developing countries.

Much can be done to improve the

education and training that we of the temperate zones give young people from developing countries, both in their own countries and abroad. This training has usually been deficient in practical or farm level experience, training in administration, training with a tropical emphasis, and in giving too much emphasis on basic rather than applied or practical research.

The young people of our society, and those of most developing countries are seriously disturbed, if not in open revolt, about the world as it is. They want to make the world a better place in which to live. Biologists have an exciting challenge — namely, to interest young people in biological science and show them a way whereby they can have a real impact on two of the world's most serious problems — hunger and malnutrition.

References

Abel, M. E., and A. S. Rojko. 1967. World food situation. Prospects for world grain production, consumption, and trade. Foreign Agr. Econ. Rept. No. 35. USDA, Washington, D.C.

Brown, Lester R. 1965. Increasing world food output. Foreign Agricultural Economic Report No. 25. USDA, Econ. Res. Serv.

Chevangeon, Jean. 1956. Les maladies cryptogamiques du manioc en Afrique occidentale. *Encycl. Mycologique,* **28:** 1-205.

Food and Agriculture Organization of the United Nations. 1966. FAO, *Production Yearbook.* Vol. 20.

Frankel, Charles. 1968. The university and the world. *Univ. Rev.,* **1:** 8-11. State University of New York.

Harrar, J. G. 1961. Socio-economic factors that limit needed food production and consumption. *Fed. Proc.,* **20:** 381-383.

Hornig, D. F. (Chairman). 1967. *The world food problem. Report of the Panel on the World Food Supply.* A report of the Presidents' Science Advisory Committee. Vol. 1. The White House, Washington, D.C.

Jones, William O. 1959. *Manioc in Africa.* Stanford University Press, Stanford, Calif. 315 p.

Moseman, A. H. 1966. National systems of science and technology for agricultural development. Proceedings of the St. Paul Conference. University Directors of International Agricultural Programs. University of Minnesota, St. Paul, Minn.

Nolla, Jose A. B. 1962. The necessarily broad view in studying tropical plant diseases. *Phytopathology,* **52:** 946-947.

Paarlberg, Don. 1968. World food: Present situation and future prospects. In: World Markets and the New York Farmer.

Paddock, W. C. 1967. Phytopathology in a hungry world. Ann. Rev. *Phytopathology,* **5:** 375-390.

Padwick, G. W. 1956. *Losses Caused by Plant Diseases in the Colonies.* Phytopathological papers No. 1. Commonwealth Mycol. Inst., Kew, Surrey, England.

Vicente-Chandler, Jose. 1967. A prospectus of natural resource use and research programs for semitropic regions. The Southeastern area meeting of the National Association of Conservation Districts, San Juan, Puerto Rico.

Editor's Comments
on Papers 17, 18, and 19

17   JORDAN and MURPHY
Excerpt from *A Latitudinal Gradient of Wood and Litter Production, and Its Implication Regarding Competition and Species Diversity in Trees*

18   ODUM
Excerpt from *The Rain Forest and Man: An Introduction*

19   JANZEN
Tropical Agroecosystems

TOTAL TROPICAL ECOSYSTEM PRODUCTIVITY AND PRODUCTIVITY USEFUL TO MAN

Although plant production in tropical ecosystems is usually higher than in temperate ecosystems, this does not mean that productivity useful to man is higher in the tropics.

In a series of papers, I have developed the idea that the high productivity in tropical forests is due to high leaf production, and that net annual increment of wood is not different from temperate zones (Jordan, 1971a; Jordan, 1971b; Jordan, in press). The two figures reproduced from Jordan and Murphy (Paper 17) are part of the evidence. Comparisons of tree growth in temperate and tropical forests by Weaver (1979) and Huston (undated) show that annual diameter increment is not greater in tropical forests than in temperate forests. Knight (1978) and Kato et al. (1978) remark on the low rates of wood production they found in tropical forests. Whitmore (1975) comments:

> The luxuriousness and appearance of unbridled growth given by the vegetation of the perhumid tropics does not therefore arise from an intrinsically higher growth rate than exists amongst temperate species. The unfamiliar life-forms of palm, pandan, the giant monocotyledonous herbs of the Scitamineae (gingers, bananas, and their allies), and the abundant climbers make a vivid impression of 'vegetative frenzy' on the botanist brought up in a temperate climate. The appearance of rapid growth of pioneer trees of forest fringes and

clearings, which forms the other part of the impression, does not result, as far as we yet know, from a particularly efficient dry-weight production or energy conversion but arises from the architecture of the tree, which results from the capacity for unrestricted elongation of internodes and production of leaves in the continually favourable climate. Temperate herbs, by contrast, die down annually, and although their annual production is impressive, growth is not cumulative and in many cases does not extend higher than eye-level to dwarf the observer.

Because the total annual solar radiation in the tropics is greater than at higher latitudes, equal annual production of wood at high and low latitudes means that wood production at high latitudes is more efficient, when efficiency is defined as the ratio of calories of wood produced to the calories of solar radiation entering the system. The question naturally arises, why should production of wood in temperate regions be more efficient than in the tropics? The reason may stem from the fact that much of what is now the temperate zone at one time had a tropical climate (Chaney, 1947). As the climate in these higher-latitude regions became colder, the annual amount of light available to plants decreased, due to lengthening of the cold-induced dormant periods. The plant species that were efficient at converting photosynthate into permanent structural biomass were better adapted, because relatively rapid increase in structural biomass increased the ability to compete for light, an important advantage during the period of decreasing length of growing season.

In contrast, in the tropics efficient production of structural biomass may be less advantageous than having a large supply of carbohydrates available for maintenance costs. One reason that maintenance costs are higher in the tropics is because respiration rates are higher in tropical trees during leafless periods than in temperate trees following leaf shedding. In tropical regions, the maintenance costs for support of the photosynthetic system are more or less constant during the whole year, while in temperate regions the respiration losses during the winter season are reduced to very low levels (Medina and Klinge, in press). Evidence comes from a comparison of carbohydrate loss among plants in tropical and temperate zones during the dormant season. Carbohydrate reserves of *Jacquinia pungens*, an understory shrub, in deciduous forests of Costa Rica were reduced about 70 percent during the dormant period (Janzen and Wilson, 1974). In contrast, in the central United States, *Robinia pseudo-acaia* showed reduction of total reserve carbohydrates in the living bark of about one percent between the winter months of December and April (Siminovitch et al., 1953).

Because productivity of native tropical forests is not high, tropical foresters often advocate conversion of native tropical forests to plantations in the belief that the rate of wood production in plantations is higher. Since foresters measure production in terms of volume, not biomass, and since many plantation species have a specific gravity considerably lower than most primary species, the seemingly high wood production of plantations often results, in part, from the system used for measurement. Plantation productivity also seems high because herbicides and pesticides are used to suppress weeds and parasites. If chemicals or other forms of energy are used to control pests and competitors, energy ordinarily lost by trees to competitors and parasites becomes available for stem growth. Odum (Paper 18) explains that controlling pests supplements the energy available to trees for biomass production. An accounting system that measures wood output but not total energy input, including petroleum derivatives, deceives us about net gain from plantations.

In certain cases, species of *Eucalyptus* established in plantations of the humid tropics are growing exceptionally well, and reports of their growth seem to substantiate the claim that tropical plantations are highly productive. However, the importation of *Eucalyptus* from Australia is recent, and herbivores and parasites in the new environment may not have adapted to exploit this new resource. If they do, productivity could decline.

The first paper in this section indicates that wood production in tropical regions is not greater than in temperate zones. It is not only wood production that is lower in the tropics than in temperate areas. Rates of food production are also not necessarily greater in the tropics than in temperate regions. Janzen (Paper 19) shows that, despite continual warmth, sunshine, and rain in many regions of the tropics, agricultural productivity is not high. He points out that an environment that is good for crops is also good for insects. In the temperate zone, a cold season reduces the insect populations. When there is no periodic dieback of insects, their exponential growth rates can quickly result in crop destruction. The influence of insect herbivores is especially great in monocultures such as cotton that are becoming more common in the tropics.

The ideas developed here result in a different impression of the productive capacity of the tropics than resulted from the papers in the preceding section, where the high productivity of the

227

tropics was emphasized, and Paper 16 left the impression that high production of natural tropical ecosystems meant that tropical ecosystems also had high potential economic productivity. In this section, the papers contrasted this impression and developed the idea that tropical ecosystems should not be looked at as a potential source of high economic production. Although total productivity of natural forests can be high, productivity useful to man, on an area basis, is not greater than in temperate zones. The fact that some areas of the tropics get lots of sun and rain does not mean that they will produce economic crops at high rates.

REFERENCES

Chaney, R. W., 1947, Teritiary Centers and Migration Routes, *Ecol. Monogr.* **17**:141–148.

Houston, M., undated, A Comparison of the Growth Rates of Tropical and Temperate Trees, Division of Biological Sciences, University of Michigan, Ann Arbor, Mich.

Janzen, D. H., and D. E. Wilson, 1974, The cost of being dormant in the tropics, *Biotropica* **6**:260–262.

Jordan, C. F., 1971a, Productivity of a Tropical Forest and Its Relation to a World Pattern of Energy Storage, *J. Ecol.* **59**:127–142.

Jordan, C. F., 1971b, A World Pattern in Plant Energetics, *Am. Sci.* **59**:425–433.

Jordan, C. F., in press, Productivity of Tropical Rain Forest Ecosystems and the Implications for Their Use as Future Wood and Energy Sources, in *Ecosystems of the World: Tropical Rain Forests*, vol. 2, F. B. Golley ed., Elsevier, Amsterdam.

Kato, R., Y. Tadaki, and H. Ogawa, 1978, Plant Biomass and Growth Increment Studies in Pasoh Forest, *Malay. Nat. J.* **30**:211–224.

Knight, D. H., 1978, Tree Growth and Mortality in a Species-Rich Tropical Forest, Research proposal submitted to the National Science Foundation.

Medina, E., and H. Klinge, in press, Tropical forests and tropical woodlands, in *Encyclopedia of Plant Physiology*, Springer Verlag.

Siminovitch, D., C. M. Wilson, and D. R. Briggs, 1953, Studies on the chemistry, of the living bark of the black locust in relation to its frost hardiness. V. Seasonal transformations and variations in the carbohydrates: Starch-sucrost interconversions. *Plant Physiol.* **28**:383–400.

Weaver, P. L., 1979, Tree Growth in Several Tropical Forests of Puerto Rico, Research Paper S0-152, U.S. Dept. of Agriculture, Southern Forest Experiment Station, Rio Piedras, Puerto Rico, 15 p.

Whitmore, T. C., 1975, *Tropical Rain Forests of the Far East*, Clarendon Press, Oxford, 282 p.

17

Reprinted from pp. 415 and 423–424 of Am Midl. Nat. **99**:415–434 (1978)

A Latitudinal Gradient of Wood and Litter Production, and Its Implication Regarding Competition and Species Diversity in Trees

CARL F. JORDAN

Centro de Ecologia, Instituto Venezolano Investigaciones Cientificas, Apt. 1827, Caracas, Venezuela

and

PETER G. MURPHY[1]

Department of Botany and Plant Pathology, Michigan State University, East Lansing 48824

[*Editor's Note:* Only Figures 1 and 2 of this article have been reproduced here.]

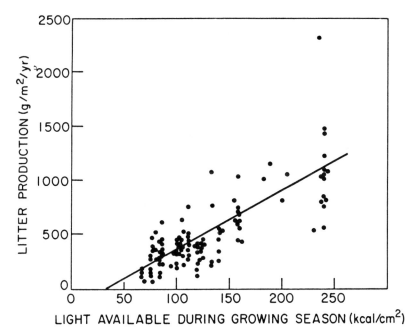

Fig. 1.—Litter production of mesic forests as a function of total possible amounts of light available during growing seasons

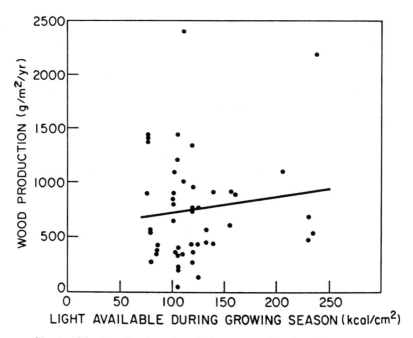

Fig. 2.—Wood production of mesic forests as a function of total possible amounts of light available during growing seasons

18

Reprinted from *A Tropical Rain Forest: A Study of Irradiation and Ecology at El Verde, Puerto Rico*, H. T. Odum and R. F. Pigeon eds., U.S. Atomic Energy Commission, Washington, D.C., 1970, pp. A8–A10

A TROPICAL RAIN FOREST: A STUDY OF IRRADIATION AND ECOLOGY AT EL VERDE, PUERTO RICO

Howard T. Odum
University of North Carolina

and

Robert F. Pigeon
U.S. Atomic Energy Commission

[*Editor's Note:* In the original, material precedes this excerpt.]

TROPICAL FORESTS, CLIMAX, AND MANAGEMENT REGIMES

The problem facing man in management of his tropical forest lands may be a choice among several alternative classes of energy use. In Fig. 2, the energy circuits are used to compare some types of overall forest management now available to man and

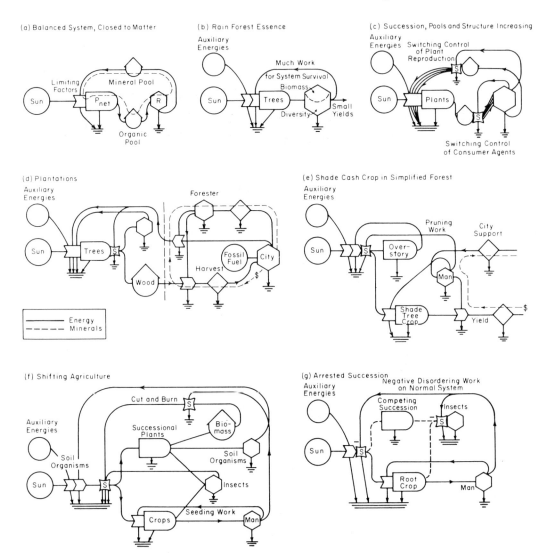

Fig. 2 — Energy flow diagrams. (a) The basic plan of many ecological systems. (b-g) Six systems of tropical land use.

nature. When an ecological system is at steady state, holding an average pattern of structure and metabolism constant, it is said to be at climax. With many kinds of input and output flows of energy and matter (called boundary conditions), many kinds of climax are possible (Fig. 2). Can quantitative data, such as some in this book, help determine the circumstances favorable for each class of use ?

The worldwide problem of tropical forest regime management concerns selection of energy type from these and other possible steady states. In all the diagrams, note the closed loop formed by energy flows and their loop-back multiplier work actions. These form a balance of payments that may be necessary to sustain any system against alternatives, because the feedback arrangement prevents an energy use from being only a drain on its energy source. The user protects the source, and hence itself, with a reward loop. Consider the models of Fig. 2 in relation to man's use of rain forest lands.

Figure 2(a) is the idealized closed-to-matter system found in some laboratory microcosms, such as balanced aquaria. Note the production and respiration centers. Two storages that serve to control the flows are the pool of total organic matter and the pool of energy containing limiting minerals that gate and release photosynthesis.

Anticipating the arguments presented in Sec. I, we suggest that the real rain forest is not so closed and has substantial auxiliary energies of wind, water, and chemical inflows [Fig. 2(b)]. In the fully developed rain forest, biomass and mineral storages tend to be assigned and divided among many specialized sites and operations and are so intimately tied to their separate species that they are less like pools. Hence the organic storage is shown only as that within the respiratory hexagon. Note that to harvest from this system is to remove some of its structure, by definition changing it to something less.

Figure 2(c) shows the successional system in a rain forest climatic regime. Much of the work of the species in the respiratory work group goes toward programmatic switching, which step by step develops a climax forest again. The steps of the program are the responses of the component species. The program is sent as a giant message by seeds and other propagation to the site from the parent climax nearby or from its storage at an earlier time. Unlike the passive cards of man's simple computer programs, the statements of the rain forest successional program have reassorting and self-organizing mechanisms. Systems (b) and (c) are useful to man where self-regulation is useful for protecting watersheds, serving recreational needs, and protecting the general planetary life-support system.

Contrasting with these examples that use their own energies is the plantation yield system [Fig. 2(d)] which requires subsidy from human energies to channel energy yield. Spectacular examples are the plantations of Asiatic exotic cadam trees first developed at the Institute of Tropical Forestry by C. B. Briscoe and F. H. Wadsworth, who tested many exotics in Puerto Rico, discovering phenomenal net growths on Tabonuco rain forest land. Notice the plantation success at 5 years of age at Catalina, P. R., in Fig. 3. Seed from Puerto Rico also have been used successfully in Central America as well (Grijpma, 1967). More on the cadam is given in Chaps. I-7, I-8, and I-10. Although herbivorous epidemic insect or microbial infestation may eventually be expected, requiring control action, there is little sign of it yet.

To the left of the dashed line in Fig. 2(d) is the ecosystem, where energies are guided into wood pools by management control work of various types, including the deletion (by switch S) of the successional and self-regulating developments that would receive the energies otherwise. A yield plantation is coupled into the economy of urban man through the forester, who has a balance of dollar payments with the urban civilization in addition to a balance of work payments. In Chaps. G-0, I-6, and I-10, more detailed diagrams are given which show the role of stress, especially from gamma radiation. At El Verde the radiation was both a drain and a programmatic releaser.

Intermediate between the yield and the climax is the simplified rain forest, Fig. 2(e), which provides some yields of specialized products, such as rubber, cacao, and coffee, but retains enough forest components, such as overstory shade trees, and diversity to maintain the soil, the microclimate, the mineral cycle, nitrogen fixing, and some of the animal regulators. The mass of the yield is small compared to a forestry plantation. The work of the partial forest structure may be regarded as part of the energetic cost of the production of the quality product.

Shifting agriculture is another of the tropical land use systems which through cutting and burning diverts much ecosystem energy as it becomes stored back into processes for keeping the ground suitable for crops. Shown in Fig. 2(f) are the results of cutting and burning, which act as a switch to couple several years of energy storages by a partly successional ecosystem to help man start the program over. In the past before the use of fossil fuel, the total energy available to man in harvests after the first year or two was not adequate to sustain him even when all his work was directed to maintaining an arrested successional state. In this system, we may regard a steady state as existing on the average in time and place. Climaxes are usually made up of continuing

Fig. 3 — Plantation of the Asiatic exotic cadam tree *(Anthocephalus cadamba)* five years after planting at Catalina, Puerto Rico. Center tree is 15 in. in diameter at breast height and 50 ft tall.

series of component successional, stages. Shifting agriculture generates a pulse whether the climate has one or not.

When work energies by man in weeding and insect control are adequate to arrest succession, an agricultural system of tropical land management results, as shown in Fig. 2(g), which is achieved by steady application of disordering stress, such as regular cutting, herbicides, grazing animals, land tilling, and other disturbances. In some climates the steady application of stress in combination with harvest does allow agricultural usage without a shifting system or without subsidy of energies from the industrialized economy, as shown in Fig. 2(g). In other climates, subsidy of energies from the outside is required.

[*Editor's Note:* In the original, material follows this excerpt.]

Reprinted from *Science* **182**:1212–1219 (1973)

Tropical Agroecosystems

These habitats are misunderstood by the temperate
zones, mismanaged by the tropics.

Daniel H. Janzen

Tropical countries (*1*) have one major problem: how to evolve a social system that is tailored to the carrying capacity of a small resource base and yet have any resources left once the experiments in setting up the system have run their course. This challenge must be met in a very harsh sociobiological environment. Some of the outstanding environmental traits of most tropical countries are (i) past and present harvest of resources by temperate zone countries at prices unrelated to the worth of the resources at their place of origin; (ii) borders established directly or indirectly by temperate zone countries that were partitioning a resource for their own use; (iii) many nearly equal and opposing pressures acting on social structures, pressures generated not so much by the immediate environment as by the hybridization of two or more social structures with radically different goals in resource use; (iv) potential and realized resources per person already lower than in most temperate zone countries; (v) current social aspirations modeled after exploitative social systems that evolved in resource-rich habitats to deal with the harvest of highly pulsed, regionally homogeneous agricultural resources; and (vi) usable productivity per unit of human effort expended that is considerably lower than that in the temperate zones.

Scientists and policy-makers in the temperate zones often express high hopes for the future productivity of tropical agriculture (*2–6*), but constructive criticism of tropical agroecosystems (*2–24*) is in a primitive state. Nearly all research in tropical agriculture is highly reductionist, parochial, and discipline-oriented. This can be

The author is associate professor of zoology. University of Michigan, Ann Arbor 48104.

quickly observed by perusal of books such as *Farming Systems in the Tropics* (*2*) and *Pests and Diseases of Tropical Crops and Their Control* (*25*), as well as tropical agricultural journals (*26*). Articles with a holistic approach (*21, 27, 28*) are a conspicuous rarity in the trade journals, with the exception of those in recent volumes of *Tropical Science*.

It is widely believed in temperate zone countries that tropical countries disregard the rules of sustained-yield agroecosystems out of ignorance. This condescending evaluation is sometimes correct for certain aspects of the decision-making process. However, there are many more situations in which a key manager is deliberately maximizing short-term returns at the expense of long-term returns. It is not an acceptable defense to point out that technological knowledge, whether that of the culture or of the world at large, is not immediately available to the persons carrying out the act. If the cost of making technological knowledge available were to be charged against the project, even short-term exploitation would often be uneconomical.

Short-term exploitation is conspicuous at all levels of agricultural sophistication in the tropics, except perhaps in those rare "primitive" cultures whose traditions of resource harvest are still intact (*29, 30*). What tropical countries so rarely grasp (*22*) is the fact that agriculture in the temperate zone countries evolved (and is still evolving) from short-term exploitation to sustained-yield agriculture while operating off a much larger natural capital than the tropical countries possess. Furthermore, this natural capital is in part obtained from the tropics (or other "undeveloped" areas) at a cost much less than its value (*31*).

Short-term exploitation is particularly easy in contemporary tropical societies. Government attitudes are generally "frontier exploitative" (*32*), and the "tragedy of the commons" (*33*) is promoted by undefined ownership of resources despite the fact that much of the land has been under stable subsistence agriculture for thousands of years. The temperate zone countries have said to the tropics, "Look at all the nice cash crops you can grow for us to buy," but have neglected to teach the tropics at the same time how to preserve the natural capital and harvest its natural interest.

By assuming that technological ignorance is the sole cause of agricultural problems in the tropics, we allow this ignorance to become the scapegoat for all ills of the agroecosystem (*8, 10, 12*). In fact, the scientific and folklore communities know quite enough to deal with most of the technological problems in tropical agroecosystems, or if not, how to get that information. As Talbot states in his analysis of deterioration of Masai rangeland, "These adverse ecological consequences of the developments were not intentional. They were, however, anticipated, predicted, and documented by some range managers, wildlife ecologists and other biologists who knew the area" (*34*, p. 695). There are many examples of a disastrous tropical agroecosystem existing side by side with a highly successful one—but under a different social system (*35*). This strongly suggests that the social rather than technological environment is at fault in problems of tropical agroecosystems.

It is a common argument that technological advance in the tropics will buy time in the war against population increase and deterioration of natural capital (*5*). However, there is little evidence that anything is being done with the time bought. It is of no use to fund a soil or natural resource survey for a major development scheme (*36, 37*) when there is a preordained number of settlers (*38*). I feel that the plea for technological advance gives the scientific community a perfect excuse to continue their reductionist and esoteric approaches (*12, 39*) rather than to put their efforts into the far more frustrating task of generating sustained-yield tropical agroecosystems and ensuring that technological advances are integrated with them. Few basic studies in tropical biology genuinely seek to adapt their technology and findings to the agroecosystem (*40*), although many of them could. A few pious sentences in the introduction (*41*), or the use of economically important animals in experiments, does not remove a study from the category of "biological art form." Some argue that a crisis is needed to alter the situation (*42*). However, like other forms of tropical change, approaching tropical crises tend to be inconspicuous and cannot be recovered from as easily as can crises in the temperate zones.

When examining the problems that confront the development of a sustained-yield tropical agroecosystem (SYTA), it is impossible to separate the biological problems of practicing agriculture in the tropics from those of inadequate education, public facilities, administration, and social aspirations. The regions under discussion are both tropical and undeveloped, and it would be a major tactical error to attribute their overall difficulties to either of these traits.

I focus on some of the areas that seem to be generally unappreciated or ignored by those in the temperate zones who influence the development of SYTA's. In most cases, there is a conflict between optimization and maximization. Reductionism is the order of the day in the contemporary forces shaping SYTA's, and descriptions and analyses of SYTA's are influenced by this philosophy. Tropical agroecosystems are characterized by attempts to maximize outcomes of single processes and the glorification of this maximization. The major challenge in the tropics today is to determine which reductionist lines of research and development should be halted or deflected in deference to optimization processes within holistically designed SYTA's.

Productivity

Net annual primary productivity may be higher in the moist, lowland tropics than anywhere else in the world (*43*), but what really matters is the difference between the cost of turning that productivity into human desiderata and the value of the output (*11, 17, 44, 45*). This difference is very poorly understood as it applies to the tropics. There is a strong tendency for tropical administrators to evaluate labor as free input, to value land only for food and fiber production, and to value products

in terms of the world market rather than national life-support systems. When people in the temperate zones say (*46*, p. 440):

> The need is universally recognized for drastic increases in production of food and fiber to feed and clothe a rapidly expanding [tropical] population, a large percentage of which is now undernourished and poorly clothed. It is also recognized that much of the increase required must come from the intensification of agricultural production in the developing nations.

and, "A continual guarantee of increasing agricultural productivity is absolutely essential for our tropics" (*27*, p. 1), they forget that tropical people are no more interested in spending all their waking hours picking beetles off bean bushes and transplanting rice by hand than they are. High-yield tropical agriculture requires immense amounts of very accurate hand care (*2, 47–49*) or tremendous amounts of fossil fuel (*50*), or both.

If agricultural production costs were determined equally and fully throughout the world, most of the lowland tropics would be classified as marginal farmland. Some researchers have come to this conclusion on the basis of weather data alone (*9, 19*). As Paddock puts it, "The hungry nations have been and are hungry because they have a poor piece of real estate" (*15*, p. 898). This is well illustrated by the very high cost and slow rate of development of tropical Australia as compared with temperate Australia. Tropical Australia lacks a large, free labor force and its products are in direct competition with those of temperate Australia (*9*). Oddly, the temperate zones accept the concept of nonagricultural use of marginal farmland at the national level, but not at the international level.

In the tropics, "optimum population size and optimal political area are almost irreconcilable: for a state to reach a reasonable size of population it must overstep the optimum-area limits; for it to remain within a reasonable area means more often than not a midget population. . . ." (*51*, p. 435). There is no biological reason that the capacity to support human life should be evenly distributed over the earth's surface, nor why it should be correlated with the primary productivity of natural ecosystems or with the biomass (standing crop) of these ecosystems. Temperate-tropical comparisons aside, as population density and cash crop-

ping for export increase, the use of marginal land within the tropics increases. In addition to being fragile and having low productivity, marginal farmlands in the tropics have greatly fluctuating productivity. Colonization of such areas may appear justified for several years, and during this time the invading population severs its cultural-economic connection with its homeland (*18*). Then, when drought (*18, 34*), hurricane (*52*), or resistance to pesticides (*8, 53*) occurs, it is termed a "natural disaster." Because one person can be sustained at a minimal standard of living more easily in the tropics than in the temperate zones, the population in the tropics is likely to have been greater before the catastrophe than it would have been in marginal farmland in the temperate zones.

Year-Round Warmth

The year-round warmth of the lowland tropics is a mixed blessing (*11*). High year-round soil temperatures lead to very rapid breakdown of litter, with subsequent leaching of soil nutrients before they can be taken up by plants (*54*). Plant diseases breed year round (*27*), and pests breed freely in stored food that is not chilled by winter cold (*53*). In addition, stored foods degenerate rapidly because of their own metabolic activity at high temperatures. Even in areas with a severe dry season, many insect species are present as active adults; they are concentrated at local moist sites or are breeding on alternate hosts (*55, 56*). Insect pests are therefore available for immediate colonization of newly planted fields, even during the harshest time of year; the same is probably true of plant diseases (*27*). Tropical herbivorous insects are highly adapted for making local migrations (*55, 56*); this makes it difficult to protect crops by introducing heterogeneity of fields in time and space.

One possible remedy is unpleasant for the conservationist. The agricultural potential of many parts of the seasonally dry tropics might well be improved by systematic destruction of the riparian and other vegetation that is often left for livestock shade, erosion control, and conservation. It might be well to replace the spreading banyan tree with a shed. The tremendous number of species of insects (*56*) and diseases (*27*) that characterizes the tropics might be severely reduced through habitat destruction. This conclusion might change the policy problem to a consideration of how much land should be set aside purely for conservation; the remaining land might not even approximate a natural ecosystem (57). Some studies even suggest that "overgrazed" pastures may have a higher overall yield than more carefully managed sites (58), especially if the real costs of management are charged against the system. If one wishes a high yield from a particular site, year-round warmth necessitates complex fallow systems to deal with the weeds and insects. However, it is possible that, over large areas, a much lower yield per acre in fields under continuous cultivation could produce the same average yield per acre as fallow systems. Social complications, rather than pests, are likely to be the major barrier to experimentation leading to SYTA's based on extensive, rather than intensive, agriculture; tropical countries are conspicuously hostile to schemes requiring tight administrative control over large areas by single sources of power.

It is not only superior nutrient dynamics of the soil that cause the seasonally dry tropics to be more productive agriculturally than the wet tropical lowlands. In the ever-warm tropics, irrigating between subtropical oases (36, 59) and between wet seasons is tempting, but it eliminates the only part of the physical environment that is on the farmer's side in his competition with animals and weeds. The less extreme the dry season (or the more thorough the irrigation), the less extreme are the seasonal dips in insect pest population, with which the farmer can synchronize his crop's growth. There are numerous parallel cases between the natural communities of the tropics and those of temperate zones (60, 61).

Ecosystem Fragility

Two very different concepts are involved in the "stability" so often attributed to tropical ecosystems. On the one hand, owing to the apparent lack of variation in the weather within each year (62) and the apparently small variations in the climate from year to year, temperate zone peoples often regard the tropics as stable. However, much of this stability is illusory (63), as any farmer on a large scale will

confirm after plowing under his third attempt to grow rice on a site in the seasonal tropics where rice can be grown only in wet years.

On the other hand, the complex biological systems of the tropical lowlands are very easily perturbed and cannot be easily reconstituted from roadside and woodlot plants and animals (20), as could many North American habitats. For this reason, the complex processes in SYTA's are likely to be highly unstable. For example, a great variety of horticultural practices and strains of common tropical food plants have accumulated over the centuries (64). They are closely adjusted to local farming conditions and coevolved with the other dietary resources of the area. When high-yield hybrids are introduced, the local strains (65) and practices (66) are quickly abandoned. This later leads to (i) expensive and complex programs to reevolve these strains when adjusting hybrid monocultures to SYTA's (65), (ii) increased dependence on pesticides and complex breeding programs to keep abreast of the pest problem in single-strain monocultures, and (iii) increased imbalance in the distribution of wealth among farmers (6, 15, 16, 22). The same may be said for the replacement of indigenous floras by foreign grasses (67) and pure stands of foreign trees (14, 68), the generation of complex irrigation systems susceptible to market perturbations (69), and the destruction of adaptive village structures by population pressure (70) or cash cropping (17, 30). As mentioned earlier with respect to the pest community, one way to remove fragility is to remove complexity. However, monocultures are clearly unstable in certain circumstances (23, 57, 71), at least with respect to the demands made on them.

Crops and Spacing

Long distances in space and time between conspecific plants in the lowland tropics are a major element in their escape from their host-specific herbivores (11, 13, 60, 61, 72–74). The monocultures or moderately mixed stands that characterize modern agriculture are thus a much greater departure from normal in the tropics than they are in the temperate zones. In this sense, modern agriculture removes a much greater proportion of the plant's defense in the tropics than in the temperate zones. However, as has been

correctly emphasized (45, 57, 71, 72, 75), crop heterogeneity is a mixed bag.

First, there is heterogeneity among monoculture fields in time and space. Here, the benefits of heterogeneity depend on whether the vegetation that is interspersed with the crop field sustains a pest community of less risk than the benefit of the entomophagus parasites and predators it also contains. The outcome has to be determined individually for each site, and in the tropics, it may well go either way (72, 76). The efficacy of letting a field lie fallow depends also on the proximity of seed sources for wild plants (30, 77) and the value of these wild plants for other uses (78). We cannot even infer that a reduction in yield after a shortened fallow period is the result of less effective pest control (79).

Second, there is heterogeneity within the field. Often viewed as the answer for the tropics, this practice has two major problems: harvesting a mixed crop requires greatly increased labor and skill, and different crops may well require mutually incompatible treatments (48, 68, 80). Furthermore, crop plants have had much of their chemical and mechanical defense system bred out of them. For many pests, a field of four or five crops may be a monoculture (13, 74).

While some of the most complex mixed cropping is in the tropics (2), the tropics also have some very successful monoculture agriculture, if human labor is not included in the cost calculation (47). Finally, in some cases in the tropics, a monoculture may have a greater productivity than mixed crops (81).

Chemical Defenses against Pests

Secondary compounds are a tropical plant's other major form of defense. However, tropical crops, perhaps even more than those in the temperate zones, have had many of their internal defenses bred out of them in man's quest for less toxic or offensive food. It is almost impossible to grow vegetables in pure stands in the lowland tropics without heavy use of pesticides (11, 82). Furthermore, when there is intense selection for higher yields and other energy- and nutrient-consuming traits, the plant probably reduces its defense outlay in order to balance its internal resource budget. "Miracle grains" may be especially susceptible to insects and disease for internal reasons, as well as their genetic and horticultural uniformity.

In the tropics, as in the temperate zones, plants' internal defenses are often replaced with pesticides. However, tropical insects should develop resistance to pesticides as fast as or faster than insects in the temperate zones. One of the classic stories of mismanagement of a tropical agroecosystem is the losing battle between large-scale cotton production with the aid of pesticides and the evolution of insects' resistance (*53, 82, 83*). The modern tropics are dotted with doomed pesticide disclimaxes requiring ever-increasing amounts of chemicals for their maintenance. Only now are the side effects being monitored for a few major crops (*84*).

There are several reasons to expect a more rapid evolution of a pesticide-resistant pest community in tropical agroecosystems than in temperate agroecosystems: (i) the coevolution of herbivores and plant chemistry has always been a major aspect of tropical community structure—if there is a biochemical defense genome in insects, this is probably where it is most highly developed (*11*); (ii) the larger the proportion of the insect community that is hit by the pesticide, the more rapidly resistance may be expected to appear (*85*), and in tropical communities it is commonplace for an insect that is rare in nature to be very common in adjacent fields—even the use of systemic pesticides against vampire bats (*86*) has this problem; (iii) if tropical insects are as localized in their geographic distributions as they appear to be, there will be less chance for dilution of resistant genotypes by susceptible genotypes from unsprayed neighboring regions (*82*); and (iv) in species-rich tropical communities (*27, 56, 87*), the pool from which resistant species may be drawn is much larger than in a temperate zone community.

Tree crops, particularly prominent in discussions of tropical agroecosystem potential (*73, 88, 89*), deserve special mention here. In contrast to annual plants, it is impossible to breed resistant tree strains each year in order to keep ahead of pests that are resistant to natural and artificial pesticides. Not only are the breeding times of pest and host disproportionate, but farming tree crops is a long-term investment, and the loss of a tree crop to a newly resistant pest is a much greater loss to the agroecosystem than the loss of an annual crop.

Soils

Soils in the tropical lowlands are often a nutrient reservoir of very low capacity (*54, 90, 91*). Plant ash from burning, ions from the very rapid litter breakdown, and chemical fertilizers are rapidly leached from the soil if not taken up by plants. There is generally a deep layer of nutrient-poor material over unweathered rock. Chemical fertilizers are a far more complex solution than they would appear to be. Because of the high rate of leaching from the soil, fertilizers must be added in far greater amounts than are actually taken up by the plant, and this creates a pollution problem. This overdose also raises the real cost of the crop. If fertilizers are added frequently, but in small amounts, the amount of work put into the crop is greatly increased. Even less appreciated is the fact that, since the soil nutrient pool is very small, a careful balance of chemical fertilizers must be added to avoid toxicity; sulfate of ammonia, the standard nitrogenous fertilizer in much of the tropics, may be doing more harm than good in that it acidifies an already acid soil (*91*).

In shifting agriculture, fields are commonly left fallow after 2 to 5 years of farming. The standard explanation for this is exhaustion of the nutrients in the soil. However, the real cause is lowered yield, and pest insects and competing weeds probably contribute as much as or more than soil depletion does to lowered yield (*11, 30, 92*). Magnificent stands of native weeds grow in the abandoned fields—and often in fields before they are abandoned. It is a very great mistake to analyze the adaptive significance of subsistence cultivation patterns in the tropics solely in terms of soil nutrient depletion. Ruthenberg's detailed description of tropical agriculture (*2*) contains not one sentence analyzing pest problems. The literature of tropical agriculture is replete with fertilizer trials, and there is almost no information on the dynamics of field colonization by insect and weed faunas (*93*).

Heterogeneity of Pest Distribution

There are at least five major kinds of pest communities that may be encountered as background to a tropical agroecosystem. As mentioned earlier, the insect community of the lowland seasonal tropics differs strikingly from that of the lowland aseasonal tropics, primarily because of the difference in intensity of the dry season in the two habitats.

The third major pest community is that of upper elevations. Cooler soils and the lower humus decomposition rates associated with them are undoubtedly partly responsible for the higher yields per acre of fixed-field agriculture at upper elevations in the tropics [and the focus of major societies on them (*94*)]. However, one cannot ignore the effect of cool weather in slowing the growth rates of insect and weed populations. The elevation at which this effect is maximal is a complicated function of the decline of plant photosynthesis with increasing elevation, the amount of photosynthate metabolized at night, and the growth rates of insect and weed populations. I have recently found that there are more species and a greater biomass in natural insect communities at elevations of 500 to 1000 meters than in the lowland tropics (*56*). This suggests that man may be able to harvest more there if he is clever about it. Ironically, it is the intermediate to high elevations that are often ignored in overall investigations of tropical productivity (*95*, figure 1, p. 47).

The fourth major pest community is that of tropical islands. In addition to having very few species, native insect populations on tropical islands have an amazingly low biomass (*56*). Aside from the obvious potential effects on natural plant community structure and decomposition (*60, 96*), this means that crops on islands should have fewer challenges from native pests than those on the mainland. Further, when a pest is introduced, it is unlikely to be fed on by a native entomophage. These observations speak poorly for the extrapolation of results from tropical island agroecosystem studies (*97*) to mainland circumstances.

The fifth major type of pest community is that produced by plants growing on very poor soils. I have recently found that animal communities in Borneo are drastically reduced when supported by tropical rain forest growing on nutrient-poor white sand soils. The conspicuous success of lowland rice monoculture in Southeast Asia may be due, in part, to a generally depauperate insect community, as compared to that of other parts of the lowland tropics.

Finally, and to put it bluntly, next to nothing is known about the losses caused by insects and weeds in tropical agroecosystems. The evaluation systems so badly needed (98) are not only difficult to develop in areas with a poorly educated population, but they may cost more in cash and complexity than the value of the crop.

Cash Cropping

One of the largest stumbling blocks to the development of SYTA's is the philosophy that cash crops, usually for export, are the best use of the land, and that subsistence agriculture [including nomadism (99)] is a nuisance that must be tolerated to feed the farmer. For example (100, p. 569):

The basic idea behind argicultural development in East Africa has been that it must increase the cash income from the land. Development has usually meant the introduction of a cash crop, such as cotton, pyrethrum, milk, coffee or tea into a subsistence economy, and the new system is expected to increase the farmer's incomes fivefold or more.

In his 1971 text *Introduction to Tropical Agriculture*, designed for junior high school students in the tropics, Sutherland states, "What is wrong with subsistence agriculture is that everything that is produced is used up by the people. The people only grow what they need" (4, p. 5). Such reductionist economics leads easily into very distorted analyses. In his detailed description of tropical farming methods in 1971, Ruthenberg provides an example (2, pp. 108–109):

Although [alternative] practices are traditionally known, they are rarely employed in farming systems where cash cropping has been introduced and where land shortage is a recent phenomenon. In many of these situations, particularly in the drier savanas, gullies increase rapidly in number and size, soil conservation usually being neglected as cash cropping and incomes per head increase, mainly because of the unfavorable short-term input-output relationship of the labor invested. The way out of this undesirable situation probably does not lie in a return to traditional agricultural methods, but in additional cash cropping which, by changing the economic setting, can make soil conservation economically worthwhile.

It is repeatedly stated that tropical staples are ignored in research programs (101), while export crops are studied extensively. Some cash cropping is necessary for a country's SYTA, but when the crop is grown for export there is often a large social cost that

is not charged against the product. When crops are grown in plantation-size stands, often to generate the crop uniformity desired by temperate zone markets, it disrupts the local agroecosystem. Farming peoples are lured from small holdings by wages and then are unable to return to their land when prices drop or disease eliminates the crop from the area. They cannot return because others have taken over their land, closely adapted seeds and stocks have been lost, sites have degenerated for lack of close care, details of farming the site have been forgotten, and the people are psychologically habituated to the things money can buy. Families on wages set the size of their families by the amount of cash coming in, rather than by the amount of land (homesteads) available and the multitude of other natural systems regulating population. This removes one of the main feedback loops in population control; large tropical families are often the result of planning rather than ignorance of birth control mechanisms.

Subsistence farming of steep slopes and other marginal farmlands is commonly the result when large commercial establishments own or control the best land (18). For example: "Whereas smallholders usually have to operate where they are settled, and adapt to the natural habitat, and are thus compelled to diversify production, the firm [engaged in cash farming] can select the most favorable economic and natural location, which is chiefly on land suitable for monoculture" (2, p. 194).

As cash cropping becomes a larger proportion of the total production of an area, there is generally a decrease in the variety of crops the farmer can grow and still mesh with the community's or world's plans for development (15, 102). The sensitivity of tropical crop monocultures to economic perturbations is well known (2, 23, 73). Demand for labor and machinery becomes highly pulsed, and production may be limited by the cost of maintaining people and draft animals between periods of maximum need (2). The more pulsed the labor demand, the less possible it is to execute the complex crop timing required to generate high yields. (Experiment stations can produce high yields by virtue of large labor, fossil fuel, and pesticide supplies for their small plots.) As the agroecosystem turns entirely to cash crop production, there is no upper limit to the

security desired by the farmer, and the tendency to mine the soil and then move elsewhere becomes overpowering (103). Ultimately, the country may find itself in the position of having very little idea of the real value of its farmland in supporting its people on a sustained-yield basis, as their incomes are set by the taste and biochemical whims of the temperate zone countries. One of the major reasons that species-rich neotropical rain forests are not harvested for export as a mixed-species sustained yield, as is done by the African Timber and Plywood Company in Nigeria, is that the North American markets are not willing to accept the large variety of wood types that European markets will accept.

Political Expedience

Although seldom openly acknowledged, much of the motive for governmental manipulation of tropical agroecosystems is political. Occupancy implies ownership; an argument for development of the Australian tropics appears to be the irrational notion that occupancy will decrease the likelihood of its being invaded (9)—and this is not an uncommon sentiment with respect to the agricultural development of the Amazon basin. Farming is a job that many administrators assume can be done well by anybody (18); agrarian resettlement programs in the tropics commonly have as a driving force the need to quiet restive slum dwellers or starving farmers on marginal land during droughts (38). Fragmentation of large landholdings after revolutions need not be the best use of that land, even for a highly nationalized agroecosystem. Experiment stations tend to be political footballs, with the maximum life of an experiment limited to the amount of time between major elections (18, 104).

When farming populations are displaced even short distances, their age-old farming traditions often do not function well, and the reeducation programs generated by governments are notorious for technological and psychological insensitivity (18). The displaced smallholders are poor farmers, and it is often concluded that the smallholder is incapable of farming the tropics. For example, to make census-taking easier, the government of Sarawak forced the Iban upland rice farmers to live in village (longhouse) units of ten or more families, which increased rates of land degradation near the village and decreased crop protection at greater distances from the village (*30*). The people displaced by hydroelectric impoundments are usually relocated in areas where their age-old riparian farming traditions are of little use; the people downstream are of even less concern (*8, 18*). The following is a representative story (*105*, p. 597):

As part of an attempt to introduce cash-cropping to the district, the Zande Scheme opened in the 1940's with the commissioner resettling five thousand homesteads in the Yambio area. The theory was that the cotton-producing scheme would be more successful if the supervision were easier. Ultimately 40 thousand families were resettled, almost the entire population. The cotton crop was a success for the first few years and the yields were high, but after three years of operation the production dropped off markedly. Force was then applied to attain the desired production levels and the Azande became plantation "peons" instead of the prime actors in a great drama of the advance of the stone age.

This would appear to be only quaint history today, but in fact it would probably be impossible to fit this population back into the tightly integrated local ecosystem they once occupied, and such settlement programs are currently in progress elsewhere (*89*).

Interference by the Temperate Zone

Can SYTA's really be developed if new traditions are constantly being bombarded by innovations from other social systems? Well-meaning persons are constantly injecting fragments of temperate zone agricultural technology into the tropics without realizing that much of the value of these fragments is intrinsic not to the technology, but rather to the society in which that technology evolved. Temperate zone countries tend to give "aid" in forms of which they have an excess, or in forms that will benefit their foreign trade (*24*). The Peace Corps, military bases, tractors, miracle grains, grain surpluses, hydroelectric dams, and antibiotics without birth control are a few examples. More often than not, these acts are simply modern versions of buying Manhattan for a few trinkets. That the tropical country "cannot resist" these gratuities is hardly justification for giving them. There appears to be no moral code for the injection of

temperate zone technology into the tropics (*106*). Although DDT is banned in the United States, it is freely exported to the tropics. American cigarettes are sold in Central and South America without cancer warning labels. By eradicating tsetse flies, we encourage the raising of cattle in preference to wild game animals, the harvest of which may have been conducive to an SYTA. In the long run, modern drugs without concomitant birth control will take more lives than they save and will lead to a long-range lowering of health and standard of living.

A major force in tropical agroecosystems is "international development," as exported by the temperate zones. It is "a nebulous term, and its meaning seems to reflect the opinion, interest and profession of the beholder" (*107*). An important aspect of international development is illustrated by the following comments on irrigation, which apply equally to other areas (*107*):

Many development projects, whether in Australia, Massiland, Saudi Arabia, or Rhodesia, fail because they do not take this question of carrying capacity into consideration. Water is provided perhaps, and the land is thus enabled to support more animals and people. But seldom is provision made to hold populations at the new levels that land can support. *In consequence, the land deteriorates, deserts spread or become more barren, and a greater number of people end up worse off than they were before development of the area took place* [italics added]. One can question whether international development agencies should continue to play this losing game.

Conclusion

I have listed some of the ways in which the lowland tropics are not such a warm and wonderful place for the farmer, some of the reasons why it may be unreasonable to expect him to cope with the problems, and some of the ways in which the temperate zones make his task more difficult. The tropics are very close to being a tragedy of the commons on a global scale (*69, 103*), and it is the temperate zone's shepherds and sheep who are among the greatest offenders (*31*). Given that the temperate zones have some limited amount of resources with which they are willing to repay the tropics, how can these resources best be spent? The first answer, without doubt, is education, and the incorporation of what is already known about the tropics into

that education. Second should be the generation of secure psychological and physical resources for governments that show they are enthusiastic about the development of an SYTA. Third should be support of intensive research needed to generate the set of site-specific rules for specific, clearly identified SYTA's.

The subject matter of youths' cultural programming is presumably determined by what they will need during the rest of their lives. A major component of this programming should be the teaching of the socioeconomic rules of a sustained-yield, nonexpanding economy, tuned to the concept of living within the carrying capacity of the country's or region's resources. Incorporating such a process into tropical school systems will cause a major upheaval, if for no other reason than that it will involve an evaluation of the country's resources, what standard of living is to be accepted by those living on them, and who is presently harvesting them. Of even greater impact, it will have to evaluate resources in terms of their ability to raise the standard of living by Y amount for X proportion of the people in the region, rather than in terms of their cash value on the world market.

For such a change to be technologically successful, it will require a great deal of pantropical information exchange. This information exchange will cost a great deal of resource, not only in travel funds and support of on-site study, but in insurance policies for the countries that are willing to take the risk of trying to change from an exploitative agroecosystem to an SYTA. For such an experiment to be sociologically successful, it will require a complete change in tropical educational systems, from emphasizing descriptions of events as they now stand, to emphasizing analysis of why things happen the way they do. This will also be very expensive, not only in retreading the technology and mind-sets of current teaching programs, but in gathering the facts on why the tropics have met their current fate.

There is a surfeit of biological and agricultural reports dealing with ecological experiments and generalities which suggest that such and such will be the outcome if such and such form of resource harvest is attempted. It is clear that human desiderata regarding a particular site are often radically different from the needs of the "average" wild animals and plants that formed the basis for such experiments and generalities. A finely tuned SYTA will come close to providing a unique solution for each region. The generalities that will rule it are highly stochastic. The more tropical the region, the more evenly weighted the suboutcomes will be, and thus the more likely each region will be to have a unique overall outcome. For example, it is easy to imagine four different parts of the tropics, each with the same kind of soil and the same climate, with four different, successful SYTA's, one based on paddy rice, one on shelterwood forestry, one on tourism, and one on shifting maize culture.

A regional experiment station working holistically toward an SYTA is potentially one of the best solutions available. As currently structured, however, almost all tropical experiment stations are inadequate for such a mission. Most commonly they are structured around a single export crop such as coffee, sugar, rubber, cotton, cacao, or tea. A major portion of their budgets comes directly or indirectly from the industry concerned. This industry can hardly be expected to wish to see its production integrated with a sustained-yield system that charges real costs for its materials. When an experiment station is centered around a major food crop, such as rice or maize, the goal becomes one of maximizing production per acre rather than per unit of resource spent; this goal may often be translated into one of generating more people. More general experiment stations tend to be established in the most productive regions of the country and, therefore, receive the most funding. Such regions (islands, intermediate elevations, areas with severe dry seasons) need experiment stations the least because they can often be successfully farmed with only slightly modified temperate zone technologies and philosophies. The administrators of tropical experiment stations often regard their job as a hardship post and tend to orient their research toward the hand that feeds them, which is certainly not the farming communities in which they have been placed.

The tropics do not need more hard cash for tractors; they need a program that will show when, where, and how hand care should be replaced with draft animals, and draft animals with tractors. The tropics do not need more randomly gathered, esoteric or applied agricultural research; they need a means

to integrate what is already known into the process of developing SYTA's. The tropics do not need more food as much as a means of evaluating the resources they have and generating social systems that will maximize the standard of living possible with those resources, whatever the size. The tropics need a realistic set of expectations.

References and Notes

1. As used here, the tropics are those regions lying approximately between the Tropic of Cancer and the Tropic of Capricorn.
2. H. Ruthenberg, *Farming Systems in the Tropics* (Clarendon, Oxford, 1971).
3. P. P. Courtenay, *Plantation Agriculture* (Bell, London, 1971); A. H. Bunting, *J. Roy. Soc. Arts* 120, 227 (1972); J. A. Tosi and R. F. Voertman, *Econ. Geogr.* 40, 189 (1964); F. R. Fosberg, *BioScience* 20, 793 (1970); M. Drosdoff, Ed., *Soils of the Humid Tropics* (National Academy of Sciences, Washington, D.C., 1972).
4. J. A. Sutherland, *Introduction to Tropical Agriculture* (Angus & Robertson, London, 1971).
5. H. D. Thurston, *BioScience* 19, 29 (1969).
6. W. D. McClellan, *ibid.* 21, 33 (1971).
7. J. Phillips, *The Development of Agriculture and Forestry in the Tropics* (Faber & Faber, London, 1961); R. F. Smith, *Bull. Entomol. Soc. Amer.* 18, 7 (1972); H. G. Wilkes and S. Wilkes, *Environment* 14, 32 (1972); F. L. Wellman, *Ceiba* 14, 1 (1969); T. B. Croat, *BioScience* 22, 465 (1972); M. U. Igbozurike, *Geogr. Rev.* 61, 519 (1971); H. O. Sternberg, in *Biogeography and Ecology in South America*, E. J. Fittkau, J. Illies, H. Klinge, G. H. Schwabe, H. Sioli, Eds. (Junk, The Hague, 1968), pp. 413–445; M. Yudelman, G. Butler, R. Banerji, *Technological Change in Agriculture and Employment in Developing Countries* (Development Centre Studies, Employment Series No. 4, Organization for Economic Cooperation and Development, Paris, 1971); L. R. Holdridge, *Econ. Bot.* 13, 271 (1959); A. I. Medani, *Trop. Agr. Trinidad* 47, 183 (1970); S. Odend'hal, *Hum. Ecol.* 1, 3 (1972); M. I. Logan, *Geogr. Rev.* 62, 229 (1972).
8. M. T. Farvar and J. P. Milton, Eds., *The Careless Technology* (Natural History Press, Garden City, N.Y., 1972).
9. B. R. Davidson, *The Northern Myth* (Melbourne Univ. Press, Melbourne, Australia, 1965).
10. D. H. Janzen, *Natur. Hist.* 81, 80 (1972).
11. ——, *Bull. Ecol. Soc. Amer.* 51, 4 (1970).
12. ——, in *Challenging Biological Problems*, J. A. Behnke, Ed. (Oxford Univ. Press, New York, 1972), pp. 281–296.
13. ——, in *Proceedings of the Tall Timbers Conference on Ecology, Animal Control, and Habitat Management* (Tall Timbers Research Station, Tallahassee, Fla., 1972), pp. 1–6.
14. H. O. Sternberg, paper presented at the 12th Technical Meeting of the International Union for Conservation of Nature and Natural Resources (1972); H. M. Gregersen, *J. Forest.* 69, 290 (1971); *Forest Prod. J.* 2, 16 (1971).
15. W. C. Paddock, *Annu. Rev. Phytopath.* 5, 375 (1967).
16. J. L. Apple, *BioScience* 22, 461 (1972); W. C. Paddock, *ibid.* 20, 897 (1970); C. R. Wharton, *Foreign Aff.* 47, 464 (1969); W. G. Peter, *BioScience* 21, 1718 (1971).
17. D. R. Gross and B. A. Underwood, *Amer. Anthropol.* 73, 725 (1971).
18. T. T. Poleman, *The Papaloapan Project: Agricultural Development in the Mexican Tropics* (Stanford Univ. Press, Stanford, Calif., 1964).
19. J. Chang, *Geogr. Rev.* 58, 333 (1968).
20. A. Gómez-Pompa, C. Vázquez-Yanes, S. Guevara, *Science* 177, 762 (1972).
21. I. G. Simpson, *Trop. Agr. Trinidad* 45, 79 (1968); J. J. Oloya, *ibid.*, p. 317; V. C. Uchendu, *ibid.*, p. 91; Anon., *ibid.* 25, 1 (1948); R. O. Whyte, *ibid.* 39, 1 (1962).
22. T. Aaronson, *Environment* 14, 4 (1972).
23. C. Brooke, *Geogr. Rev.* 57, 333 (1967).
24. C. Clark and M. Haswell, *The Economics of Subsistence Agriculture* (Macmillan, London, 1964).

25. G. Fröhlich and W. Rodewald, *Pest and Diseases of Tropical Crops and Their Control* (Pergamon, Oxford, 1969).
26. For example: *Tropical Agriculture, Experimental Agriculture, Dasonomia Interamericana, Ceiba, Turrialba, Indian Journal of Agricultural Science, Malayan Forester, Caribbean Forester, Tropical Ecology, Journal of the Rubber Research Institute of Malaya, Tropical Agriculturalist*.
27. F. L. Wellman, *Ceiba* 14, 1 (1969).
28. P. Foster and L. Yost, *Amer. J. Agr. Econ.* 51, 576 (1969).
29. G. Reichel-Dolmatoff, *Amazonian Cosmos* (Univ. of Chicago Press, Chicago, 1971); W. M. Denevan, *Geogr. Rev.* 61, 496 (1971); D. R. Harris, *ibid.*, p. 475.
30. D. Freeman, *Report on the Iban* (London School of Economics, Monographs on Social Anthropology No. 41, Athlone, New York, 1970).
31. G. Borgstrom, in *The Careless Technology*, M. T. Farvar and J. P. Milton, Eds. (Natural History Press, Garden City, N.Y., 1972), pp. 753–774.
32. J. V. Fifer, *Geogr. Rev.* 57, 1 (1967).
33. G. Hardin, *Science* 162, 1243 (1968).
34. L. M. Talbot, in *The Careless Technology*, M. T. Farvar and J. P. Milton, Eds. (Natural History Press, Garden City, N.Y., 1972), pp. 694–711.
35. D. Lowenthal, *Geogr. Rev.* 50, 41 (1960).
36. J. H. Stevens, *Geogr. J.* 136, 410 (1970).
37. A. Smith, *Matto Grosso: Last Virgin Land* (Michael Joseph, London, 1971).
38. J. C. McDonald, *Foreign Agr.* 10 (No. 13) 8 (1972).
39. W. Meijer, *BioScience* 20, 587 (1970).
40. L. J. Webb, J. G. Tracey, W. T. Williams, G. N. Lance, *J. Appl. Ecol.* 8, 99 (1971).
41. W. A. Williams, R. L. Loomis, P. de T. Alvim, *Trop. Ecol.* 13, 65 (1972).
42. R. F. Smith and R. van den Bosch, in *Pest Control: Biological, Physical and Selected Chemical Methods*, W. W. Kilgore and R. L. Doutt, Eds. (Academic Press, New York, 1967), pp. 295–340.
43. F. B. Golley and H. Leith, in *Tropical Ecology with an Emphasis on Organic Productivity*, P. M. Golley and F. B. Golley, Eds. (International Society of Tropical Ecology, Athens, Ga., 1972), pp. 1–26; H. Leith, *Trop. Ecol.* 13, 125 (1972).
44. J. A. Bullock, *Malayan Nature J.* 22, 198 (1969); P. Wycherly, *ibid.*, p. 187.
45. D. Gifford, *Bull. Ecol. Soc. Amer.* 53, 9 (1972).
46. L. D. Newsom, in *The Careless Technology*, M. T. Farvar and J. P. Milton, Eds. (Natural History Press, Garden City, N.Y., 1972), pp. 439–459.
47. P. H. Haynes, *Trop. Agr. Trinidad* 44, 215 (1967).
48. G. C. Wilken, *Geogr. Rev.* 62, 544 (1972).
49. R. van den Bosch, *Environment* 14, 18 (1972).
50. H. T. Odum, *Environment, Power, and Society* (Wiley-Interscience, New York, 1971).
51. G. Hamdan, *Geogr. Rev.* 53, 418 (1963).
52. F. R. Fosberg, *Biol. Conserv.* 4, 55 (1971).
53. A. P. Kapur, in *Problems of Humid Tropical Regions* (Unesco, Paris, 1958), pp. 63–85.
54. P. H. Nye and D. J. Greenland, *The Soil Under Shifting Cultivation* (Technical Communication No. 51, Commonwealth Agricultural Bureaux, Harpenden, England, 1962).
55. Y. Gillon, in *Proceedings of the Annual Tall Timbers Fire Ecology Conference* (Tall Timbers Research Station, Tallahassee, Fla., 1972), pp. 419–471; D. Gillon, in *ibid.*, pp. 377–417; D. H. Janzen, *Ecology* 53, 351 (1972).
56. ——, *ibid.*, in press.
57. J. S. Kennedy, *J. Appl. Ecol.* 5, 492 (1968).
58. T. H. Stobbs, *Trop. Agr. Trinidad* 46, 187 (1969).
59. E. Rivnay, in *The Unforeseen International Ecologic Boomerang*, M. T. Farvar and J. Milton, Eds. (Conservation Foundation, Washington, D.C., 1968), pp. 56–61.
60. D. H. Janzen, *Amer. Natur.* 104, 501 (1970).
61. ——, *Annu. Rev. Ecol. Syst.* 2, 465 (1971).
62. ——, *Amer. Natur.* 101, 233 (1967).
63. G. W. Leeper, *J. Aust. Inst. Agr. Soc.* 11, 188 (1945).
64. C. Geertz, *Hum. Ecol.* 1, 23 (1972).
65. D. H. Timothy, in *The Careless Technology*, M. T. Farvar and J. P. Milton, Eds. (Natural History Press, Garden City, N.Y., 1972), pp. 631–656.

66. J. Rutherford, *Geogr. Rev.* **56**, 239 (1966).
67. J. J. Parsons, *Tubinger Geogr. Studien* **34**, 141 (1970); *J. Range Manage.* **25**, 12 (1972); A. J. Oakes, *Trop. Agr. Trinidad* **45**, 235 (1968).
68. G. R. Conway, in *The Unforseen International Ecologic Boomerange*, M. T. Farvar and J. Milton, Eds. (Conservation Foundation, Washington, D.C., 1968), pp. 46–51.
69. G. C. Wilken, *Geogr. Rev.* **59**, 215 (1969).
70. R. K. Udo, *ibid.* **55**, 53 (1965).
71. T. R. E. Southwood, in *Proceedings of the Tall Timbers Conference on Ecology, Animal Control, and Habitat Management* (Tall Timbers Research Station, Tallahassee, Fla., 1971), pp. 29–51.
72. J. E. Edmunds, *Trop. Agr. Trinidad* **46**, 315 (1970).
73. G. Gottsberger, *Phytologia* **22**, 215 (1971).
74. D. J. Pool, unpublished manuscript.
75. R. F. Smith and G. R. Conway, in *The Careless Technology*, M. T. Farvar and J. P. Milton, Eds. (Natural History Press, Garden City, N.Y., 1972), pp. 664–665.
76. B. Gray, *Annu. Rev. Entomol.* **17**, 313 (1972); K. F. S. King, *Agri-silviculture: The Taungya System* (Department of Forestry, University of Ibadan, Nigeria, 1968).
77. A. K. Khudairi, *BioScience* **19**, 598 (1969).
78. N. Meyers, *ibid.* **21**, 1072 (1971).
79. R. K. Udo, *Geogr. Rev.* **61**, 415 (1971).
80. S. Y. Peng and W. B. Size, *Trop. Agr. Trinidad* **46**, 333 (1969); C. E. Yarwood, *Science* **168**, 218 (1970).
81. R. E. Johannes *et al.*, *BioScience* **22**, 541 (1972).
82. S. Parasram, *Trop. Agr. Trinidad* **46**, 343 (1969).
83. T. Scudder, in *The Careless Technology*, M. T. Farvar and J. P. Milton, Eds. (Natural History Press, Garden City, N.Y., 1972), p. 664.
84. J. H. Koeman, J. H. Pennings, J. J. M. De Goeij, P. S. Tjioe, P. M. Olindo, J. Hopcraft, *J. Appl. Ecol.* **9**, 411 (1972).
85. R. L. Benson, *BioScience* **21**, 1160 (1971).
86. R. D. Thompson, G. C. Mitchell, R. J. Burns, *Science* **177**, 806 (1972).
87. E. O. Pearson and R. C. M. Darling, *The Insect Pests of Cotton in Central Africa* (Empire Cotton-Growing Corporation and Commonwealth Institute of Entomology, London, 1958).
88. P. Grijpma, *Turrialba* **20**, 85 (1970); F. P. Ferwerda and F. Wit, Eds., *Outlines of Perennial Crop Breeding in the Tropics* (Miscellaneous Papers No. 4, Landbouwhogeschool, Veenman & Zonen, Wageningen, Netherlands, 1969).
89. R. Wikkramatileke, *Geogr. Rev.* **62**, 479 (1972).
90. R. Wetselaar, *Plant Soil* **16**, 19 (1962); W. A. Williams, *Trop. Agr. Trinidad* **45**, 103 (1968).
91. J. Phillips, in *The Careless Technology*, M. T. Farvar and J. P. Milton, Eds. (Natural History Press, Garden City, N.Y., 1972), pp. 549–566.
92. R. Daubenmire, *Trop. Ecol.* **13**, 31 (1972); M. C. Kellman and C. D. Adams, *Can. Geogr.* **14**, 323 (1970); D. G. Ashby and R. K. Pfeiffer, *World Crops* **8**, 227 (1956); R. J. Shlemon and L. B. Phelps, *Geogr. Rev.* **61**, 397 (1971); G. P. Askew, D. J. Moffatt, R. F. Montgomery, P. L. Searl, *Geogr. J.* **136**, 211 (1970).
93. See, for example, the 1962 through 1972 issues of *Tropical Agriculture*.
94. D. R. Dyer, *Geogr. Rev.* **52**, 336 (1962).
95. A. L. Hammond, *Science* **175**, 46 (1972).
96. D. H. Janzen, *Ecology* **53**, 258 (1972).
97. E. Hacskaylo, *BioScience* **22**, 577 (1972); D. D. MacPhail, *Geogr. Rev.* **53**, 224 (1963); H. T. Odum and R. F. Pigeon, Eds., *A Tropical Rain Forest* (Atomic Energy Commission, Washington, D.C., 1970).
98. R. F. Smith, in *Proceedings of the Tall Timbers Conference on Ecology, Animal Control, and Habitat Management* (Tall Timbers Research Station, Tallahassee, Fla., 1971), pp. 53–83; L. Chiarappa, H. C. Chiang, R. F. Smith, *Science* **176**, 769 (1972).
99. M. J. Mortimore, *Geogr. Rev.* **62**, 71 (1972).
100. E. W. Russell, in *The Careless Technology*, M. T. Farvar and J. P. Milton, Eds. (Natural History Press, Garden, City, N.Y., 1972), pp. 567–576.
101. C. L. Burton, *Trop. Agr. Trinidad* **47**, 303 (1970).
102. D. R. Hoy and J. S. Fisher, *Geogr. Rev.* **56**, 90 (1966).
103. D. A. Preston, *Geogr. J.* **135**, 1 (1969).
104. W. Popenoe, in *Plants and Plant Science in Latin America*, F. Verdoorn, Ed. (Chronica Botanica, Waltham, Mass., 1945), vol. 16, pp. 1–11.
105. M. McNeil, in *The Careless Technology*, M. T. Farvar and J. P. Milton, Eds. (Natural History Press, Garden City, N.Y., 1972), pp. 591–608.
106. R. E. Train, in *ibid.*, pp. xvii–xix.
107. G. K. Myrdal, in *ibid.*, pp. 788–789.
108. This study was supported by grants GB-25189 and 350-32X from the National Science Foundation. I appreciate the facilities made available by the Hope Department of Entomology and the Department of Zoology, Oxford University, and the constructive criticism of the manuscript by R. Carroll, J. Vandermeer, C. M. Pond, and D. B. McKey.

Editor's Comments
on Papers 20 Through 24

MAINTENANCE OF PRODUCTIVITY IN NATURAL AND HUMAN-MODIFIED SYSTEMS

In the previous sections I have shown that despite high total productivity of tropical ecosystems, productivity useful to man is not necessarily higher than that of ecosystems at other latitudes. Not only is economic productivity not higher, productivity following cultivation in tropical systems often decreases more rapidly than productivity in temperate systems (Sanchez, 1976). In this section, the selections explain, for two types of tropical ecosystems, how the systems maintain their productivity in the natural state and why they rapidly lose productive potential when cultivated. Two of the papers describe agricultural systems that were designed to circumvent the natural limitations and maintain low but sustained yield.

The first type of tropical ecosystem is typified by the forest of the central Amazon Basin, which grows on soils very low in nutrient content. Despite the infertile soils in this region, the undisturbed forest ecosystems are able to maintain production at rates similar to tropical forests growing on more fertile soil (Jordan and Herrera, 1981). Jordan and Herrera discuss reasons why the undisturbed forests on nutrient-poor soils can sustain production but why production decreases when the forests are converted to agricultural areas. There are nutrient-conserving mechanisms in the undisturbed forest that prevent serious nutrient losses.

One of the mechanisms to sustain production in the undisturbed forests is hypothesized by Richards (Paper 20) in his classic book *The Tropical Rain Forest*. He postulated direct cycling of nutrients from decomposing humus to roots in nutrient-poor tropical soils. Went and Stark (Paper 21) discuss mycorrhizae as one of the mechanisms for direct nutrient cycling in nutrient-poor tropical ecosystems.

Other nutrient-conserving mechanisms on infertile soil in undisturbed ecosystems may be production of secondary plant compounds, which make leaves less palatable to herbivores, and mast fruiting, which results in less seed loss to predators than annual fruiting (Janzen, Paper 22). Additional mechanisms include superficial roots, which are very efficient in transferring nutrients released by decomposing organic matter (Jordan and Stark, 1978; Stark and Jordan, 1978); rapid turnover of small roots (Jordan and Escalante, 1980); sclerophyllous leaves, which are resistant to insect attack and to leaching (Sobrado and Medina, 1980); epiphylls, which adsorp cations and fix nitrogen (Witkamp, 1970); thick bark, which insulates the phloem from leaching (Jordan and Uhl, 1978); and large pore spaces in the soils, which result in rapid drainage and decreased opportunity for leaching (Nortcliff and Thornes, 1978).

The most significant feature of these nutrient-conserving mechanisms is that they are all associated with the living forest. When the structure of the forest is destroyed, these mechanisms are destroyed, nutrients are leached out of the system, and the productive potential of the system declines.

Sioli (Paper 23) recommends that the infertile soils of the Amazon region be used primarily for selective forestry or perennial tree crops, which will not drastically alter the basic nutrient-cycling functions of the forest. Only the varzea areas along the riverbanks that are replenished with nutrients by annual flooding should be used for annual food crops.

A second type of tropical ecosystems that is increasingly being converted to agricultural use is the tropical montane forest where soil erosion rather than nutrient leaching is a principal cause of declining crop production when the native forests are converted into large-scale monocultures. Exponential growth of pest populations also is a problem in the monocultures. Native species within the forest are protected to some extent from exponential growth of parasites and herbivores by the high species diversity of the forest. The parasites and herbivores, which are often species specific, multiply more slowly under primary forest than when they encounter a monoculture of host species.

Wilken (Paper 24) describes how some small-scale and traditional farmers in mountainous areas of Central America partially recreate forest conditions in their fields to stabilize soil, shade crop seedlings, and increase spatial heterogeniety, which reduces species-specific parasites and predators. Farmers also carry litter from nearby forest to gardens and mix it with the soil to enrich it with nutrients and improve its structure.

Although these practices of land management produce low yields, the yields are sustainable. In the long term, such systems are better both ecologically and economically than systems that intensively exploit an area and yield a high profit for a few years but ultimately degrade the ecosystem's productive capacity.

REFERENCES

Jordan, C. F., and G. Escalante, 1980, Root Productivity in an Amazonian Rain Forest, *Ecology* **61**:14–18.

Jordan, C. F., and R. Herrera, 1981, Tropical Rain Forests: Are Nutrients Really Critical? *Am. Nat.* **117**:167–180.

Jordan, C. F., and N. Stark, 1978, Retencion de nutrientes en la estera de raices de un bosque pluvial Amazonico, *Acta Cient. Venez.* **29**:263–267.

Jordan, C. F., and C. Uhl, 1978, Biomass of a "Tierra Firme" Forest of the Amazon, Basin, *Oecol. Plant.* **13**:387–400.

Nortcliff, S., and J. B. Thornes, 1978, Water and Cation Movement in a Tropical Rain Forest Environment, *Acta Amazonia* **8**:245–258.

Sanchez, P. A., 1976, *Properties and Management of Soils in the Tropics*, Wiley, New York, 618 p.

Sobrado, M. A., and E. Medina, 1980, General Morphology, Anatomical Structure, and Nutrient Content of Sclerophyllous Leaves of the "Bana" Vegetation of Amazonas, *Oecol.* **45**:341–345.

Stark, N., and C. F. Jordan, 1978, Nutrient Retention in the Root Mat of an Amazonian Rain Forest, *Ecology* **59**:434–437.

Witkamp, M., 1970, Mineral Retention by Epiphyllic Organisms, in *A Tropical Rain Forest*, H. T. Odum, ed., U.S. Atomic Energy Commission, Washington, D.C., pp. H177–H180.

20

THE ENVIRONMENT

P. W. Richards
University College of North Wales, Bangor

[*Editor's Note*: In the original, material precedes this excerpt.]

THE CYCLE OF PLANT NUTRIENTS

In the rain-forest climate, as in all climates in which the movement of water in
the soil is predominantly downwards, the trend of soil development is always
towards impoverishment. Soluble substances are continually being washed
down into the deeper layers of the soil and removed in the drainage water. The
most important common characteristic of all rain-forest soils, whether of the red
earth or the podzol type (except perhaps some very immature soils) is thus their
low content of plant nutrients. This being so, it seems paradoxical that rain-
forest vegetation should be so luxuriant. The leached and impoverished soils of
the wet tropics bear magnificent forest, while the much richer soils of the drier
tropical zones bear savanna or much less luxuriant forest. This problem has
been considered by Walter (1936) and Milne (1937) for African forests, and by
Hardy (1936a) for those of the West Indies, and all these authors reach a similar
conclusion. In the Rain forest the vegetation itself sets up processes tending to
counteract soil impoverishment and under undisturbed conditions there is a
closed cycle of plant nutrients. The soil beneath its natural cover thus reaches
a state of equilibrium in which its impoverishment, if not actually arrested,
proceeds extremely slowly.

Fresh plant nutrients are continually being set free by the decomposition of
the parent rock. Provided the horizon in which they are set free is not too deep
for the tree-roots to reach, a part of these nutrients is taken up by the vegetation
in dilute solution. Some of these substances are fixed in the skeletal material of
the plant—the cell walls—others remain dissolved in the cell sap. Eventually
all of them are returned to the soil by the death and subsequent decomposition

of the plant or its parts. The top layers of the soil are thus being continually enriched in plant nutrients derived ultimately from the deeper layers. The majority of the roots, including nearly all the 'feeding' roots, are in the upper layers of the soil. Most of the nutrients set free from the humus can be taken up again by the vegetation almost immediately and used for further growth. The loss, if any, must be very slight; Milne (loc. cit. p. 10) has shown that in the Usambara Rain forest the electrolyte content of the streams is very low. It can thus be seen that in a mature soil the capital of plant nutrients is mainly locked up in the living vegetation and the humus layer, between which a very nearly closed cycle is set up. The resources of the parent rock are only necessary in order to make good the small losses due to drainage.

The existence of this closed cycle makes it easy to understand why a soil bearing magnificent Rain forest may prove to be far from fertile when the land is cleared and cultivated. When the forest is felled the capital of nutrients is removed or set free in the soil and the humus layer is often destroyed at the same time by burning and exposure to the sun. As Milne (loc. cit. p. 10) says: 'The entire mobile stocks are put into liquidation and, as is usual at a forced sale, they go at give-away prices and the advantage reaped is nothing like commensurate with their value.' Crops planted where rain forest has been cleared may thus do very well for a few seasons, benefiting from the temporary enrichment of the soil, but before long, unless special measures are taken, a sterile, uncultivable soil may develop. On the ordinary system of native cultivation practised in rain-forest areas it is rare for more than two or three harvests to be obtained in succession without a long intervening period of 'bush fallow'. Even in British Honduras, where the annual rainfall is not more than about 180 cm., the yield of maize on forest clearings falls from about 800–1000 lb. (350–450 kg.) in the first year to about 400–600 lb. (180–270 kg.) in the third (Charter, 1941, p. 16). On very poor porous rain-forest soils such as the Wallaba podzol in British Guiana (p. 242) it may be impossible to obtain even one crop. The changes in the soil due to the clearing or thinning of the forest emphasize the delicate equilibrium of soil and vegetation in a natural rain forest.

Since the poverty of rain-forest soils in plant nutrients has been realized, some authors have been inclined to see great significance in the fact that a large proportion of rain-forest plants, like those of many other types of vegetation, contain mycorrhizal fungi. Janse (1897) examined seventy-five species from primary forest in West Java and found that sixty-nine of them possessed mycorrhiza. Data are not available for other regions, but there is little reason to doubt that a careful examination of the root systems of rain-forest plants would show that a very large percentage of them are associated with mycorrhizal fungi. The role of mycorrhiza is so little understood and the subject as a whole so controversial that there can be little use in discussing it here. The mycorrhiza of rain-forest plants may be of great ecological significance, but at present little can be said with certainty as to what that significance may be.

In this connexion reference should also be made to the abundance of leguminous trees in many Rain forests. Since the soils are so poor in nutrients the possession of root nodules containing nitrogen-fixing organisms is probably of considerable advantage to rain-forest plants. In five forest types at Moraballi Creek, British Guiana (p. 236), Davis & Richards (1933–4) found great differences in the proportion of Leguminosae. Among trees 4 in. in diameter and over the sample plot figures were as follows:

Forest type	Mora	Morabukea	Mixed	Greenheart	Wallaba
Percentage of Leguminosae (individuals)	59	33	15	14	53

Leguminosae are thus much more abundant in the Mora and Wallaba forest types than elsewhere. The Mora community grows on a swamp soil subject to frequent flooding, the Wallaba community on a highly leached podzol; the other forest types are found on red or brown soils of the red earth group (see Chapter 10), in which nitrification is possibly more active and leaching less intense. There is perhaps some significance in these facts, and in the abundance of Leguminosae in rain forest in general. In rain-forest consociations the single dominant tree species is frequently leguminous (Chapter 11).

THE SOIL AS A DIFFERENTIAL ECOLOGICAL FACTOR

It has already been stated or implied in several places that in the moist tropics, as elsewhere, differences of soil manifest themselves in differences in the composition of the plant communities. Edaphic factors are not only responsible for variations in floristic composition within primary Rain forest itself, but there is evidence that soil conditions may even result in the development of scrub, e.g. the padang and muri communities referred to above (p. 211), or some types of savanna (Chapter 15), as edaphic climaxes in a rain-forest climate. This is contrary to the view which was for long generally accepted—that in the tropics soil differences are of little ecological importance, certainly of less importance than in the Temperate zone. Only recently has enough knowledge been obtained of the composition of tropical plant communities to dispel this misconception.

The subject will be further discussed in later chapters; meanwhile two examples may suffice to show how close is the correlation between the soil and the climax vegetation.

Among the five primary forest communities at Moraballi Creek, British Guiana, two show a particularly striking correlation with soil conditions. The Wallaba forest (consociation of *Eperua falcata*) is confined to bleached ('white') sand capping plateaux and ridges; the Greenheart forest (consociation of *Ocotea rodiaei*) occurs on brown sand, which is usually found in belts along the contour of the ridges. The two communities are sometimes found in proximity and when this is so the boundary between them is extraordinarily sharp, coinciding exactly with the change from the bleached to the brown sand. In one transect the transition was complete within a distance of less than 150 m. (see Davis & Richards, 1933–4, fig. 1, p. 121). Elsewhere in British Guiana an even more abrupt boundary between these types of Rain forest has been observed; Wood (1926, pp. 4–5) says of the Bartica-Kaburi district:

'In most cases it was possible to step in one stride from the white sand where the greenheart never occurred to the brown sand where it did, generally at some point on the slopes down from the flat ridge to the creeks. The most striking instance was seen...where a nest of *akushi* [leaf-cutting] ants had thrown up soil over an area about 30 ft. [9 m.] square. On the upper half of the nest the soil was white, on the lower half brown, and the dividing line was sharp enough to lay a hand across. Two trees were growing out of the nest so close together that a man would have to go sideways to pass between and they occurred one just on the white soil and the other just on the brown soil below. The upper tree was the last wallaba and the lower the first greenheart on that slope.' Though the boundaries between the other forest types at Moraballi Creek were generally not as sharp as that between the Wallaba and Greenheart consociations, a close relation between forest type and soil was evident in all of them.

The other example is the distribution of the Ironwood forest of Borneo and Sumatra, a peculiar consociation dominated by *Eusideroxylon zwageri* to the almost complete exclusion of other species of trees (p. 258). Gresser (1919) observed that this type of forest is confined to pure or loamy sand. The later investigations of Witkamp (1925) showed that *Eusideroxylon* may actually be used as a geological indicator. In the Koetai district of Borneo the rocks consist of Tertiary sands, sandstones, clays and clay-shales, with occasional limestone and marl. The ironwood disappears or becomes rare on passing from sand or sandstone to clay, shale, marl or limestone, and the boundary is sharp, except where the rocks themselves merge gradually. Most of the district is an anticline, and sandstone, clay, etc. occur as parallel bands. Exactly correlated with this, the ironwood communities occur in long narrow bands, marking the outcrops of sand and sandstone.

These instances are by no means isolated and clearly illustrate the decisive part often played by soil in determining the distribution of rain-forest communities. The effect of soil differences is often overlooked because tropical vegetation types may be very similar superficially and detailed floristic analysis is needed to demonstrate their distinctness.

Though the effectiveness of soil differences in determining the composition of tropical plant communities can no longer be questioned, the edaphic factors concerned are not necessarily the same as those which chiefly operate in the ecology of temperate vegetation. The edaphic factors usually regarded as of most importance in temperate regions (apart from maritime vegetation) are soil acidity (pH), content of calcium carbonate and other bases, soil texture, water-supplying capacity and aeration, while recently the redox potential of the soil has been claimed as ecologically significant. In the wet tropics it is doubtful if acidity or base status are often of great importance, since nearly all of the soils are acid and more or less markedly deficient in bases. The differential edaphic factors, as far as our present very imperfect knowledge extends, seem to be most often water-supplying capacity, aeration and soil depth; they thus seem to be mainly physical rather than chemical, though Hardon's (1937) demonstration, that plants growing on the extremely leached lowland tropical podzols of Malaysian padangs (p. 211) have a much lower base content than the same species growing on a more fertile soil, suggests that different degrees of base deficiency may sometimes be an important ecological difference. The exceptional deficiency of tropical podzols in plant nutrients of all kinds (which is much greater even than that of normal rain-forest soils) may be one of the factors responsible for their highly characteristic vegetation. The presence in many, but not in all, tropical soils of large quantities of aluminium (see below) is possibly another chemical edaphic factor of differential value.

Whether calcicole and calcifuge species exist in the flora of wet tropical regions is a question which has been variously answered. Van Steenis (1935a) has discussed it briefly. Climax rain-forest communities characteristic of calcareous soils are unknown because in a humid tropical climate mature or even moderately mature soils never have a high lime content. In some parts of the tropics, e.g. Trinidad and British Honduras, the forest on limestone ridges differs considerably from that on the non-calcareous rocks of the surrounding lowlands, but it is clear that the difference is mainly due to excessive drainage and the thin covering of soil over the porous limestone rather than to any chemical factor. Thus in the moister parts of Trinidad the limestone ridges bear Semi-evergreen Seasonal forest, unlike the Evergreen Seasonal forest of the sand and clay soils in the same climate, but where the rainfall is lower similar Semi-evergreen forest occurs on non-calcareous clay not overlying limestone (Beard, 1946b). Here the water-supplying capacity of the soil is certainly the crucial factor. Charter (1941) has also pointed out that species almost confined to calcareous soils in British Honduras occur in Trinidad in a drier climate on non-calcareous sands and clays.

In the Malay Peninsula, Java and Borneo there are steep-sided limestone crags with a highly characteristic flora. These hills are almost bare of soil, except in hollows, and do not support high forest. Klein (1914) drew attention to certain species which appeared to be peculiar to such hills and regarded these as true calcicoles, i.e. as dependent on the chemical rather than the physical properties of the limestone. Van Steenis (loc. cit.) is however of the opinion that most of these plants are cremnophytes (rock-crevice plants) rather than calcicoles and, according to him, some of them also occur in similar rocky habitats on non-calcareous rocks. The present author's own observations on the flora of limestone rocks and of non-calcareous sandstone cliffs in Sarawak (Borneo) tend to confirm the views of Klein rather than those of van Steenis; the limestone rocks at Bidi and Bau have a highly characteristic flora quite unlike that of the sandstone, including the remarkable *Moultonia singularis* and the fern *Phanerosorus* (*Matonia*) *sarmentosus* which occurs on limestone rocks in several widely separated localities and seems never to occur on non-calcareous rocks. The problem needs a more critical investigation than it has yet received.

If the existence of calcicole species in the wet tropics is still doubtful, it is fairly well established that some rain-forest plants are calcifuge. The avoidance of limestone by *Eusideroxylon zwageri* has already been noted. Though it is sometimes stated that the Dipterocarpaceae as a family avoid limestone (van Steenis, 1935, p. 59), Symington (1943) states that Dipterocarpaceae are in fact absent on the limestone hills in the south of the Malay Peninsula, but probably only because these hills are inaccessible to species of dipterocarps suited to xerophilous conditions; in the north of Malaya, nearer the semi-deciduous and deciduous forests of Burma and Thailand, several dipterocarps are found on the limestone hills and at least one of these (*Hopea ferrea*) seems to show a distinct preference for limestone.

The possibility that the abundance of aluminium in tropical soils acts as a differential ecological factor is raised by Chenery's work (1948 and unpubl.) on aluminium-accumulating plants. Many tropical species, both woody and herbaceous, belonging chiefly to large cosmopolitan or pan-tropical families (Rubiaceae, Melastomaceae, several families of pteridophytes) have been shown to accumulate large quantities of aluminium in their tissues. These plants are abundant in both primary and secondary rain-forest communities and Chenery believes they play a fundamental part in succession and soil-forming processes in the tropics. It is possible that some of the differences between the vegetation of tropical soils rich in aluminium, such as lateritic red earths, and siliceous soils relatively poor in aluminium (e.g. podzols) may be partly due to the differences in the abundance of aluminium.

In spite of some instances of the apparent importance of chemical edaphic factors it seems probable, as has already been said, that physical edaphic factors, especially those affecting the water supply of the plant and the supply of oxygen to its roots, play a larger part in the ecology of tropical vegetation. The occurrence of characteristic plant associations on river margins and land liable to flooding, of others on sites with exceptionally free drainage, the close correlation observed in British Guiana between soil texture and the climax forest communities (Chapter 10), all these are ecological phenomena for which physical edaphic factors must be mainly, though not necessarily exclusively, responsible. The precise nature of these factors awaits further investigation.

A similar view of the relative importance of chemical and physical edaphic factors in the tropics has been put forward by Hardy (1936b) with reference to the climax forest communities of Trinidad. In the south-central district of the island Hardy studied the distribution of two varieties of Semi-evergreen Seasonal forest and three of Evergreen Seasonal forest. When a map of the vegetation was superimposed on a soil map, a certain coincidence became obvious. A number of individual species were also noted as occurring only on sand, clay, silt, marl or mud-flow soils. Hardy considers that the main factor controlling the distribution of the species and forest communities in this region is the water-supplying power of the soil, which depends on its physical properties as well as on topography and climate. Soil, climate and topography interact in a complex manner, sometimes tending to counteract one another, sometimes acting in the same direction.

'Soil type appears to be significantly important mainly with respect to those physical features that decide its moisture relations. Its chemical properties and attributes appear to exert little or no influence, except in so far as they affect the behaviour of soil water. Among the chemical features, lime and magnesia contents and humus content alone appear to be important in this connection. Other chemical factors (nitrogen supply, phosphate content, potash content, etc.) seem to be quite subsidiary in deciding the broad distribution of vegetation types, although they assume important roles when forest lands are utilized for the growing of commercial crops.' (Hardy, loc. cit. p. 28.)

In the more recent work of Beard (1946 b), the prime importance of the physical edaphic factors is equally strongly emphasized, and the author concludes that it is the moisture relations of the habitat, which are the resultant of the three factors, soil, topography and climate, which chiefly govern the distribution of vegetation in tropical countries such as Trinidad.

In British Honduras there are two main groups of forest communities: those on soils derived from limestones and those on soils derived from non-calcareous rocks such as granodiorites, slates, schists, quartzites and sandstones. Though the two groups of communities are found on soils differing in base content and other chemical properties, Charter (1941) believes their distribution is largely controlled by moisture relationships.

There is some evidence that aeration as well as the water-supplying power of tropical soils is often a potent ecological factor. This is especially likely to be true in the wettest climates and under conditions of topography or soil texture leading to impeded drainage. Chenery & Hardy (1945), in their interesting description of the 'moisture profiles' of certain Trinidad soils, showed that of the soils they examined, only three free-draining sandy soils were adequately aerated down to the full depth of the profile at all seasons of the year. The others were saturated with water at least during the wet season and inadequately aerated except in the surface layers. Different plant species undoubtedly vary very greatly in their tolerance for poorly aerated soils. Coster (1935 b) has suggested that the trees of the Java lowlands have lower soil oxygen requirements than those of the more porous and better drained upland soils. His view that competition for oxygen by the tree-roots is of great ecological importance in tropical forest has been referred to above (p. 190).

Another factor closely bound up with aeration is soil depth. Tropical trees, though broadly speaking shallow rooted, vary in the depth of penetration of their root systems. In Chapter 4 it was shown that there is a close relation between the buttressed habit and the depth of the root system, buttresses developing on superficial lateral roots and being seldom present in trees with a well developed tap-root; it was also shown that the frequency of buttressing in different types of Rain forest shows a definite relation to the effective depth of the soil. Where the soil is actually shallow or, owing to a high water-table, 'effectively' shallow, the root systems tend to be superficial and buttressing is common. Sykes (1930) has described a *Gossweilerodendron-Cylicodiscus* type of Mixed Rain forest in Nigeria which is confined to deep sandy soils; he suggests that the reason for this is that the dominant species need room for the development of their long tap-roots. Though it is possible that soil depth may sometimes act in this purely mechanical way, it must clearly interact with aeration in limiting the thickness of the soil layer which the tree-roots can exploit.

REFERENCES

[*Editor's Note*: Only the references cited in the preceding excerpt are reproduced here.]

Beard, J. S. (1946b). The natural vegetation of Trinidad. *Oxf. For. Mem.* no. **20**.

Charter, C. F. (1941). *A reconnaissance survey of the soils of British Honduras.* Government of British Honduras, Trinidad.

Chenery, E. M. (1948). Aluminium in plants and its relation to plant pigments. *Ann. Bot., Oxford*, N. S., **12**, 121–36.

Chenery, E. M. & Hardy, F. (1945). The moisture profile in some Trinidad forest and cacao soils. *Trop. Agriculture, Trin.*, **22**, 100–15.

Coster, C. (1935b). Wortelstudiën in de Tropen. IV. Wortelconcurrentie. *Tectona*, **28**, 861–78.

Davis, T. A. W. & Richards, P. W. (1933-4). The vegetation of Moraballi Creek, British Guiana: an ecological study of a limited area of Tropical Rain Forest. Parts I and II. *J. Ecol.* **21**, 350–84; **22**, 106–55.

Gresser, E. (1919). Resumeerend rapport over het voorkemen van ijzerhout op de olieterreinin Djambi. I. *Tectona*, **12**, 283–304.

Hardon, H. J. (1937). Padang soil, an example of podsol in the Tropical Lowlands. *Verh. Akad. Wet. Amst.* **40**, 530–8.

Hardy, F. (1936a). Some aspects of cacao soil fertility in Trinidad. *Trop. Agriculture, Trin.*, **13**, 315–17.

Hardy, F. (1936b). Some aspects of tropical soils. *Trans. 3rd. Int. Congr. Soil Sci. (Oxford)*, 1935, **2**, 150–63.

Janse, J. M. (1897). Les endophytes radicaux de quelques plantes javanaises. *Ann. Jard. bot. Buitenz.* **14**, 53–201.

Klein, W. C. (1914). Kalkplanten in Nederlandsch-Indie. *Trop. Natuur.* **3**, 133–5.

Milne, G. (1937). Essays in applied pedology. I. Soil type and soil management in relation to plantation agriculture in East Usambara. *East Agr.
Agric. J.* (July 1937).

Steenis, C. G. G. J. van (1935a). Maleische Vegetatieschetsen. *Tijdschr.
ned. aardrijksk. Genoot.* Reeks 2, **52**, 25–67, 171–203, 363–98. (The
map also appears in *Atlas van tropisch Nederland*, 1938, Weltevreden,
Java.)

Sykes, R. A.(1930). Some notes on the Benin forests of Southern Nigeria.
Emp. For. J. **9**, 101–6.

Symington, C. F. (1943). Foresters' manual of Dipterocarps. *Malay. For.
Rec.* **16**.

Walter, H. (1936). Nährstoffgehalt des Bodens und natürliche Waldbest-
ände. *Forstl. Wschr. Silva.* **24**, 201–5, 209–13.

Witkamp, H. (1925). De ijzerhout als geologische indicator. *Trop. Natuur,*
14, 97–103.

Wood, B. R. (1926). The valuation of the forests of the Bartica-Kaburi
area. *Report by the Conservator of Forests to the British Guiana Combined Court, Second Special Session,* 1926. Georgetown, British
Guiana.

Mycorrhiza

F. W. Went and N. Stark

Two of the most interesting cases of symbiosis in the plant kingdom, lichens and mycorrhiza, both involve fungi. For about 100 years it has been recognized that lichens are an association of a specific fungus with a specific alga forming a third organism which resembles neither of its components. Where no other organisms can grow, this symbiotic entity, the lichen, is amazingly successful. Of the hundreds of plants living in the severe climate of Antarctica, all but a few mosses and flowering plants are lichens. This association has become so specific and perfect that it is the lichen unit which produces and disperses, not the separate alga and fungus. Thus, by joining forces, the symbiotic unit can live and function where each partner by itself would be unsuccessful.

Less than 100 years ago it became recognized that fungi were involved in another important symbiosis, namely, with roots (Kelley, 1950). In this case fungi live in and around the roots of higher plants without injuring them. Frank gave this association the name of mycorrhiza or "fungus-root." The main difference between lichens and mycorrhiza is that the latter involves not an alga and a fungus, but a higher plant root and a fungus joined symbiotically.

Fungal hyphae frequently are found inside or outside of plant cells, but in most of these cases the plant cells are dead or sickly because of the saprophytic or parasitic nature of those fungi. Examples of the damaging association of fungi with higher plants are the soggy mess of plant cells killed by *Botrytis* infection or the weakened wheat plant cells attacked by rust.

Roots harboring mycorrhizal fungi, on the other hand, are healthy, vigorous and long-lived. A mycorrhizal root can be recognized, in addition to the presence of fungi, by the large number of thick, short, lateral roots, abundant

branching, and in some cases a sheath of hyphae serving to increase the absorbing areas. In fact, some pine roots on which mycorrhiza was first recognized lack root hairs, the specialized root cells for absorption. Mycorrhiza is best-developed in trees, orchids, and saprophytes. It is found mainly in soils rich in humus and frequently poor in nutrients, and under those conditions it is essential for the growth of orchids and saprophytes, and highly beneficial to the growth of trees (Rayner, 1927). How essential mycorrhiza is for growing orchids is clear because noninfected orchid seedlings cannot grow (Meyer, 1966) at all.

And the essentiality of mycorrhiza for trees has been repeatedly observed in the laboratory and in the field. When pines were first introduced into Puerto Rico and Australia, they grew very poorly until soil containing mycorrhizal fungi was introduced (Briscoe, 1960). After the addition of the right kind of fungi to the soil, mycorrhiza developed and the pines started to grow rapidly.

Theories on Mycorrhizal Fungi

In the case of lichens, the contribution of each of the partners of the symbiosis is obvious. The fungus is unable to synthesize its own energy food, and derives this from the algae which carry out photosynthesis. The algal cells in their turn are protected by the surrounding fungal mycelium which also supplies the essential inorganic nutrients.

In the case of mycorrhiza, the mutual benefits of the two partners are not so clear-cut. Consequently, there are a number of different theories about the basic significance of the fungus to the roots and to the plant as a whole.

We have already disposed of the theory that the fungi in mycorrhiza are parasitic. Mycorrhizal fungi may take substances from living roots in normal symbiosis, or under extreme conditions they may be too virulent and become parasitic, but the normal mycorrhizal relationship is primarily symbiotic (Har-

ley, 1959). Early workers thought that fungi brought nitrogen and other materials to the root and numerous studies have shown that this is partly true (McDougal, 1943). Work by Knudson (1922) on sugar in orchid seed germination suggests that one of the main functions of the fungus is to supply sugars which the seed must otherwise obtain from decomposing organic matter. Some orchids become independent of the fungus when they begin photosynthesis in excess of respiration. Similarly, the saprophytic higher plants which lack chlorophyll must depend on mycorrhiza to bring in their food supply (Harley, 1959; Bierhorst, 1953). The hyphae which penetrate the forest litter apparently obtain nutrients and food substances from the digestion of dead organic matter, or from living tree roots in the case of *Monotropa* (Meyer, 1966) and transfer these to the roots of orchids or saprophytes. It has been suggested that *Monotropa* is not a saprophyte, but a parasite on a saprophytic fungus.

In some poor soils, ectotrophic fungi which envelop the root cells may take sugars and other substances from the roots (Harley and McCready, 1950; Melin and Nilsson, 1957). Endotrophic fungi which do penetrate the root cells directly may take some materials from these cells, as well as give up a wide variety of nutrient materials to the cell (Harley, 1959). Some authors claim that the fungi obtain growth substances from the plant roots (Ulrich, 1960). Others feel that they aid mineral uptake by creating an excess of hydrogen ions near the root (Routien and Dawson, 1943).

The consensus of opinion seems to be that mycorrhizal fungi vary greatly in their relation to the host plant; without sufficient organic matter, the fungus takes sugars and growth substances from the roots and passes phosphates and minerals from the soil to the root, especially in soils low in minerals (Melin and Nilsson, 1957). The other

Dr. F. W. Went is Head of the Laboratory of Desert Biology of the Desert Research Institute, University of Nevada, Reno. Dr. N. Stark is an ecologist with the Laboratory of Desert Biology.

extreme is apparent in the nongreen orchids which obtain all of their food through fungal hyphae which digest dead organic matter.

The Importance of Mycorrhiza to Trees

Evidence suggests that mycorrhizal fungi are more than casual associates with tree roots (Slankis, 1958). If mycorrhiza is so important to the temperate forests that some species will not grow without them (Kessell, 1927), how much more important it must be to the tropical rain forests where mycorrhiza is so abundant and the soils are often poorer. A few early studies mention mycorrhiza in the rain forest (Janse, 1897; Alexander, 1961), but only the orchid mycorrhiza and a few crop trees have been studied in any detail in the tropics.

Janse (1897) described the endotrophic mycorrhiza of coffee in humus-rich soils in Java and postulated an external source of carbon used by the fungi. Studies by Reed and Fremont (1935) in southern California on semitropical citrus trees growing outside the tropics showed that citrus roots receiving fertilizer developed a resistance to invasion by mycorrhizal fungi. In unfertilized soils the mycorrhizal fungus invaded the roots, resulting in a condition close to parasitism.

Recent Observations from Amazon Forests

We were recently members of the Alpha Helix Expedition to the Amazon. The Alpha Helix, research vessel of the Scripps Institute of the University of California at San Diego, is a most excellent floating laboratory, equipped with standard and specialized laboratory equipment. The vessel was anchored offshore and supplied power to the sister shore camp research laboratories, making it possible to bring the scientist directly to the problems in the field or in the water.

We made our initial observations on the abundance and importance of mycorrhiza in the tropical rain forest east of Manaus, Brazil. In addition, we traveled through many other parts of the Amazon forest, observing and collecting materials from the periodically inundated forests at the confluence of the Rio Negro and Rio Branco, the forests behind Borba on the Rio Madeira, the forests of the upper Amazon in Peru, near Tingo Maria (Aucayaco),

Fig. 1

One striking thing about the rain forest soils is that the fine feeder roots occur mainly in the well aerated upper 2-15 cm of humus where waterlogging does not occur (see Fig. 1). Below the thin humus layer in the eastern and central Amazon soils, we found white or gray sterile sand or poor, yellow clays, which support lush rain forest. In temperate soils, feeder roots are found deeper in the soil, which is normally richer in humus and available minerals.

Another prominent character of these tropical soils is the thin litter layer. Rarely do more than 1-2 cm of litter accumulate on the tropical forest floor, although not because of sparse leaf fall. Denudation plots for leaf fall show that 4.5 to 12.6 g of dry material are added to each square meter of forest floor per day during July through September. Careful examination of the humus and older litter shows that feeder roots, with largely endotrophic mycorrhiza, weave between and through leaf litter, fallen branches, and the heavy, woody fruits. Most of the feeder roots never reach mineral soil, but lie in layers between the leaves and are attached to the dead organic matter by hairs and fine hyphae. A few feeder roots which penetrate below the humus layer are mostly non-mycorrhizal. The flattening of their tips and the attachment between root hairs and dead organic matter resembles the attachment between root hairs and soil particles. The mycorrhizal roots encircle and penetrate fallen fruits and branches. Roots may grow through the pith of dead stems and even into termite galleries and under the bark of dead

branches forming thick mats. In all cases the roots are snugly bound by root hairs and hyphae to the decomposing organic matter. Ultimately, all organic matter in contact with roots and hyphae becomes soft and decomposes. The root hairs do not penetrate the dead organic matter directly, but are connected to it by hyphae and rhizomorph tissue.

Many woody tropical fruits contain much more material than is normally used by the germinating seed directly. Where roots contact dead organic matter, the woody portion softens under the influence of fungi while the unattached part remains firm and woody. It is not unusual to find a seedling with its roots and fungi wrapped around its own heavy woody fruit, digesting it and transferring material to the seedling. Organic debris may also soften through the activities of termites.

Equally poor soils with highly developed mycorrhiza were found west of Manaus near the juncture of the Rio Negro and Rio Branco. The forests around Borba on the Rio Madeira had slightly better soils with deeper humus layers and feeder root mats, and abundant mycorrhiza. The forests of the Alto da Serra on the south coast of Brazil, those near San Ramon and La Merced on the east side of the Andes in Peru, and those near Tingo Maria in the western upper Amazon Basin were generally much richer than the soils of the central Amazonian lowlands. The former had thick humus layers with mycorrhiza as well as feeder roots into the mineral soil.

Although many of these observations

256

agree with what was known about mycorrhiza in temperate zones, there is one fundamental addition. That is the close tie-in of the fungal hyphae in and around roots, and the mycelium pervading the litter layers of the soil. Thus, we completely disagree with the statement by Alexander in a recent textbook on soil microbiology (1961, p. 82), "The [mycorrhizal] fungus is not a soil microorganism in a strict sense, and its ecological niche is properly within the root association."

When mycorrhiza is destroyed by cutting, burning, and cropping, the infertility of the mineral soil soon becomes evident, and a once-lush climax forest is replaced by poor brush (Sioli, 1966). The difference between the richness of the climax forest and the low-productivity of the cropped land and second growth appears to be tied to the presence or absence of mycorrhiza. This difference has been blamed on leaching and washing of soluble minerals from the soil after cutting, burning, and cropping. This constitutes a drain from the mineral working capital of the system.

Direct Mineral Cycling Theory

Richards, in his excellent review on the rain forest said, ". . . there is a delicate balance in the primary forest between soil and vegetation, plant nutrients circulating in an almost closed cycle between the two, the small amount of soluble nutrients lost by leaching being replaced from unweathered soil minerals" (1952). He did not mention mycorrhiza as a part of the "almost closed cycle." Sioli (1966) described the mineral distribution in tropical soils, but he did not attribute the recycling to mycorrhiza. Stone (1950) felt that some types of mycorrhiza pass materials through the hyphae to the root cells.

We suggest that the "almost closed cycle" referred to by Richards (1952) is indeed closed, and that the closing link which is essential to maintain the mineral working capital of the system is through mycorrhiza, where fungi cycle nutrients (organic substances containing minerals or free minerals) directly from dead organic matter to the living roots with only a minimum of leakage into the mineral soil. The minerals lost from the closed organic system are partly reclaimed by roots and partly leached. There is probably a very slow replenishing of minerals from the sand or clay

soil. In bacterial decay, minerals are made soluble, released into the soil, and then taken up by roots. In direct mineral cycling, the minerals remain tied up in living or dead organic matter, they are transferred through hyphae from dead branches or leaves to living roots and very little mineral is made soluble and moved into the soil. This would explain why the feeder roots are mainly restricted to the humus layer, and why there are more mycorrhizal roots in the poorer soils. Also, humus and roots give off substances which stimulate fungal growth (Meyer, 1966).

In the temperate forests, the ectotrophic fungi appear to be largely unable to digest cellulose and lignin. The rapid disappearance of the forest litter in the tropical soils suggests that the microorganisms there can digest these materials readily. We find no reason to believe that tropical tree mycorrhiza behaves in exactly the same way as temperate tree mycorrhiza. The two areas may have mycorrhiza with differing metabolisms. The tropical mycorrhiza may act more like orchid mycorrhiza than conventional tree mycorrhiza. Many studies exist on temperate and orchid mycorrhiza, while that of the tropical forest has been neglected.

Our observations show that tropical forests have mainly endotrophic mycorrhiza which digests cellulose and lignin, and is able to break down woody fruits, pods, branches, and leaves on the forest floor.

Went's (1957) studies with tomato showed that where root systems were totally submerged in a nutrient solution, growth was poor; while growth was good if a portion of the root system was above the solution. The difference was attributed to the leaching of a soluble growth factor from roots submerged in water. If the upper parts of the roots were above the water, this growth factor and another factor related to iron utilization was not leached, and healthy and optimally growing plants resulted. Perhaps the rain forest maintains its surface feeder root system above zones of waterlogging as a means of retaining factors needed for growth and iron utilization. Results obtained by growing tomatoes and other plants with their whole root systems in a nutrient fog box (Went, 1957) support this interpretation. In the forest we have to consider these surface roots as mycorrhiza. Therefore,

poor growth and the chlorotic appearance of coffee and other forest plants with poor surface root development may result from lack of hormone production and deficient iron uptake.

The Role of Fungi in Soil

In the previous review, emphasis was laid on the fungus-root interactions, as is customary in mycorrhizal discussions. Let us now shift the emphasis on soil-fungus interactions. In most fertile soils, fungi are common, as is observed microscopically. No matter whether one looks at a forest, meadow, steppe, or desert soil (except those of absolute deserts), hyphae and rhizomorphs are ubiquitous. Especially in the more organic and sandier soils, hyphae bind soil particles. When root systems are carefully lifted from such open soils, all roots are surrounded by soil particles which are only partly attached by root hairs. Often the predominant attachment is by mycelium which not only connects most soil particles, especially the organic ones, but also surrounds the younger and older roots of both annual and perennial plants. Such roots do not have the appearance of mycorrhiza, and probably these hyphae are only part of the rhizosphere without penetrating the roots. Whereas in moister soils bacteria may play a dominant role, in arid regions and in rapidly draining soils fungi seem to be the main litter decomposers. It is generally accepted that they have a greater biomass.

In all desert soils where we have worked, wooden sticks decompose in about 2 years through fungal degradation, as far as they are inserted in the soil. They cannot be pulled out of the soil without a several millimeter thick layer of sand, which is attached to the sticks through a dense mycelium. Also in other light soils, and certainly in the upper horizons of forest soils, particles hang together through the dense network of mycelium, and thus fungi contribute much to the cohesiveness of soils, more so than generally accepted by soil scientists. And in all cases this mycelium is directly tied in with plant roots.

Another indication of the magnitude of fungal activity in soils is found in the appearance of mushrooms, the fruiting stage of the fungi. In forests, *Boletus, Russula,* and *Cantharellus* are common; in meadows, *Coprinus* and *Psalliota* are seen; in steppes, *Bovista* are noted; and in deserts, the fruiting bodies of *Podaxon, Tulostoma,* and *Battarrea* are regularly seen, especially where soil has been disturbed, where more organic material was buried, or where roots were severed.

All dead roots of desert plants disappear within a few years from the desert soils, apparently by fungal activity. One can see dead brown roots attached to living roots of small desert plants by a network of mycelium. Since there is little fungal mortality of desert annuals, these fungi are definitely not parasitic. The mycelium which penetrates the desert soil much deeper than the roots of most desert annuals (which are often restricted to only the upper few centimeters of the soil) perhaps even supplies water to these tiny annual plants.

Another phenomenon common in deserts is the lusher growth of so many annual plants near shrubs. For instance, *Phacelia* and *Rafinesquia* plants are two to ten times larger when growing around shrubs or even dead plants (Went, 1942). This might also be associated with fungi, which can feed on the organic material produced by and collected under such shrubs.

All the previous considerations seem to set us back 125 years, since Sprangel and Liebig so effectively dismissed the organic nutrition theory, which had held sway until then. It had been previously accepted that it was the organic matter in the soil which was necessary for the nutrition of crops. Liebig (1840) showed that crop plants grew proportionally to the amounts of inorganic nutrients — phosphates, nitrates, and potassium — present in the soil. And now we find, that after all, the organic content of certain soils may be significant for plant growth, ranging from rain forest trees to desert plants. A quantitative interpretation will have to await further research. Although growth of certain plants may be enhanced by the organic content of the soil, most crop plants presumably follow Liebig's Law of the Minimum — based on the inorganic nutrients in the soil.

Practical Aspects of Direct Mineral Cycling

If the direct nutrient cycling theory proves to be correct, we have the beginnings of an explanation of the enigma of lush forests on poor Amazonian soils. The native Indians have never achieved a greater population density than one person per 5 square miles, with life largely based on hunting. By contrast, the rich soils of Java, originally covered by rain forest, with efficient agriculture can support 2000 persons per square mile of arable land.

Many of the past attempts at agriculture in the Amazon and other tropical South American countries have met with failure because of a lack of knowledge of the basic ecology of the forests. Everyone thought, quite naturally, that here was the most lush, extensive rain forest in the world, which should be replaceable by rich crops.

Because the pattern of cut, burn, and crop quickly depletes the organic matter and thus removes the major source of minerals, the cleared land is abandoned after a few years by a largely nomadic riparian population. As a result of this frequent clearing, enormous areas formerly covered with rich rain forest now support poor, straggly second growth or scrub, and it may require hundreds of years of careful management to nurse them back to the original high production levels.

In the upper Rio Negro, there are extensive areas covered by what is called "caatinga" or thorn scrub vegetation of extremely low productivity. More land is being cleared daily to meet the same fate.

According to our observations, the solution for the Amazon lies in developing the detailed ecological knowledge necessary to use the land without depleting its natural mineral reserves This seems to require the testing of crop plants with mycorrhizal roots which can recycle the minerals in the same manner as the climax forest without producing a serious drain from the mineral working capital of the organic system. Tree crops with mycorrhiza around Manaus do well, but techniques are needed which are specifically suited to these tropical ecosystems. The solution to the problems in the Amazon cannot come soon enough to avoid further deterioration of the land, before it becomes essential to meet population expansion. Many Amazonians live on low protein diets and low cash crops. There is a definite need for better management of the Amazonian resource.

Summary

We have advanced a theory, based on observations of soils over a wide area of the Amazon Basin, which explains how minerals are recycled. The theory is called the direct mineral cycling theory. It is discussed in light of our observations and in relation to the great body of research on temperate climate mycorrhiza. The theory is based on the fact that the bulk of minerals available in the tropical rain forest ecosystems is tied up in the dead and living organic systems. Little available mineral ever occurs free in the soil at one time. Mycorrhiza which is extremely abundant in the surface litter and thin humus of the forest floor is believed to be capable of digesting dead organic litter and passing minerals and food substances through their hyphae to living root cells. In this manner, little soluble mineral leaks into the soil where it can be leached away. This is contrary to the conditions which result when bacteria are the main agents of decay, where the minerals are directly released into the surrounding soil.

Unlike the ectotrophic mycorrhiza of temperate forests, tropical trees appear to have largely endotrophic mycorrhiza in which fungi have actually penetrated into living cells. The endotrophic mycorrhiza is most generally described from orchids and, until now, was thought to be rare in trees. The endotrophic mycorrhiza of the jungle is thought to function more like orchid mycorrhiza than like temperate tree mycorrhiza. This report points to the scarcity of research on the tropical forest soil organisms and to the urgent need of such research to aid the declining agriculture of the Amazon.

References

Alexander, M. 1961. *Introduction to Soil Microbiology.* John Wiley & Sons, New York.

Bierhorst, D. W. 1953. Structure and development of the gametophyte of *Psilotum nudum. Am. J. Bot.,* **40:** 649-658.

Briscoe, C. B. 1960. The early results of mycorrhizal inoculation of pine in Puerto Rico. *Caribbean Forester,* **20:** 73.

Harley, J. L., and C. C. McCready. 1950. Uptake of phosphate by excised mycorrhiza of Beech. I. *New Phytologist,* **49:** 388-397.

Harley, J. L. 1959. *The Biology of Mycorrhiza.* Plant Science Monographs, Leonard Hill, Ltd., London. 233 p.

Janse, J. M. 1897. Les Endophytes radicaux de quelques plantes Javanaises.

Ann. Jardin Botan. Buitenzorg, **14:** 53-201.

Kelley, A. P. 1950. *Mycotrophy in Plants.* Chronica Botanica Co., Waltham, Mass. 223 p.

Kessell, S. L. 1927. Soil organisms: the dependence of certain pine species on a biological soil factor. *Empire Forestry J.,* **6**(1): 70-74.

Knudson, L. 1922. Non-symbiotic germination of orchid seeds. *Botan. Gaz.,* **73:** 1.

Liebig, J. von. 1840. *Die Chemie in ihrer Anwendung auf Agricultur und Physiologie.* Braunschweig Verlag Friedrich Vieweg und Sohn.

McDougal, D. T. 1943. Study of symbiosis of Monterey pine with fungi. *Yearbook Am. Phil. Soc.,* p. 170-174.

Melin, E., and H. Nilsson. 1957. Transport of C¹⁴ labelled phosphate to the fungal associate of pine mycorrhiza. *Svensk Botan. Tidskr.,* **51:** 166-186.

Meyer, F. H. 1966. In: *Symbiosis, Vol. I,* M. S. Henry (ed.), Academic Press, Inc., New York. 478 p.

Rayner, M. C. 1927. *Mycorrhiza.* New Phytologist Reprint No. 15, Weldon & Wesley, Ltd., London. 246 p.

Reed, H. S., and T. Fremont. 1935. Factors that influence the formation and development of mycorrhizal associations in citrus roots. *Phytopathology,* **25**(6): 645-647.

Richards, P. W. 1952. *The Rain Forest.* Cambridge University Press, London. 450 p.

Routien, J. B., and R. F. Dawson. 1943. Some interrelationships of growth, salt absorption, respiration and mycorrhizal development in *Pinus echinata. Am. J. Botany,* **30:** 440-451.

Sioli, H. 1966. In UNESCO. *Scientific Problems of the Humid Tropic Zone Deltas and Their Implications.* Proceedings of the Dacca Symposium, Paris. 422 p.

Slankis, V. 1958. Mycorrhiza of forest trees. Forest Soils Conference, Sept. 8-11, 1958.

Stone, E. L. 1950. Some effects of mycorrhiza on the phosphorus nutrition of Monterey pine seedlings. *Proc. Soil Sci. Soc. Am.,* **14:** 340-345.

Ulrich, J. M. 1960. Auxin production by mycorrhizal fungi. *Physiol. Plantarum,* **13:** 429.

Went, F. W. 1942. The dependence of certain annual plants on shrubs in Southern California Deserts. *Bull. Torrey Botan. Club,* **69:** 100-114.

Went, F. W. 1957. *The Experimental Control of Plant Growth.* Chronica Botanica Co., Waltham, Mass. 343 p.

22

Reprinted from *Biotropica* 6:69–103 (1974)

Tropical Blackwater Rivers, Animals, and Mast Fruiting by the Dipterocarpaceae

Daniel H. Janzen

Department of Zoology, University of Michigan, Ann Arbor, Michigan 48104, U.S.A.

ABSTRACT

It is proposed that tropical nutrient-poor white sand soils produce blackwater rivers, rivers that are rich in humic acids and poor in nutrients, because the vegetation growing on these soils is exceptionally rich in secondary compounds. The humic acids (= tannins and other phenolics) may even be only the more conspicuous of the secondary compounds that leach out of the living vegetation and the litter. While the water and the soil (including litter) may be expected to have a low productivity and animal biomass solely on the basis of its low nutrient content, it is quite possible that large amounts of secondary compounds are also debilitating to the animal community. An exceptionally high concentration of secondary compounds is expected in the vegetation growing on white sand soils for two reasons. First, this is an expected outcome in habitats where the loss of a leaf to an herbivore or through deciduous behavior is relatively a much greater loss than on nutrient-rich soils. Second, the plants growing there belong for the most part to families exceptionally rich in secondary compounds, a characteristic which is in turn selected for by the chemical defense requirements of plants growing in low diversity stands. The small amount of data that is available from Sarawak white sand habitats shows that the carrying capacity for animals is very greatly reduced. The postulated cause is reduced primary productivity and/or much of the productivity being used by the plant for secondary compounds (unharvestable productivity), or stored for seed crops at very long intervals (unavailable productivity). It is proposed that mast fruiting at the community level, as displayed by trees in the Dipterocarpaceae, is a mechanism of escape from seed predators that is unique to this part of the tropics (S.E. Asia) because this area has reduced animal communities (both on white sand soil sites and in general), and because the climate is sufficiently uniform for such an intra- and inter-population cueing system to evolve. Without experimentation, it is impossible to know, however, if the animal community is reduced solely due to overall lowered primary and harvestable productivity, or as well to the inevitable reduction in animal numbers when many of the trees in a habitat wait more than a few years for their highly synchronized seed crops. The occurrence of numerous tropical habitats with a very low diversity of trees inviolates the currently popular dogma that diversity is mandatory for stability in tropical habitats. I propose that the trees in such monotonous habitats are exceptionally well-protected chemically with respect to foliage, and have either very toxic seeds or well-developed mast cycles.

THE TROPICAL ECOLOGIST is confronted with four related questions:

I) Why do some tropical habitats produce blackwater rivers?

II) Why do these habitats contain drastically reduced numbers of animals?

III) How is it that mast (= gregarious) fruiting has evolved at the population and community level in a group of lowland tropical trees, the southeast Asian Dipterocarpaceae?

IV) How is it that certain lowland tropical trees persist in forests of very low species richness?

I wish to explore the idea that these are expected and ecologically related properties of habitats based on soils that are exceptionally deficient in nutrients. Each of these four questions is discussed as though it were a somewhat independent line of inquiry. However, I relate them in an attempt to generate a holistic evolutionary ecological viewpoint of tropical habitats with low productivity. It must be recognized from the start that this inquiry is largely intended to generate testable hypotheses and show that sufficient circumstantial evidence exists to make their proposal reasonable.

I. TROPICAL BLACKWATER RIVERS

Some lowland tropical river basins contain tributaries with water that is clear yet brown to nearly black in color. The Rio Negro of the northern Amazon basin is the largest and best-known tropical blackwater river (Ducke and Black 1953; Fittkau 1967; Klinge and Ohle 1964; Marlier 1965, 1973; Sioli 1955, 1964, 1967b, 1968a,b; Sioli and Klinge 1962; Sioli et al. 1969), but smaller bodies of blackwater occur in the southern Amazon (Sioli 1967a), various tributaries of the Congo (Clerfayt 1956; Doubois 1959; Marlier 1973; Rougerie 1958), Nigeria (Clayton 1958), Sarawak (Anderson 1963, 1964b; Ashton 1971; Inger and Chin 1962; Janzen 1974; Richards 1936, 1963), Malaya (Johnson 1967a,b, 1968; Mizuno and Mori 1970), East Indies (Polak 1933), Guyana (Carter 1934; Davis and Richards 1933, 1934), Suriname (Heyligers 1963; Stark, pers. comm.), coastal Brazil (Anonymous 1960), Guatemala (L. G. Brinson 1973; M. M. Brinson 1973), northern British Honduras (pers. obser.), and other sites (see Klinge 1968 for a thorough bibliography).

Blackwater rivers usually flow from podzolized white quartz sand soils (regosols, bleached sands) or peat swamps (Handley 1954; Klinge 1965, 1967, 1969; Richards 1941). Klinge (1965, 1968) estimated that at least 4475 million hectares of the tropics are podzolized. If we may infer from the known cases (Klinge 1967), most of this soil is sandy and may produce blackwater rivers. The white sands are usually eroded from ancient aeolian or alluvial sandstones (Hardon 1936; Johnson 1968; Richards 1941, 1952, 1973; Sioli 1964; Sombroek 1966), though Heyligers (1963) feels that Suriname white sands may be formed *in situ* by leaching. For example, the white sand soils in the watershed of the Rio Negro are alluvial deposits eroded from the Roraima sandstone in interior Venezuela, northern Brazil, and the Guianas (Zinke, pers. comm.). In view of their secondary or even tertiary origin, high porosity (leading to rapid leaching by heavy tropical rain) and low ion retention properties, it is not surprising to find that they are probably the most nutrient-poor soils in the world when compared with tropical latosols and other soil types (Arens 1963; Hardon 1937; Herrera 1972; Heyligers 1963; Klinge and Ohle 1964; Mohr 1944; Mohr and van Baren 1954; Richards 1963; Sioli 1954, 1967b; Sioli *et al.* 1969; Stark 1970, 1971a,b; Webb 1968). In a recent review of Amazon river basin soils at several hundred sites, Zinke and Castro (1973) found white sand soils to contain half or less as much carbon as nearly all other soils, and consistently to be at the low end of the gradient of phosphorus, nitrogen, calcium, potassium, and sodium concentration. Ungemach (1969) concluded that except for iron, the concentration of nutrients in the water of the Rio Negro was equal to that in the rainwater of the region; which suggests that there are no nutrients being weathered from parent material in white sand soil ecosystems.

Tropical blackwaters are usually very acidic (pH between 3 and 4.5), contain many fewer inorganic ions than do clear, white, or muddy waters in the same drainage basin, have low oxygen content, have low light penetration, and contain high concentrations of dark brown "humic acids" (Anonymous 1972; Black and Christman 1963; Carter 1934; Fittkau 1967; Flores *et al.* 1972; Foldats 1962; Ghassemi and Christman 1968; Hardon 1937; Hutchinson 1957; Joachim 1935; Johnson 1967a,b, 1968; Klinge and Ohle 1964; Marlier 1965, 1973; Mizuno and Mori 1970; Mohr and van Baren 1954; Richards 1941; Roberts 1973; Schnitzer and Khan 1972; Shapiro 1957; Sioli 1954, 1955, 1964, 1967a,b, 1968a,b; Sioli *et al.* 1969; Waksman 1938; Williams 1968). However, I must stress early on that there are undoubtedly many kinds of blackwaters; and this situation may account for many of the exceptions to the statements to be made in the following pages. The low nutrient content of blackwater rivers is partially explainable by the observation that the soil they drain is not being formed by the weathering of base-rich rock. The polyphenolic humic acids are also well-known chelating agents for inorganic ions, and may be preventing their uptake by plants. The low oxygen concentration is probably due to the lack of aquatic plants (see below), though it must be remembered that the surface of the water and water at rapids should be well oxygenated. The low light penetrance is due to the humic acids. While only circumstantial evidence is available, it appears that the high acidity of the water is solely due to organic (humic) acids. However, the folklore of the lowland tropics suggests that soil temperatures are so high that litter decomposition occurs too rapidly for the accumulation of humic material. However, blackwater rivers also occur at cool tropical high elevations, and there, usually issue from peat bogs just as they do in the mid-latitudes. We are then left with the question of why should there be a high concentration of humic acids in these rivers?

Before dealing directly with this question, let us examine the circumstantial evidence for the effect of humic acids on the organisms in blackwater rivers. While the humic acids are organic compounds and therefore surely susceptible to degradation by decomposers, they are phenolics and belong to that special ecological class of toxic organic compounds known generally as "secondary compounds" or "defensive compounds" (Levin 1971; Whittaker and Feeny 1971). They are thus expected to be generally toxic, difficult to degrade, and persistent to a greater degree than other chemical plant debris. We should also expect them to have a negative effect on the organisms in blackwater rivers.

What suggestion is there that the humic acids, and other secondary compounds, in the blackwater rivers are detrimental to the organisms in blackwater rivers? Blackwaters have long been suspected to be toxic to aquatic organisms by local people and field naturalists. The humic acids in blackwaters are polyphenolic compounds, compounds that are well known to be toxic to a wide variety of organisms (see below) owing to their propensity for forming insoluble complexes with proteins (Bate-Smith 1973a,b; Feeny 1968, 1969; Handley 1954; Ribereau-Gayon 1972). Johnson (1967b) found that 9 of the 15 species of fish in Malayan blackwaters are air-breathers or surface swimmers. Many of the fish in the upper Rio Negro and in Florida blackwaters are obligatory air-breathers (McNab, pers. comm.), and Carter's (1935) work on Guyana air-breathers was with fish from blackwater streams. M. M. Brinson (1973) and L. G. Brinson (1973) have found that blackwater, forced out of the peat swamps at the west end of Guatemala's Lake Izabal by rainy season flooding, is highly toxic and repellent to the fish that live in the ordinary water of the lake. In Malayan blackwaters, almost all species of insects are air-breathers (Johnson 1968). The standard explanation for air-breathing by these aquatic organisms is the low oxygen content of the water, but it seems equally likely that protein-complexing humic acids should play havoc with the oxygen-exchange surfaces of gills. Levanidov (1949) reports that water slaters (*Ascellus aquaticus*) placed "in water from the upper layer of a peat bog" died within 24 hours. Dunson and Martin (1973) report strong toxicity to fish of the organic acids in blackwaters from a Pennsylvania bog. In both these cases death was attributed to the low pH, but that conclusion may be viewed as just another way of describing the toxic effects of the humic acids. Geisler *et al.* (1971) found that Amazonian clear-water Characidae and Cichlidae were able to survive in acidic blackwater once the humic materials had been filtered out, suggesting toxicity went beyond mere acidity. Lisk and his associates at Cornell University (pers. comm.) have found that the saponins washed from birch bark are responsible for the heavy spring fish mortality in the Black River in the New York Adirondack Mountains. In traveling on the Rio Negro, Spruce (1908: 270) pointed out that there is "the great advantage of voyaging on black waters, that no carapaná (or zanuedo, as the Spaniards call them) interrupts one's repose . . . and this I could do undisturbed [at night] by the insects which are the greatest torment to the traveler on the [white water] Amazon." While the absence of biting insects may be due to a shortage of prey for aquatic larvae, it may be due as well to blackwater being an unfavorable chemical medium for their larvae. It has been suggested by numerous people that in addition to humic acids, blackwater rivers may contain other phenolics, such as the phytoanalogues of insect hormones. The trees of blackwater river drainage basins are likely suspects to contain these compounds (see below), and phytoanalogues of insect hormones have been shown to be highly effective against mosquito and black fly larvae (Cumming and McKague 1973; Spielman and Williams 1966; Williams 1970), and acorn barnacle larvae (Gomez *et al.* 1973). In speaking of Amazonian blackwaters, Fittkau (1971) says "the whole region is free of Culicidae, except for those few forms which live in pockets of water trapped by epiphytic bromeliads";

however, he also says that these same waters are very rich in chironomid species.

The productivity of blackwater rivers is extremely low. This characteristic is usually attributed to the low nutrient level, or lack of trace elements, acting by themselves or in concert with the low oxygen and low light penetrance. Low pH is often included in this array, but that may in itself be viewed as a toxic effect of the humic acids. However, there are enough cases where one or more factor is nullified and productivity still stays low to make the humic acids (and perhaps other secondary compounds) quite suspect. For example, Carter (1934), on Guyana blackwater streams, says "the only characteristic in which the forest [black] waters differ from the [temperate zone] moor waters is their relatively high content of nutrient salts, which are typically present in small amounts in moor waters." He also noted that flowing blackwater streams are well oxygenated. In comparing an Amazonian white water lake with a blackwater lake, Marlier (1965) found a very small number of fish with few species and low plankton productivity in the blackwater lake; however, his tables show that the oxygen content is even higher in the blackwater lake than in the white water lake.

"The Rio Negro might be called the Dead River —I never saw such a deserted region" (Spruce 1908: 268). "For the human populations living along the banks of black-water rivers, these rivers always have the fame of being 'hunger rivers'" (Sioli 1968a). "Turtle are very rarely met with in the Rio Negro, but only on some of its lower branches" (Spruce 1908:275). "Leaving San Fernando the party took their famous short cut down the 'black' or coffee-coloured, mosquito-less rivers past Javita and the portage of Pimichin, entered Rio Negro . . . The return journey was made up the 'white' Rio Casiquiare, which connects the Rio Negro with the Orinoco, and is remarkable for the wildness of its banks and the blood-thirstiness of its mosquitoes" (Sandwith 1925). Roberts (1973) and Marlier (1973) report that the fish and invertebrate biomass is greatly reduced in Amazonian blackwater rivers as compared with other lakes and rivers in the Amazon basin. Patrick (1964) states that in the non-blackwaters of the Amazon basin "the number of species [of animals] were greater than in the acid [blackwater] streams as is the case in temperate zone streams." Gessner (1964) states that phytoplankton productivity is very low in the Rio Negro. Malayan blackwaters have "very low standing crops of animals and plants . . . and must be regarded as unproductive" (Johnson 1967a). "Collections [of fish] from 24

[Malayan] blackwater habitats . . . comprise only about one tenth of the Malayan fauna. Thus the fish of blackwaters are very restricted in variety" (Johnson 1967b). In Malayan blackwaters, the standing crop of fish may be as low as 0.5 gm/m²; they feed almost entirely on material falling into the water, yet in this detritivore community, "microphagous organisms are mostly rare or absent" (Johnson 1968); Cladocera, annelids, rotifers, gastrotrichs, nematodes, and protozoans are rare (Johnson 1968). Algae are rare, except for a few species which are locally common, and "higher plants are likewise uncommon and often absent" (Johnson 1968). In a Malaysian nonblackwater stream, Bishop (1973) estimated the fish standing crop to be 18 gm/m² and the invertebrate fauna to be very diverse. Fish are "slow growing and stunted" in Wisconsin blackwater lakes fed from peat bogs, and fertilization does not completely eliminate the effect (Johnson and Hasler 1954; Stross and Hasler 1960). In Tasek Bera, a blackwater lake in Malaya, "detritus on the bottom was thick, but no benthonic animal was found on it"; it had reduced numbers of zooplankton, virtually no molluscs and only 5 species of fish, whereas 5 white water lakes in Thailand and Cambodia had 13 to 29 fish species (Mizuno and Mori 1970). When describing Amazonian blackwater rivers, Sioli (1968a) states that floating meadows, "the biotope of probably the richest community of aquatic animals which exist in all Amazonian waters," are very common on Amazonian white water rivers, yet are completely absent from blackwater rivers. "The black-water rivers, even the big black-water rivers with wide mouth bays, exposed with their whole surfaces to the solar radiation, have very little primary production of phytoplankton and of submerged higher water plants. . . ." ". . . many of those river-lakes are more or less blackwaters with very little phytoplankton development and . . . the true great clearwater rivers as e.g., *the Tapajos and the Xingu develop even water blooms, in spite of the chemical poorness of the water*" (my italics).

Sioli's comment on the Tapajos and Xingu emphasizes that it is not merely the low nutrients in blackwater rivers that lead to their low standing crops. Johnson (1967a) has pointed out that "somewhat surprisingly the calcium concentration in most of the [Malayan] blackwaters proves to be higher than in non-blackwater tree-country habitats and the highest value of all is for a blackwater stream." The low pH cannot be the only factor, as some Malayan blackwater streams flow from peat bogs and then through limestone areas that raise the pH above 6.0; these streams have snails, somewhat more insects

and more fish, but the fish are highly restricted to these waters rather than occurring in other basic waters as well (Johnson 1968). To attribute the effects in the previous paragraphs to low oxygen content and low light penetration does not answer why plants and animals are absent from the surface, and ignores the fact that they are abundant in tropical turbid rivers.

To return to our original question, I propose that the rainwater runoff into blackwater rivers is exceptionally rich in "humic acids," and probably other toxic organic compounds, because 1) the leachate from fresh vegetation and decomposing litter on these soils is exceptionally rich in phenols and other plant defensive chemicals, and 2) the poor soil leads indirectly, and high input of phenols directly, to a litter and soil community relatively incompetent at degrading these same secondary compounds. In short, this appears to be a situation like that described for the podzolization of Russian taiga soils. Here, the phenol-rich lower plants, conifers, and ericaceous shrubs produce litter out of which leaches a dilute but highly acidic organic solution, whose free acid groups are gradually paired with basic ions as they move down through the sand soil, resulting in a clean sand layer over a layer of precipitated polyphenols ("humus," Ponomareva 1969: 260). The same process is clearly the case with the formation of mor under phenol-rich plants at more temperate latitudes (Dimbleby 1962; Handley 1954). On tropical white sand soils, originally sand rather than derived by weathering ordinary rock, it is quite possible that there is even less basic material. In addition to precipitating out in a "giant podzol" or very hard ortstein (e.g., Heyligers 1963; Klinge 1967), most of the phenolics are then simply being leached into the streams in a highly active state, continuing to complex with proteins, minerals, and other compounds as they move through the ecosystem. There is no *a priori* reason to expect that the secondary compounds in the litter are produced *de novo* by the soil flora, even though they may produce some for the same protective reasons as do the higher plants, and alter the composition of secondary compounds while degrading or avoiding them.

HYPOTHESIS 1.   There are no studies directed at the subject, but there are three areas of strong circumstantial evidence that the leachate from litter and vegetation over tropical lowland white sand podzols is exceptionally rich in toxic phenolics and other secondary compounds.

A). In a habitat that has extremely low primary productivity (see section II), yet a climate favorable to animals year round, there should be very strong selection for plants that are exceptionally rich in chemical defenses. This is so for at least three reasons. a) The loss of a leaf or other part to an herbivore should lower the plant's fitness proportionately more than in more fertile habitats, because of the greater cost of replacing both the nutrients consumed and the damaged part in infertile habitats. Proportionately more of the plant's resources are thus expected to be spent on defenses. b) In a habitat with lower productivity, the plant should produce better-protected leaves as a consequence of selection for leaves with a longer half-life. The vegetation on tropical white sand soils is commonly evergreen even when there is a severe dry season (e.g., Brünig 1973; Heyligers 1963; Huinink 1966; Takeuchi 1961). A lower turnover rate of leaves lowers the inevitable losses of nutrients each time one leaf is replaced with another (and see Small 1972b). The more slowly the herbivore damage accumulates on a leaf, the longer it will be before the leaf must be replaced for internal economic reasons. The evergreen habit of trees on white sand soils, to be discussed later, may be viewed as an outcome of this adaptive behavior. As Ponomareva (1969) puts it when speaking of the Russian taiga, "coniferous tree species, each year shedding only part of their needles, perennial subshrubs, evergreen plants, etc., are all adaptive forms of live nature to conditions under which mineral nutrients are strongly leached out of the plant's environment." We may also note that evergreen leaves tend to have heavy waxy cuticles, material that may on the one hand be viewed as a contribution to the organic leachate in blackwaters and on the other hand as adaptive in minimizing leaching from living leaves in a habitat poor in nutrients. c) In habitats with low productivity and containing vegetation whose seeds escape from predators by extreme toxicity or mast fruiting (= gregarious fruiting), we find low species richness of trees and high aggregation (see below). This spacing pattern should be associated with selection for those species that are exceptionally rich in secondary chemical defenses, as the plant has no escape in space (Janzen 1970).

B). The plants that grow on tropical white sand soils (usually podzols) belong to families that are particularly rich in defensive chemicals. On Malaysian white sand soils (and their overlying peat swamps) the commonest woody plants (species and individuals) are in the Anacardiaceae, Annonaceae, Apocynaceae, Araucariaceae, Burseraceae, Casuarinaceae, Dipterocarpaceae, Ebenaceae, Ericaceae, Euphorbiaceae, Fagaceae, Guttiferae, Lauraceae, Leguminosae, Melastomataceae, Moraceae, Myrtaceae, Podocarpaceae, Rubiaceae, Sapotaceae, and Thymelaeaceae (Anderson 1961, 1963, 1964b; Ashton 1964; Brünig 1965, 1968, 1969a,c, 1973; Fox 1967; Meijer 1970; Richards 1936, 1952, 1973). A list of the families of trees over 12 inches in girth in a Sarawak peat swamp forest (Anderson 1961) could hardly be more foreboding to an herbivore (numbers of species in parentheses): Dipterocarpaceae (5), Thymelaeaceae (1), Leguminosae (1), Sapotaceae (3), Ebenaceae (1), Apocynaceae (1), Crypteroniaceae (1), Anisophylleaceae (1), Anacardiaceae (3), Annonaceae (5), Rubiaceae (1), Sterculiaceae (1), Myrtaceae (6), Rutaceae (1), Aquifoliaceae (1), Sapindaceae (2), Xanthophyllaceae (2), Guttiferae (6), Icacinaceae (1), Fagaceae (1), Euphorbiaceae (1), Oleaceae (1), Burseraceae (1), Chrysobalanaceae (1), Myristicaceae (1), and Meliaceae (1). On the extensive white sand podzol described in New Guinea by Hardon (1936), the vegetation was almost entirely of gymnosperms (*Araucaria, Dacrydium, Phyllocladum, Podocarpus*), Ericaceae, ferns, mosses, and lichens (1860 to 2400 m elevation). With the exception of Melastomataceae and Annonaceae, the wood and roots of plants in all these families are well known to contain large amounts of latex, essential oils, resins, tannins, and other phenolic or terpenoid defensive compounds (Ashton 1964; Bate-Smith 1962; Browne 1955; Burkill 1935; del Moral and Cates 1971; del Moral and Muller 1969; Foxworthy 1927; Harshburger 1916; Hon 1967a,b; Jackson 1957; Mors and Rizzini 1966; Ohtaki *et al.* 1967; Quisumbling 1935, 1951; Richards 1936; Senear 1933; Slooten 1952; Standley 1920-1926; Thapa 1968; Webb 1968; Whitford 1906; Whittaker 1970; Williams 1960; Willis 1966), and other toxic secondary compounds such as juvenile hormones (e.g., Mansingh *et al.* 1970; Ohtaki *et al.* 1967; Retnakaran 1970; Williams 1970 and included references), and high silicon concentrations (Burgess 1965; Jones and Handreck 1967; Menon 1956).

In Guyana, sandy podzols support forests where more than 40 percent of the large trees are Lauraceae (greenheart; *Ocotea = Nectandra*) or Leguminosae (*Eperua*; Davis and Richards 1933, 1934; Richards 1941, 1952). On the Rio Negro and other Brazilian podzolized white sand soils, *Eperua* spp. are among the most common trees (Ducke 1940; Sioli 1960; Sioli and Klinge 1962; Takeuchi 1961, 1962a). The *Ocotea* trees are renowned for being "more resistant than any other known timber to the attacks of insects and marine borers" (Davis and Richards 1933; Menon 1956), and "old logs and stumps of greenheart (*Ocotea*) decay extremely slowly, lasting unchanged for many years" (Davis and Richards 1933). *Ocotea venenosa* also contains a very toxic vertebrate poison (Schultes 1969). *Ocotea* wood is rich in alkaloids (Menon 1956) and tannins (Gonggryp and Burger 1948), and greenheart has an alkaloid in the bark (Burkill 1935). The fallen leaves of *Eperua* are nearly black (presumably from tannins) and stain the soil brown to a depth of 25 cm (Davis and Richards 1934). *Eperua* resin (gum?) is produced in a quantity sufficient for commercial harvest (Ducke 1940; Mors and Rizzini 1966). The freshly cut wood "becomes sticky with exudations from ducts similar to those in pine," the "odor of fresh wood is rather acrid, suggesting creosote," and "the average life of an untreated [fence and telephone] pole is 20 years" (Aitken 1930). Sawdust of *Eperua* and *Ocotea* has a very depressant effect on fungal growth (Ernest 1936). On Guyana, Carter (1934) says "these black-water streams are popularly supposed to derive their colour from the decaying leaves of the Wallaba (*Eperua* spp.), and it is certainly true that many streams of this type flow from Wallaba forest." After *Ocotea*, the next most abundant trees in the *Ocotea* forest are *Pentaclethra macroloba* (Leguminosae), *Eschweilera sagotiana* (Lecythidaceae), and *Licania venosa* (Rosaceae). In Costa Rica, *P. macroloba* has extremely poisonous seeds, and the tannin-rich logs persist for many years on the forest floor (Hartshorn, pers. comm.). Its roots have a conspicuous layer of tannin-rich cells just inside the epidermis (D. Janos, pers. comm.); the roots of *Eperua*, *Mora* (see below) and *Peltogyne* (purpleheart, another South American legume with very resistant wood) are dark brown (Norris 1969) and probably for the same reason. *Eschweilera longipes* has wood so rich in siliceous deposits that marine wood borer mandibles are worn down when they attempt to bore into it (Menon 1956). There is no information on *Licania venosa*, but Rosaceae produce large amounts of cyanogenic glycosides, alkaloids, and saponins (Pammel 1911), and *Licania* is one of the few trees in Costa Rican deciduous forest with evergreen leaves.

In the Suriname *Dimorphandra* (= *Mora*) forests described by Stark (1970), Lindeman and Moolenaar (1959), and Heyligers (1963), on white sand soils, the leaves are highly resistant to decay and accumulate to depths of over 200 cm, while on adjacent latosols litter accumulates only to a depth of a few cm. *Mora* is well known for dark-reddish wood extremely resistant to decay and termites—thus its use for fence posts, railway ties, and bridges—

(Allen 1956; Gonggryp and Burger 1948). Stark's (1970) *Dimorphandra* forest generates a blackwater river (Stark, pers. comm.) as does Heyligers' (1963). Spruce (1908: 304) states that the vegetation on a Brazilian white sand soil was primarily a caesalpinaceous legume tree (*Mora*, *Eperua*, *Copaifera*, *Hymenaea*, and other legumes famous for heavy resin and tannin production are caesalpinaceous legumes, Standley (1920-1926)) and a vegetation composed in great part of Myrsinaceae and Rutaceae, with the latter being used as fish poison. Ericaceae are frequently mentioned in the descriptions of South American vegetation on white sand soil (e.g., Ducke and Black 1953; Heyligers 1963; Huinink 1966; Takeuchi 1961, 1962a). Ducke and Black (1953) point out that the forest along white water rivers is "characterized by a high percentage of trees without distinct heartwood [(wood impregnated with secondary compounds)] or with softer wood than that of their congeners growing in the 'varzea' of the rivers of black water." Ashton (1964) states that on Brunei yellow podzols "the litter, whether dense or not, was largely composed of leaves showing signs of resistance to rotting . . . some tree species here, such as *Dipterocarpus globosus* and *Dryanobalanops aromatica* (Dipterocarpaceae), showed marked accumulation of litter below their crowns owing to their resistant fallen leaves."

C). At Bako National Park, a white sand and blackwater site in Sarawak (Anderson 1963; Ashton 1971; Brünig 1965, 1968, 1969a, 1973; Janzen 1974), I found virtually all the woody plants, from early second growth of kerangas forest ("Padang") to mixed dipterocarp forest to peat swamp (all on white sand), to have leathery leaves with a conspicuous resinous, acrid or aromatic odor or taste when crushed. In my experience in the tropics, foliage with such omnipresent potential toxicity is not characteristic of insolated foliage of tropical forest on latosols, and especially atypical for plants of secondary succession. As Brünig (1969c) stressed, the "fast-growing pioneer species" of good soils "are notably lacking in the secondary vegetation on podzols." These sclerophyllous leaves (Ashton 1971; Brünig 1965, 1968, 1969a,b) are smaller than expected in a lowland rainforest site and appear to have a life span of well over a year. Arens (1963), Ashton (1964, 1971), Brünig (1969a,b,c), Ferri (1960), and Richards (1952) have suggested that this morphotype of foliage is characteristic of tropical white sand podzols, and Ashton (1964) demonstrated a high concentration of tannin in the leaves of 18 tree families on sandy soils in Brunei (northern Borneo). The leaves at Bako closely resembled the leaves of gymnosperms and *Magnolia* (Magnoliaceae), *Persea* (Lauraceae), *Gordonia* (Theaceae), *Myrica* (Myricaceae), *Lyonia*, and *Leucothoe* (Ericaceae) and *Ilex* (Aquifoliaceae), which constitute the cedar, cypress, and evergreen peat swamp forest vegetation of Florida and southern Georgia on white sand soils drained by blackwater rivers (e.g., Collins *et al.* 1964; Monk 1966, 1968). Foliage of these and similar plants is well known to be long-lived and slow to decompose (e.g., Monk 1971), and to be poisonous to vertebrates, and presumably insects, owing to alkaloids, tannins, and (presumably) other phenols (Forbes and Bechdel 1931; Harshberger 1916; Muenscher 1970; Pammel 1911; Southwood 1972). Heinselman (1963) described peat bog litter as "a variable mixture of complex organic compounds including celluloses, lignins, autins, waxes, resins, alkaloids, and pectins. . . ." Furthermore, "species of *Sphagnum* are unique among the peat building plants . . . in the strongly acid reactions of their remains (pH 3.0 to 4.5), and in their consequent resistance to microbial attack" (Heinselman 1963). The list of medicinal plants gathered by drug companies from the New Jersey pine barrens (a blackwater peat-bog swamp on coastal sand soil) reads like a medicine man's warehouse: emetic, purgative, stimulant, febrifuge, astringent, acrid stimulant, sialagogue, errhine, bitter tonic, diuretic, nephritic, poisonous, demulcent, emmenagogue, irritant, rubefacient, carminative, nervine, insecticide, diaphoretic (Harshberger 1916).

HYPOTHESIS 2. Why should a high input of phenolics lead to a litter community of reduced competence, thereby allowing a high rate of organic movement into the streams? A variety of litter microorganisms can degrade phenolics ("humic acids") in soil and water (Alexander 1964; Evans 1947; Foster 1949; Henderson 1957; Hurst *et al.* 1962; McConnell 1968). However, they do it slowly, require a well-oxygenated substrate, and require accessory energy sources (Burges 1965). As Burges (1967) put it, "the decomposition of the phenolic polymers, lignin and humic acids, is not well understood. As yet no one has demonstrated an enzyme system capable of decomposing these substrates," and "humic acids under North American prairie soils have a mean residence period of about 1000 years." Fungal degradation of associated material is a common way of obtaining pure lignin (e.g., Brown *et al.* 1967). Handley (1954) concludes that mor soil (characteristically having phenol-rich litter that decomposes slowly and has a very reduced litter fauna) is "normally associated with a vegetation relatively rich in phenolic compounds and of low base content which gives rise to a relatively acid litter layer." When speaking of the Russian taiga (coniferous forest with an understory of mosses, lichens, and ericaceous shrubs), Ponomareva (1969) stressed that the plant residues "have a high content of compounds which do not decompose readily—lignin, wax, resin, etc., and which include the 'inhibiting' substances (tannins, terpenes, etc.). The latter do not decay readily, and inhibit the decay of all other organic substances . . . there is a low level of microbiological activity, and the biological cycle of elements is slow. The main result is an accumulation of forest litter at the surface, and its being leached by atmospheric waters of mobile organic compounds." She even says (*ibid.*: 257) that like tropical soils, "the northern podzols of Russia, which under natural conditions are quite often overgrown with high-grade coniferous forests . . . after deforestation require systematic fertilizing to grow annual crops."

Since the selective pressures that result in secondary compounds favor highly toxic chemicals that are difficult for animals to digest, it is not surprising that they appear to be difficult to degrade. Newly fallen tannin-rich leaves (e.g., Fagaceae such as beech, oak, chestnut) have to be leached by rainfall for many weeks to the point where the litter community will readily feed on them (Anderson 1973; Burges 1967; Cartwright and Findlay 1943; Heath and King 1964; Lutz 1928; Nykvist 1963; Shanks and Olson 1961). It is not surprising to find these compounds in soil (e.g., Burges 1965; Burges *et al.* 1964; Whitehead 1964) and to find them having a negative effect on the soil and litter organisms. They can have substantial negative effects on mycorrhizae (Brian *et al.* 1945; Gimingham 1972; Harley 1952), bacteria (Alexander 1964; Burges 1965; Corbet 1935; Kendrick and Burges 1962), whole plants (Yardeni and Evanara 1952), roots (Grümmer and Beyer 1960; Sherrod and Domsch 1970; Waksman 1938; Wang *et al.* 1967a,b), free-living fungi (Cartwright and Findlay 1943; Frankland 1966; Lutz 1928; Nierenstein 1934; Nykvist 1959), vertebrates (Arnold and Hill 1972; Glick and Joslyn 1970; Longhurst *et al.* 1968; Tamir and Alumot 1970), soil animals (Anderson 1973; Heath and King 1964), insects (Feeny 1968, 1969; Miles 1969), worms (Burges 1965; Shanks and Olson 1961), and organisms in general (Levin 1971; Nierenstein 1934; Whalley 1959; Whittaker 1970; Whittaker and Feeny 1971). The blackish-brown polyphenol-rich ortstein deep in tropical white sand podzols characteristically lacks roots (e.g., Heyligers

1963; Huinink 1966; Joachim 1935). Heyligers (1963) notes that "on the [white sand] savanna it seems to take 50 years before the charcoal is decomposed, under forest conditions perhaps shorter"; in Costa Rican deciduous forest, charcoal after a fire is generally gone in much less than 10 years. One wonders if the slow growth rate of plants on African and Australian termite mounds and the slow rate of disappearance of these mounds (Lee and Wood 1968, 1971; Nye 1955) might be due to concentrations of secondary compounds in and below the nest. Leaf litter in temperate zones takes longer to decompose under evergreen tree species than under deciduous ones (Shanks and Olson 1961). Burges (1967), Corbet (1935), Farmer and Morrison (1964), Handley (1954), Mohr and van Baren (1954), and Ponomareva (1969) stress that tannins, lignins, waxes, and resins are difficult to decompose (witness amber, Langenheim 1973), and their accumulation may make the soil so acid (down to 2.7 in the A_0 horizon) that it is effectively sterilized.

The effect of these compounds will also be enhanced by 1) the initial low nutrient quality of the soil (making the microorganisms more dependent on the organic litter itself for nutrients), 2) the high acidity of the soil (making life difficult for nitrifying bacteria and causing the immediate loss of basic ions to them through competitive uptake by fungi and higher plants), 3) the low nutrient quality of the litter (owing to selection favoring more intense removal of nutrients from a leaf before discarding it when growing on low fertility sites), 4) the removal of proteinaceous nutrients by complexing with phenols (e.g., Alexander 1964; Corbet 1935; Feeny 1969; Handley 1954), 5) a low rate of input of litter owing to low primary productivity, and 6) the high water content of the soils (owing to flooding or continuous rain in some tropical sites), which minimizes the chance for aerobic metabolism. It is ironic that the soil type least able to deal with a high input of toxic phenolic compounds should be the one to receive the highest input of them. This soil-litter substrate probably receives a very high input of other defensive compounds as well.

Finally, the relative ineffectiveness of the litter community is also suggested by the accumulation of deep leaf layers and even peat bogs (submerged and elevated) on some tropical sandy podzols (Anderson 1959, 1961, 1963, 1964a,b; Ashton 1964; Bailey 1951; Browne 1955; Brünig 1964; Clayton 1958; Coulter 1950; Heyligers 1963; Klinge 1967; Mohr 1944; Mohr and van Baren 1954; Richards 1941,

1963; Stark 1970). To accumulate litter, as in temperate zone peat swamps, podzols, and mors, should require exceptionally toxic litter at the high temperatures and rainfall of the lowland tropics.

II. REDUCED ANIMAL COMMUNITIES

I have asked why blackwaters have high concentrations of organic compounds, and proposed the answer that they drain vegetation that is exceptionally rich in secondary compounds and growing on very poor soil. Further, I suggested that the high secondary compound concentration in these plants is the outcome of selection by herbivores in habitats with exceptionally low productivity. The short answer, then, to the question of why are terrestrial animal communities reduced in blackwater drainage basins is that the productivity is low and that what *is* produced is so rich in toxins as to be largely unharvestable. This section draws together the available information on productivity and animal communities on white sand soils.

I have been unable to locate published information on the gross primary productivity of plants on tropical white sand soils, on the amount of that production that can be harvested by animals, or on the biomass or turnover of animals supported by them. There are, however, some very suggestive cues from agriculture. Richards (1952) and Aitken (1930) state that agricultural attempts on British Guiana white sand soils occupied by *Eperua* forest have completely failed. The forest on Sarawak white sand soil is called "kerangas" which roughly translated means "soil on which you cannot grow rice" (Richards, pers. comm.). Stark (1971b) points out that "the [white] sands of the Amazon Basin do not appear to hold promise as areas rich for agricultural development," that "the Brazil sands had a few species of ants and termites, but no earthworms" (Stark 1971a), and that ". . . mammals are sparse in these forests . . ." (Stark 1971a). Speaking of Suriname Amerindians, Heyligers (1963) says "for their fields they burn mainly forest growing on red loamy sands and heavier soils, rarely on white sand, on which they preferably build their houses." He then comments that in the white sand areas "no earthworms were found." In a study of Welsh earthworm distribution by habitat, Pierce (1972) found none in peat soils and 3 to 7 species in soils less rich in phenolic compounds. Vegetation on white sand soils that has been subject to disturbance is invariably described as stunted, scrubby, and poorly developed (Arens 1963; Eiten 1963; Ferri 1960; Heyligers 1963; Richards 1952; Sioli 1967a; Spruce 1908; Stark 1970, 1971a,b), implying a very slow rate of regeneration. Nye and Greenland (1960) do not even mention white sand soils in their thorough review of tropical shifting agriculture, with the implication being that indigenous peoples have not used white sand soil for agriculture.

It is well known in agricultural practice that plants grown on nutrient-poor soils commonly contain less minerals than those grown on nutrient-rich soils. In Java, Hardon (1937) has shown that the foliage of *Dacrydium elatum* (a gymnosperm, Podocarpaceae) contains more than three times as much ash and about seven times as much calcium when grown on andesitic laterite soils as when grown on white sand soils. *Crudia amazonica* is a caesalpinaceous legume tree common to sites flooded by either black or white water rivers; where the trees are flooded by black water their leaves contain less phosphorous and potassium than where they are flooded by white water (Williams *et al.* 1972). Benzing (1973) found the rainwater stem flow off of *Taxodium* (a gymnosperm growing in pure stands in blackwater swamps on white sand soils in the southeastern United States) to be the lowest in nutrient content of a number of tree species, and noted that the nutrient content of epiphyte leaves (which are obviously growing in a very nutrient-poor site) is exceptionally low. Canadian bog evergreen plants have substantially less minerals in their leaves than do leaves of deciduous forest trees on upland sites (Small 1972). If such data turn out to be representative of wild vegetation, it may well be that, secondary compounds aside, the foliage of vegetation on white sand soil sites is of lower nutrient value per bite to a herbivore than from more fertile sites.

Aside from considerations of primary productivity, what the plant does with what it makes is likewise important here. Mast fruiting at long intervals and production of exceptionally toxic seeds (see section III and discussion) lower the productivity from the animals' viewpoint. In the case of mast fruiting, a major source of foods for rainforest animals is absent for intervals of several years. There are two reasons why foliage and seeds that are rich in secondary compounds should be of lower value even to the specialist that can feed on them. First, in detoxifying the compounds, the specialist will have to expend some of the nutrients obtained (Freeland and Janzen 1974); and witness the requirement of bacteria for an additional energy source if they are to decompose tannins (Burges 1965). The "conventional wisdom" that pesticide-resistant strains of insects do not survive in competition with wild types is highly suggestive of this

phenomenon. Second, a higher proportion of the plant's primary productivity is going into defensive compounds rather than material that can be easily assimilated. As Kira and Shidei (1967) have emphasized in their review of tropical primary production, two well-developed rainforests of the same biomass may be produced by widely different rates of primary production; even the tall and mixed dipterocarp forest growing on the white sand hillsides at Bako (see below) can be produced and maintained by very low primary productivity and therefore have very little harvestable productivity for animals.

In addition to the above indirect evidence, three recent studies in Sarawak partly document the very low animal biomass and suggest low harvestable productivity of vegetation growing on white sand soils.

BAKO: The habitat that I examined ranged from kerangas forest (with its secondary, or "padang" vegetation) on the top of 100 m tall and flat-topped sandstone hills, down to peat swamp forest in the valley bottoms near sea level (Bako National Park, on the coast of the South China sea, near Kuching, Sarawak, Malaysia; see Anderson 1963; Ashton 1971; Brünig 1965, 1968, 1973; Janzen 1974 for a more detailed description). On the equator (1°6' N, 110°5' E), Bako receives 3 to 4 m of rain, which is distributed quite evenly throughout the year. The scrubby and obviously disturbed (see below) "padang" vegetation on the hilltops was 1 to 4 m tall and contained nearly pure stands of the small tree *Ploiarum alternifolium* (Bonnetiaceae), with taller scattered shrubs and trees such as *Tristania, Whiteodendron,* and *Rhodamnia* (Myrtaceae), *Calophyllum* (Guttiferae), *Cratoxylum* (Hypericaceae), and *Dacrydium* (Podocarpaceae). When undisturbed, the hilltops supported a dense forest of 10 to 20 m tall trees in these families and those listed in the previous section for vegetation on white sand soil (Brünig 1968, 1973). Interspersed with the secondary vegetation were large areas (several were more than one-half hectare) of bare black hardpan, which was a B horizon left after erosion of the sand; this black ortstein was more than 1 m thick in places (fig. 1). The hillside forest looked like ordinary lowland tropical rainforest (= mixed dipterocarp forest of Brünig 1973) and was rich in Casuarinaceae, Dipterocarpaceae, Euphorbiaceae, Guttiferae, Lauraceae, Myrtaceae, and Podocarpaceae, and graded into the peat swamp forest described by Anderson (1963) in the valley bottoms.

1) Despite the total (and apparently effective

FIGURE 1. (A) Interface between hardpan (ortstein) exposed by erosion of white sand soil, and kerangas forest on the tops of the hills at Bako National Park. (B) 60-cm-tall ortstein boulder left behind following erosional fragmentation of a continuous hardpan such as that in A above. The vegetation around the boulder represents 30 years of undisturbed secondary succession following clearing by fire and cutting.

269

prohibition of hunting in the Park for at least 15 years (Anderson, pers. comm.)), small vertebrates were extremely rare. In 11 days of working in the forest and along its edges, I saw only one arboreal (Geckonidae) and one terrestrial (Agamidae) insectivorous lizard, despite intensive search for 4 days. In structurally comparable vegetation and weather in Central America, the count would have been in the hundreds. During the same period, each morning I saw only 3 to 5 birds of all sizes in the forest and along its edges; birds were so scarce that each one caused special notice. At a similar site in Central America, the bird count would easily have been in the hundreds. The absence of bird calls was deafening; Ashton says that kerangas forest can be defined by its silence (Richards, pers. comm.). "Most of the forest areas of Bako appear to have a rather impoverished bird fauna, particularly so the *kerangas* areas" (Rothschild 1971).

Each day at the site, I walked about 5 km through secondary and primary forest on a wet white sand path (fig. 2A); in 10 days I saw the tracks of only one large felid and no other mammals. No rodent runways were found in the old field mentioned below, and the only sign of small mammals in the forest was two nuts that had been chewed open by a rodent. One squirrel was seen in the forest in 11 days. Harrison (1965) says of Bako, "the observer may find less than one mammal per 5 acres, and only perhaps 1 bird per acre, with frogs and reptiles correspondingly rare." For 3 days a 200-gram piece of canned ham remained untouched on the side of the creek through the old field. Aside from the pseudobulbs of orchids, in the Central American tropics, fleshy (starchy) storage tubers are found above ground on plants only in habitats where rodents are very rare: tops of high mountains (e.g., *Macleania* (Ericaceae) in Costa Rican cloud forest), and on Caribbean islands. At Bako, such tubers were conspicuous on three genera of epiphytic ant-plants and several species of vines (Janzen 1974).

Pieces of canned meat placed as ant bait in the hillside and secondary forest sometimes lasted as long as 24 hours without being found by ants (Janzen 1974); in any lowland Central American forest, such baits would remain undiscovered only a few minutes (unpub. field notes). In Suriname, Heyligers (1963) says "in the white sandy soils, especially in those of the open savanna and of the savanna scrub, animal activity is very restricted . . . in savanna wood and in savanna forest sometimes low broad sandheaps are found round the entrances of the nests of leaf-cutting ants, *Atta* spp. These ants,

however, are more often met in forest on red sandy loam."

Though no sweep samples of foliage-inhabiting insects were taken, my past experiences following a sweep net about in tropical vegetation (Janzen 1973a,b; Janzen and Schoener 1968) lead me to believe that the biomass of foliage-inhabiting insects would be about the same as on a Caribbean island or a 3000 m Costa Rican mountain top, i.e., 1 to 10 percent of that in the Costa Rican lowlands (assuming that comparisons are made between similar ages and types of the vegetation). At night, insects on the foliage were almost non-existent at Bako. I found only two termite nests in the vicinity of the site; similar searching in lowland Costa Rica would have yielded hundreds of termite nests. Flower-visiting insects were conspicuously rare even on the large flowers of *Ploiarum* (the commonest shrub in the secondary regeneration). I saw only one *Xylocopa* and two *Nomia* bees in 11 days. Only one social wasp nest was seen in the same time period, and that was in a mangrove swamp. At least two of the most common trees were wind-pollinated gymnosperms (*Casuarina, Dacrydium*). The prevalence of wind pollination, coupled with supra-annual flowering and fruiting by the common Dipterocarpaceae, indicates a distinct shortage of sugar and edible plant parts for both litter organisms and larger animals, as compared to an African or neotropical forest where the annual fruit input can be very high (Klinge and Rodriguez 1968; Smythe 1970). For example, *Spondias mombin* (Anacardiaceae) may produce as much as 5 tons (fresh weight) per hectare of fruit and seed in its annual fruiting in Panamanian rainforest (Hladik and Hladik 1969).

However, very large vertebrates were present, at least near the coastal river mouths. Brush-lipped pigs (*Sus*) rooted up the disturbed vegetation on delta clay soils near the Park buildings. A troop of macaques (*Macaca*) foraged intensely around the Park buildings, and a troop was seen twice in the hillside forest. Proboscis monkeys (*Nasalis*) have been seen in the mangrove swamps of the Park, and orangutangs (*Pongo*) have been trapped in the Park in the past (Anderson, pers. comm.). I saw one large deer (Cervidae) and one hornbill (Bucerotidae) near the beach. It is noteworthy that these are animals that can move long distances between local excesses of food and have a reputation for doing so (see section III), and probably can survive for long periods with little food (Bourliere 1973). A large varanid lizard foraged in the Park building garbage, and another was seen on the beach scavenging among beach litter. The above account would

FIGURE 2. (A) In front of Dr. Morrow is kerangas forest lightly disturbed a long time ago and in the foreground is 30-year-old secondary succession following clearing by burning and cutting. The (Lintang) trail was last scraped clean of vegetation four years earlier and has no regeneration despite use by no more than one or two persons a month. The site is about 500 m east of the site in figure 1A. (B) Looking to the right a few meters behind Dr. Morrow through 30 years of undisturbed secondary succession. The X is two meters above the ground.

be far richer in species and individuals in a similarly undisturbed rainforest site in the Central American lowlands.

2) The rate of regrowth of destroyed vegetation was extremely low. Patches totalling about 30 hectares of hilltop forest had been cut and burned to ground level, but not plowed, planted, or grazed, about 1942-1944 (information from local inhabitants near the Park); by 1972, the regeneration was only 1 to 2 m tall, with scattered trees as tall as 3 to 4 m (fig. 2). Heyligers (1963) describes a dirt road built in the 18th century across a white sand savanna: "Where this ancient road passes through the savanna, it is still easily recognizable, but where it penetrates into the forest, [on loams and latosols] it is entirely overgrown."

The low rate of secondary succession is probably due to a low immigration rate of tree seeds (owing to a lack of dispersal agents) as well as to a low growth rate of established plants. In African and neotropical evergreen lowland forests on latosols, there are numerous fruit- and seed-bearing shrubs and vines in the understory and forest edges. They fruit through much of the year, e.g., Heliconiaceae, Marantaceae, Melastomataceae, Palmae, Passifloraceae, Piperaceae, Rubiaceae, Solanaceae, Tiliaceae, etc. (and see Snow 1962, 1965), and one can find a couple of dozen species of forest trees in fruit at any time of the year, to say nothing of the peak fruiting times such as described by Foster (1973); Frankie *et al.* (1974); Janzen (1967); and Smythe (1970). At Bako there were almost no plants in fruit or flower in the forest and on its edges, implying an overall lower rate of seed generation by the forest, epitomized by the highly intermittent one of the Dipterocarpaceae. This absence of fruit and seed crops is probably one of the causes for the general absence of birds and small mammals in the forest, since fruits and seeds intended for proper dispersal agents constitute a large part of the diet of seed predators.

3) Human use also implies a low productivity by the habitat. Prior to establishment of the Park about 1959, paddy rice was grown only on the deltaic river bottoms near the ocean. While the art of growing dry-land rice on forest soil is well developed in Borneo (Freeman 1970), the farmers at Bako never cleared the forest on white sand hills for agriculture.

4) Several aspects of the extant vegetation also suggest low primary productivity. Lowland evergreen rainforest in the new world tropics is generally dotted with trees just putting out an entire new leaf crop or with flushes of new leaves in crowns already filled with older leaves. The evergreen forest in the Park

had the lowest incidence of such leaves I have ever seen in lowland tropical rainforest, and this was true for the understory as well as the canopy.

The life forms of the plants also suggest slow growth by the vegetation. The success of the vine life form clearly depends on its ability to grow rapidly in length. The biomass of vines is disproportionately reduced in tropical habitats where the absolute growth rate is reduced for all plants, such as at elevations above about 2000 to 2500 m, and in deserts. Vines are likewise extremely scarce at Bako, even in the secondary vegetation and forest edges. Their general absence on white sand soil sites has also been noted by Richards (1952). Furthermore, annual plants were almost entirely missing from the secondary succession. These small plants likewise depend on fast growth for survival in the face of competition with woody plants (usually in the early stages of succession). The absence of annuals on the white sand soils was especially conspicuous because at least 5 percent of the soil surface in the old cleared area was free of vegetation, despite the long time since the last disturbance. It is interesting in this context to note that while tree foliage commonly contains tannins, the foliage of annuals and other herbaceous plants is generally free of tannins (Bate-Smith 1973b; Feeny, pers. comm.).

5) The only fern and angiosperm epiphytes that could survive on plants growing on white sand soil at Bako were those with an exceptional method for obtaining food. Almost all epiphytes were associated with an *Iridomyrmex myrmecodiae* ant colony (fig. 3): four of the epiphytes (*Dischidia, Hydnophytum, Myrmecodia, Phymatodes*) were grown and fed by the ants, and the remaining epiphytic associates parasitized the system by rooting in the myrmecophytes' food (Janzen 1974). I found only 5 individuals of non-mutualistic fern epiphytes. These (e.g., *Drynaria*) had leaves modified for capturing falling leaves and other debris, and/or for harboring ants.

I interpret the shortage of ant-free epiphytes to be a consequence of a low rate of inorganic ion input to tree crowns from bird feces, insects, rain, dust, and falling leaves and twigs. Ungemach (1969) postulated that the extraordinarily low amount of plant nutrients in the rainwater falling in the Rio Negro river basin in Brazil was due to the low nutrient content of the soil. The postulated high concentration of phenols and other secondary compounds in leaf and bark leachate may also be of importance in regard to absence of epiphytes. The tannin-rich young bark of mangroves (e.g., Burkill 1935) is extraordinarily free of epiphytes in the neotropics even when the soil adjacent to the mangrove swamp is of high quality. While not epiphytic, the extreme abundance of the insect-capturing pitcher plants (Nepenthaceae) and sundews (Droseraceae) also implies a strong competitive advantage for terrestrial plants with special nutrient-gathering devices.

FIGURE 3. (A) *Hydnophytum formicarium* Jack, myrmecophytic epiphytes growing on a tree in the kerangas forest in figure 2A. (B) *Myrmecodia tuberosa* Jack, another myrmecophytic epiphyte in the kerangas forest in figure 2A, and its ant occupants (*Iridomyrmex myrmecodiae* Emery). These ants appear essential for epiphyte survival in this nutrient-poor habitat (see text and Janzen 1974).

The conspicuous paucity of legumes at Bako is somewhat perplexing in this respect, except that acid soils appear to be particularly harsh for root-nodule bacteria (Alexander 1964). Brünig (1969c) states that "mycorrhiza and root nodules are conspicuously more frequent in [Sarawak] tropical podzols than in red-yellow podzolic or lateritic soils." Ashton (pers. comm.) disagrees with Brünig on this point, but Stark's (1971a, b) studies that emphasize the importance of mycorrhizae in tropical forest were done on Brazilian podzols (and see Huinink 1966; Klinge 1973). Stark (1970) also states that mycorrhizae appeared to be more abundant in Suriname white sand soil that in adjacent lateritic soils. However, Norris (1969) found no root nodules on *Eperua* and *Mora* (Leguminosae) on sand soils (but they did have well-developed mycorrhizae). As well as becoming increasingly important as soil nutrients become rarer, an association with phenol-resistant mycorrhizae might be one of the specializations possible for a tree with phenol-sensitive root hairs. Plants that shed large amounts of phenols and other secondary compounds onto their roots may need mycorrhizal associations to avoid self-intoxication. The mycorrhizal fungi, which are functionally serving as root hairs for the plant, may be specialists at resisting or avoiding these compounds. We know that some fungi do live on tannin-rich substrates. In this connection it is of interest that not only do *Mora, Eperua,* and *Pentaclethra* have thick roots with extensive mycorrhizal associations, but *Hymenaea courbaril,* which is famous for heavy resin deposits beneath the single trees (see section IV), does likewise (Janos, pers. comm.). It is probably not an accident that Stark developed her direct nutrient-cycling hypothesis in white sand with a phenol-rich litter.

6) Rain pools and creeks running off the sandstone hills and hardpan were virtually devoid of animal life. No fish were seen in the creeks at Bako, and no riparian fishing birds such as herons were seen near them. I found 2 dragonfly larvae (Odonata) and a few caddisfly larvae (Trichoptera) in hours of turning rocks and searching stream bottoms. When it rained, the streams produced thick mats of white foam at the bases of waterfalls, suggesting a high saponin content in addition to whatever organic compounds turned them dark brown.

7) The very large numbers of individuals of a few plant species (low species richness and low equitability of species abundances) in the secondary succession (and see Brünig 1973 for a description of equally low values for undisturbed forest) suggest a very harsh environment (and see below with

respect to reptiles and amphibians). Shortage of food for both plants and animals is probably the cause. In the secondary succession, there were many acres where more than 80 percent of the vegetation biomass was tied up in one species of sedge (Cyperaceae), *Tectaria borneensis* (Aspidiaceae), *Ploiarum alternifolium* (Bonnetiaceae), *Nepenthes gracilis* (Nepenthaceae), and ant-plants (primarily *Myrmecodia* and *Dischidia*), and the remaining vegetation contained a maximum of about 25 species of plants. *Iridomyrmex myrmecodiae* was the only arboreal ant found in the secondary succession by baiting and collecting, and its colonies occupied or foraged over at least 90 percent of the woody vegetation. On the ground, the same ant and a large black *Crematogaster* occupied 96 percent of the foraging area (as determined by their harvesting of baits placed on the ground, Janzen 1974). In the forest, the distribution of ant biomass was almost as badly skewed in favor of *Iridomyrmex myrmecodiae* and *Camponotus* as in the open.

Numerous workers (e.g., Ducke and Black 1953; Heyligers 1963; Klinge 1968; Richards 1952; Takeuchi 1961, 1962a,b) have stressed the same low plant species richness and equitability on South American white sand podzols, even in apparently undisturbed forest.

FOURTH DIVISION, SARAWAK: As part of herpetological studies in Borneo (Lloyd *et al.* 1968 and included references) Robert Inger collected frogs, lizards, and snakes in forest on white sand soil with blackwater creeks (Nyabau), hill forest (Pesu), and alluvial forest (Labang). The podzol site has "low forest with a thin canopy . . . typical heath forest as described by Richards (1952)" while the others are "tall with a dense canopy" and produce streams with clear or turbid water (Inger, in litt.). Working with a team of collectors, Inger obtained an average of 20.5 and 24.1 reptiles and amphibians per day at Labang and Pesu during 120 and 131 days of collecting, respectively; at the white sand site, they caught 17.8 per day. While these differences in numbers of individuals do not appear dramatically different, analysis at the species level is much more revealing. On the white sand site, there were 24 frog, 13 lizard, and 13 snake species, while at Labang and Pesu, the comparable figures are 33, 33, and 38 (plus 2 turtles) and 53, 31, and 35 (plus 1 turtle), respectively. Labang and Pesu, with 106 and 120 species of herps respectively, are more than twice as species-rich as Nyabau (50 species). It is probably significant that snakes, representing tertiary consumers for they eat almost entirely vertebrates (e.g., Arnold

1972), are the most reduced of all on the poor site. Frogs were the least severely reduced in numbers of species. If I may use my own unpublished experience with gut analyses of Central American frogs and lizards for documentation, frogs normally eat much smaller amounts of insects per unit body mass than do lizards, suggesting that frogs may maintain a higher biomass or species richness on less food than lizards.

The habitat specificity of frogs, lizards, and snakes was also instructive with respect to productivity. Of the 50 species in the white sand habitat, 92 percent are found in one or both of the other two habitats. Of the 106 and 120 species at the other sites, only 37 and 33 percent were found in the white sand habitat. This is precisely the trend of habitat-fidelity that would be expected of a carnivore if the white sand site has a much lower productivity than the other two sites. The different trophic levels show the same trend; while 74 and 77 percent of the snake species at the good sites were not found in the poor site, only 64 and 65 percent of the lizards and 48 and 60 percent of the frogs at the good sites were missing from the poor site. Again, the direction of this difference between frogs and lizards is expected if we recognize that frogs probably need fewer insects per animal per day than do lizards. It is of interest in this context that many of the trees found on white sand podzols are also found growing on yellow podzol and latosols, but the reverse is not true for many of the species that grow on the more fertile sites (Ashton 1964; Richards 1936).

PEAT SWAMP FOREST: The peat swamp forests in Brunei and Sarawak are notorious for their lack of conspicuous animal life, and in the center of a large swamp, there is almost complete silence. The overstory trees, primarily the dipterocarp *Shorea albida* (Anderson 1964a; Brünig 1964), are stunted. Dead snags are conspicuous, suggesting a slow rate of dead wood turnover, and when lightning strikes in the center of this vegetation the dead trees may stand for many years (Anderson 1964a; Brünig 1964; Richards, pers. comm.); a photograph in Anderson (1961) shows that one third of the *Shorea albida* trees killed by insect defoliation in a peat swamp are still standing 10 years later.

At Bako, there were small peat swamp forests slightly inland from the river deltas. A night collecting trip in this forest for animals of any sort or size was generally a waste of time. Calling frogs were very rare, and insects were seldom encountered on the foliage. In Central American swamps lacking peat accumulation, animal life is very abundant. It seems safe to assume that a peat swamp represents hardly more than an extremely concentrated black-water river, where the breakdown of secondary compounds is further impeded by lack of oxygen. Such a site should display extremely low productivity harvestable by animals.

III. MAST FRUITING BY DIPTEROCARPACEAE

I have proposed in the previous two sections that leaves with exceptionally high concentrations of defensive chemicals are a consequence of the strong selection for anti-herbivore defenses in habitats with a climate favorable to animals yet with very low primary productivity, which in turn leads to low numbers of animals. In this section I wish to explore the hypothesis that mast fruiting by Dipterocarpaceae may be an indirect consequence of low productivity and low animal numbers. I will restrict my discussion for the most part to Sarawak and Malaya, as these are the only areas with reliable information on fruiting phenology.

Mast fruiting (called "gregarious fruiting" in much southeast Asian literature) by Dipterocarpaceae is behaviorally very similar to the supra-annual sexual phenology displayed by many North American and European canopy-member trees (e.g., *Abies, Carya, Fagus, Pinus, Quercus, Tsuga*). The latter trees have population- and community-level fruiting that is synchronized at intervals of greater than one year. The adaptive significance of being synchronized to the individual tree is clearly the advantage of fruiting when seed predators have been satiated (Janzen 1971). In Sarawak, Brunei, and southern Malayan lowland and hill evergreen rainforest, it is well known that most individuals in many dipterocarp genera (e.g., *Cotylelobium, Dipterocarpus, Dryobalanops, Hopea, Parashorea, Shorea, Vatica*) fruit synchronously over tens of square miles at 5- to 13-year intervals (Ashton 1964, 1969; Burgess 1968, 1969; Fox 1967, 1968; Medway 1972a; Meijer 1970, 1973; Wood 1956). For example, Wood (1956) reported that more than 100 species of Dipterocarpaceae flowered, and presumably fruited, in Sabah, Brunei, Sarawak, and Indonesia in 1955. While detailed data are not available, Ashton (1969) stresses that "along with them many other families flower unusually heavily."

Individual temperate-zone mast-fruiting trees that fruit out of phase are heavily selected against (Janzen 1971), and the same happens to dipterocarps that are out of phase. "Sporadic flowering and fruiting of economic species [of Dipterocarpaceae] occurs

somewhere in the [Malay] peninsula every month, but it is seldom of much value and the fruit is likely to be almost completely destroyed by insects, birds or animals" (Wood 1956). Ashton (pers. comm.) has stressed that insects and vertebrates take a very heavy toll of dipterocarp seeds except in mast years. I should point out, however, that just as we can no longer examine the coevolution of the passenger pigeon (the primary acorn predator) and mast fruiting by oaks in North America (Janzen 1971), the reduction of big mammal populations in Malaya and Borneo by hunting makes this question very difficult to examine directly.

Physiologically, mast fruiting probably involves little more than storage of photosynthate until it accrues to the amount needed to produce a seed crop of a size that accomplishes seed predator satiation. Once this level is reached, we may expect the tree to become physiologically sensitive to an external weather event that will be likewise perceived by other dipterocarps. This flowering cue appears to be a period of several rainless weeks in the case of Sarawak, Brunei, and Malayan rainforest dipterocarps (Ashton 1964, 1969; Medway 1972a; Meijer 1970; Poore 1964, 1968; Wood 1956). On an annual basis, a milder version of this same cue is used by some Malaysian bats to synchronized reproduction (Medway 1972b). With such a system we may expect cyclic fruiting at long intervals, even when notable dry spells occur at more frequent intervals. It is not surprising, then, to find, as Wood (1956) did, that fruiting years are not drier than most dry years. Ecologically and evolutionarily, the question is much more complicated than physiologically. We are confronted with four interrelated questions. 1) Why is population- and community-level mast fruiting an effective reproductive strategy for trees in these particular tropical forests? 2) Why do these trees store photosynthate for such a long time before responding to the flowering cue of a dry spell? 3) How did this behavior evolve? 4) Why did this defense mechanism appear in the Dipterocarpaceae, rather than in some other family of trees?

1) Why is mast fruiting effective in Malayan and Bornean rainforest? The short answer has to be that the animal community is in some sense smaller there than in other lowland tropical forests, such as those of Africa and the neotropics. In these latter areas, there are no reported cases of population- or community-level mast fruiting and no suggestion of their existence from the field experiences of tropical biologists whom I have consulted. Seed predator satiation occurs at the level of the individual tree's seed crop and annually at the population level in all

tropical areas, but southeast Asia is unique in having a large number of tree species involved in predator satiation by fruiting synchrony at greater than one-year intervals. The critical trait of a mast crop is that there must be so much seed that all the seed predators that find it cannot kill the seeds before some are dispersed or grow to a relatively immune stage. In temperate-zone forests, this excess of seed over animal destruction capacity occurs because inclement weather takes a heavy toll of the animal populations between large seed crops. Furthermore, because there are a small number of tree species, most individual trees are involved in mast fruiting, and therefore there are few seeds and fruits available for animals in the years between crops (Janzen 1971). In most lowland tropical forests the situation is reversed. Perhaps most important, no one tree species makes up so much of the community that it could satiate the animal community even if it were synchronized on a supra-annual basis. However, in most dipterocarp forests, either a few species, or many species in one family, constitute much of the community of adult trees.

In the temperate zones, trees on the adaptive peak of community-level predator satiation tend to have not only large seed crops, but highly edible seeds (Janzen 1971). Presumably this situation is because their fitness is higher when they produce a large number of edible seeds rather than a smaller number of slightly toxic seeds, as opposed to being on the other adaptive peak of a small number of very poisonous large seeds and satiating only specialist seed eaters that can detoxify them, as is the case with neotropical lowland forest trees on poor soil (see below). The situation is also complicated by the fact that for many mast-cropping trees in the temperate zone, some major seed predators are also dispersal agents.

Malaysian dipterocarps likewise have highly edible seeds (Browne 1955; Meijer 1969). Browne reports that on mast years in the 1940's and 1950's, 5 to 16 thousand tons of dipterocarp seeds were exported from Sarawak after being hand-gathered from the forest by indigenous peoples. They are also eaten locally, and high-quality nuts contain as much as 50 percent oil; I have found no suggestion that they contain a dramatic toxic principle, though they have a slightly resinous pericarp and may have a slight tannin content (as do acorns). Dipterocarp seeds are fed on by a wide variety of vertebrates and insects (Poore 1968; Wood 1956). However, Medway (1972a,b) very enigmatically reports that the seeds of many Malayan dipterocarps were not eaten by vertebrates at one rainforest site during

two different mast crops in Malaya. This finding suggests that either predator satiation was very effective, due perhaps to the fact that major seed-eating vertebrates such as elephants have been severely decimated, or the dipterocarp seeds are low on the vertebrates' preference list. Bearing in mind what an elephant, rhinoceros, tapir, or herd of pigs or cattle could do to the seed crop of a single dipterocarp tree, I prefer the former explanation. Harrison (1965) stresses that some of the "commonest and largest animals are predominantly or wholly frugivorous." I predict that if the oil-rich dipterocarp seed crop were dropped into an undisturbed African or neotropical rainforest growing on latosols (Bourliere 1973 reports a substantial mammal biomass in these forests), there would be no seed survival.

Is there evidence that the biomass of the seed predator community is smaller in the Malaya-Sarawak-Brunei rainforest than in African and neotropical lowland rainforests? The very fact that numerous large and highly edible seeds can lie on the forest floor long enough to generate the dense stands of dipterocarp seedlings characteristic of these forests is very suggestive. The fact that a considerable portion of the forest's seed production occurs at long intervals also implies that the animal community must be reduced, owing to the lower harvestable productivity in the "off" years. Direct evidence for low productivity would be more convincing, but except on white sand soils of Sarawak, there is none. Much of the dipterocarp rainforest is not on white sandy podzols or peat bogs. However, it is of interest that all but one of Brünig's (1973) sites used to calculate the overall species richness of trees for Borneo are sandy podzols. Therefore, on the plant side of the ledger, the evidence is very weak indeed that overall Malaysian harvestable productivity is low. About the only indication is that I and several other observers (Anderson, pers. comm.; Kellman, pers. comm.) have noted that the biomass of vines in secondary vegetation is conspicuously reduced in Malaya and Sarawak as compared to the neotropics. There is little doubt that Malaysian soils with dipterocarp forest are as poor as the latosols under other tropical lowland forests, but we don't know if they are poorer.

However, there are several suggestions that animal communities in Malaysian rainforest overall have reduced biomass as compared with large areas of African or neotropical rainforest. McClure (1966) reports a large number of species of seeds of Dipterocarpaceae and other families falling in small amounts on the Malayan rainforest floor without being eaten by vertebrates; there is no mention of the degree to which the local animal community has been decimated by humans, but in Costa Rica, at least, even with hunters present, it would be impossible to report as McClure did. Inger (Lloyd *et al.* 1968 and included references) found much less biomass of cold-blooded vertebrates in all types of Sarawak rainforest litter than Scott (pers. comm.) found with the same methods in Costa Rican rainforest on latosols and volcanic soils. Among Sarawak amphibians and reptiles, Lloyd *et al.* (1968) also found low overall inequitability and increasing inequitability in the species abundance in the progression of frogs to lizards to snakes; this result is expected in an environment with low productivity. Borneo and Malaya have areas of about 780,000 and 215,000 square kilometers and have 91 and 80 species of frogs (Inger 1966). For contrast, Costa Rica with only 55,000 square kilometers has 100 species (Scott, pers. comm.). It is my opinion that the number of kinds of habitats in Costa Rica is not conspicuously larger than in Malaysia. Malaya has only 169 species of mammals and 306 species of birds (Harrison 1963) while Costa Rica has 175 species of mammals (Wilson, pers. comm.) and 758 species of birds (Slud 1964). Malaysia contains no rodents analogous in abundance and behavior to the major large seed eaters encountered very frequently on the ground in neotropical lowland rainforests (*Cuniculus, Dasyprocta, Proechimys*), though it does have a large number of species of *Rattus* and some other small rodents. Richards reports that wild game was a much larger part of the expedition's diet in British Guiana white sand sites (Davis and Richards 1933, 1934) than in Sarawak (Richards 1936; Richards, pers. comm.). Wild pigs (*Sus* spp.) in Borneo are locally highly migratory (see below), implying that there are periods when they are absent from large areas; this situation should result in an average low density of these major seed predators.

2) Why do dipterocarps wait so long between fruiting? The length of time between fruit crops is a very important parameter in satiation of seed predators. With only a couple of years between fruiting, the majority of insects that might become specialists on dipterocarp seeds are probably avoided and vertebrate populations are reduced about as far as they can be. Medway (1972a) and others have stressed that dipterocarp seeds have no host-specific insects that feed on them. However, the longer the wait between crops the larger will be the crop, and the more likely that the tree will satiate the local seed-eating animals and those that can migrate into the habitat at the time of fruiting (e.g., deer, wild ox, pigeons, pigs, people). It is of interest

to note in this context that there are 11 species of large fruit- and seed-eating pigeons in Borneo and only 5 parrots (Smythies 1960), while Costa Rica has 6 pigeons and 16 parrots (Slud 1964). Tropical pigeons commonly fly very long distances to feed on mature fruits and seeds, while parrots tend to feed locally on a very wide variety of immature seeds and mature fruits and would appear to require a steady high level of fruit availability on a local basis. These differences are reflected in the role of the passenger pigeon and European wood pigeon as major fagaceous seed predators, while the Carolina parakeet penetrated U.S. temperate-zone forests to only a small extent. Pigs in Sarawak and Malaya were famous for mass migrations prior to their exploitation by hunting (Anonymous 1953; Medway 1969; Shelford 1916). Pig migrations "occur in different places, and further information may link them with Illipe nut or some other dipterocarp seven year cycle" (Anonymous 1953). Ironically, the lower the primary productivity of the site, the longer the tree will have to store photosynthate before it has enough for a large seed crop. The variation in mast cycles from area to area may well reflect this phenomenon. In Malaya, where white sand podzols appear to be a relatively small part of rainforest soil, the period is about 6 years (e.g., Medway 1972a; Wood 1956). In Sarawak and Brunei, where white or red-yellow podzols and peat swamps are dominant soil types (e.g., Brünig 1973), periods of 9 to 11 years between heavy mast crops appear standard (Ashton 1964; Meijer 1970) though they can be shorter in some areas (Browne 1955) and may be as long as 21 years (Wood 1956).

Throughout this discussion we should note that the selection pressure for community-level synchrony is operating on fruiting time rather than flowering time. Viewed in this way, the several-month range in flowering time (Ashton 1969; Wood 1956) becomes more understandable. Different species probably have different seed-maturation times, and flowering should be synchronized only within the species to enhance outcrossing. In fact, there may well be selection for slight interspecific separation of flowering times through competition for pollinators and through competition between pollinators. Wood (1956) states that "despite the disparity in flowering times in most of the genera, the tendency was for the fruits of those that flowered late to develop faster than those that flowered earlier, so that except in a few species of *Vatica* the fruits mostly fell between mid-August and mid-October."

However, there appears to be a fair amount of

error in timing of flowering by dipterocarps (Mc-Clure 1966; Medway 1972a; Wood 1956). I suspect that this is a straightforward expression of the fact that it is impossible to have an external environmental cue that will not be damaging, yet will be sufficiently dramatic to be sensed equally by all individuals growing on sites of variable drainage, and by individuals of variable age, health, competitive status, past history, and genetic programming. Since the latter kinds of variation also occur among the branches within a single tree crown, we may also expect flowering errors to be erratically distributed in the crowns themselves. We may also expect species to have different ways of dealing with a flowering "error," such as abortion of flower buds or flowers, or even sterile fruits. All of the above responses have been reported for dipterocarps (Medway 1972a; Wood 1956).

3) How did mast fruiting evolve in a tropical rainforest? As hinted earlier, we can recognize two quite distinct possible starting points in the evolution of mast fruiting in a lowland tropical rainforest. a) It could start with any one of the many species in a species-rich forest. In this unlikely case, the site would have to have an extraordinarily low biomass of potential seed predators. In view of the active role that seed and seedling predators probably play in maintaining habitats that are rich in tree species but do not have mast-fruiting species (Janzen 1970), such a habitat does not seem likely in the real world. However, if a mast-fruiting species were to immigrate into such a habitat, it has the potential of becoming common at the site owing to its increased numbers of seedlings relative to its neighbors. It should reduce the species richness of the site and increase the inequitability of distribution of trees among the species. I would also expect that this tree would become mutually entrained with other tree species, accentuating the process. Ironically, the larger the animal community, the faster this community-level synchrony should appear, yet the effect of synchrony should be to reduce the animal community by lowering the amount of food available to it for long periods. b) Mast fruiting could start with a species that already occurs in a nearly pure stand (owing to very severe edaphic conditions), such as *Shorea albida* forest on peat swamps (Anderson 1961, 1964a; Brünig 1964, 1968, 1969a) or a mangrove species. I feel that this is the most likely type of starting point, as supra-annual synchrony at the species level would automatically produce community-level satiation of seed predators if the habitat was large relative to other habitats from which animals could mi-

grate. Physiologically, the evolution of mast fruiting by a species would seem to involve little more than the appearance of a mutant strain of tree that requires an annual cue that is slightly harsher than usual to trigger production of flower primordia. Such a tree then produces larger seed crops in the years that it does flower and the system is in operation (Janzen 1971).

Once a mast-crop life form exists in a general region, irrespective of its origin, we may expect its ecological and evolutionary radiation into a wide variety of other habitats. It should be able to invade habitats that are richer in animal biomass than those in which it evolved, and as mentioned above it should result in the reduction of animal biomass in the invaded habitat. The latter result should be the consequence of progressively more intense selection for intra- and inter-specific synchronization as more and more of the individual trees at the site become synchronized. This observation adds significance to Ashton's (1969) discovery that trees other than dipterocarps fruit more heavily in years of dipterocarp fruiting. It is also noteworthy in this connection that fruits with small seeds (e.g., *Ficus*) that are dispersed by passing through the guts of vertebrates are not mast fruiters in the same forest (Medway 1972a).

A community rich in mast-fruiting species should display a number of interesting traits. For example, we should expect some individuals of mast-fruiting trees to act as though they were males. There may be two causes. On some occasions, a tree may respond to a flowering cue yet have only enough stored photosynthate after flower production to produce a small seed crop; it may then "make" an internal physiological decision not to use its reserves for seeds (its seeds might be eaten, for example, by even a single pig that locates the tree) but to save them for the next fruiting cycle. However, if we can assume that some outcrossing has occurred, then even a seed-barren tree has in fact reproduced in the current mast cycle. On other occasions, a particular tree may be growing on a site that is so nutrient poor that it can never make a large seed crop, or it is permanently in bad competitive straits. Here we expect it simply to flower in each mast cycle, hopefully pollinate other trees, and then abort the flowers. Facultative dioecy can very easily evolve in such circumstances. It is interesting that Ashton (1969) reports that dipterocarps occasionally flower without fruiting and that southeast Asian forests contain 20 to 25 percent dioecious tree species. It might be of value to point out that were these trees self-pollinated, there would be no adaptive value to flowering and then aborting the flowers. It is of interest in this connection that the majority of neotropical deciduous tropical forest trees are obligate outcrossers (Bawa 1974). There is no information on the degree of facultative or obligatory outcrossing among Dipterocarpaceae, but one does wonder where the pollinators come from in a mast-fruit year. We can predict that dipterocarp flowers should last more than a day, be accessible to a wide variety of pollinator life forms, and be extremely conspicuous from a pollinator's viewpoint. Temperate-zone mast-fruiting trees are almost without exception wind-pollinated.

A second characteristic expected of forests rich in mast-fruiting species is that they should have an understory with relatively few species of shrubs and small trees that reach reproductive maturity there. This result is because most tree juveniles in such a forest will grow until they die through competition, rather than serving as food when they are seeds for a large animal community (cf Janzen 1972 for a discussion of this phenomenon in pure stands of *Euterpe globosa* (Palmae) in Puerto Rico). The number of dipterocarp seedlings in Malaysian forests fluctuates greatly (e.g., Barnard 1956), and since there is no evidence of major dipterocrap seedling herbivores, we must conclude that deaths are due to competition. If the canopy-member seedlings and saplings die through competition rather than being consumed as seeds, fewer resources will be available for longer-lived and shade-tolerant understory species.

A third characteristic expected of habitats rich in mast-fruiting species is that the adults, and especially juveniles, will have clumped distributions. Ashton (1969) reports this trait in mixed dipterocarp forest, and Brünig (1973) stresses it. It was very conspicuous at Bako that adult dipterocarps produce large stands of their own seedlings around their bases. Fox (1967) reported 30,000 dipterocarp seedlings per hectare (3 per m^2) in a Sabah forest, and Poore (1968) reports 10,000 dipterocarp seedlings per hectare in the year after a mast year in Malaya. Fox (1967) states that in a dipterocarp forest, "seeds fall in abundance irregularily, at intervals of 3 to 6 years, normally allowing minimum development of 4,000 seedlings per acre in the subsequent year" and "70% of counted seedlings in a virgin area . . . consisted of *Shorea, Parashorea* and *Dipterocarpus* . . . these three species accounted for 40 percent of all trees 5 feet girth and over." Since dipterocarp seeds are gravity-wind dispersed (as expected in a forest rich in mast-fruiting species), these seedlings should be concentrated around their

parents. Once a parent is established on a site, we expect new adults to appear in her immediate vicinity, rather than at far distances as expected in an animal-rich community (Janzen 1970). I also expect this spacing pattern in habitats that have one or a few dominant tree species, but yet there are enough animals in nearby areas so that selection has favored seed escape by very toxic seeds rather than mast fruiting (see below). Large clumps of seedlings occur beneath *Eperua, Mora, Ocotea,* and *Pentaclethra* in the neotropics on white sand or other poor soils (Beard 1946; Davis and Richards 1933, 1934; Hartshorn 1973; Janzen unpub.; Richards 1952). Such aggregation also occurs when seeds escape for other reasons while still beneath the parent, such as on mammal-free Caribbean islands where it is commonplace to find dense stands of seedlings and saplings beneath parent trees even in apparently undisturbed forest.

A fourth characteristic of areas rich in mast-fruiting species should be that many of the tree species should be conspicuously site-specific. When parental replacement is set solely by competition among juveniles, the best competitors should appear to "own" any given site, and each species should become a more extreme specialist for its particular edaphic circumstances than in forests where the density of a tree species is also set by herbivory. Evolutionarily this condition should come about by the plant channeling all its resources into a maximally competitive phenotype in the context of the habitat in which it is best (e.g., raise the average seed weight without lowering the number of seeds), and by plants therefore reaching adult status only on the site where they are the very best competitors. Furthermore, as will be discussed below, the mast-fruiting tree on a different soil from that occupied by most of the population may not receive the appropriate cues and therefore may lose its seed crop by flowering out of synchrony. Ashton (1964, 1969) and Brünig (1973) have stressed and demonstrated strong site-specificity among Sarawak and Brunei dipterocarps. Kwan and Whitmore (1970) do not agree, but their analysis is based on presence/absence data rather than relative density of adult plants (and see Brünig 1973).

Strong site specificity should also be important in the physiology of mast fruiting. The larger the fraction of an adult tree population that is on one soil type, the larger will be the fraction that accumulates photosynthates at the same rate and perceives a weather event as the same cue. Viewed from the other direction, the smaller the fraction of the population that responds heavily to a fruiting cue, the lower will be the adaptive value of intra- and inter-population synchronization for a single tree. For increased inter-population synchrony, we may note that the cue will have to be progressively more distinctive for it to be simultaneously and accurately sensed by a number of different tree species growing on different substrates over a wide area. It is therefore not surprising to find that an occasional dipterocarp species or population is out of phase with the majority of species or populations (Ashton 1969; Wood 1956). We may even expect that synchrony should evolve most easily among species on poor soil because a smaller environmental perturbation may be required there for a tree to respond than on a good soil site, where the individual trees may be more capable of buffering external perturbations with internal food reserves.

Malaysian rainforest climates are particularly suited for the evolution of mast fruiting, as they are overall very uniform yet have occasional rather severe dry spells. "Within the region of tropical rain forest, the lowlands of parts of the northern half of Borneo possess the most uniform climate" (Brünig 1969b). This appears to be the case from the viewpoint of a wide variety of organisms (Berry 1964; Berry and Varughese 1968; Foxworthy 1927; Harrison 1955; Holttum 1941; Inger and Bacon 1968; Inger and Greenberg 1963, 1966, 1967; Lloyd et al. 1968; Marshall 1970; Medway 1972a; Richards 1973; Seal 1958; Ward 1969; Wycherly 1968). In neotropical and African rainforests, with their conspicuous and often erratic (even when short) dry seasons, the cue for supra-annual synchrony would have to be much more dramatic (and therefore likely to be overall physiologically detrimental) than the couple of weeks of rain-free weather that dipterocarps use as a cue. A severe dry spell of the type that occurs only every couple of years in Malaysia should be an adequate cue for dipterocarps if intra- and inter-specific rates of photosynthate storage are sufficiently synchronized. A uniform climate may also lead to fairly constant rates of annual photosynthate storage, resulting in all the trees of the population responding equally to a cue.

4) Why the Dipterocarpaceae? We may start from two rather different base points in postulating why the Dipterocarpaceae are the primary mast fruiters in Malaya-Sarawak-Brunei rainforest, rather than representatives from other families. a) There may have been a number of dipterocarp species prior to the evolution of mast fruiting. Once it appeared in one species, something about the family as a whole preadapted it for entrainment to a suite of characteristics correlated with mast fruiting, for ex-

ample, commonly growing in close proximity to other trees within the same family, being chemically well protected, being especially good at living on nutrient-poor soils. b) The first species to evolve the mast-fruiting syndrome may have been especially successful on low productivity sites with uniform climates, leading to an adaptive radiation of the new life form into a variety of habitats and the production of the "family" Dipterocarpaceae. In such a reconstruction, I am in essence stating that the Dipterocarpaceae are trees that have specialized on the Malaysian physical environment coupled with bad soil. It is of interest in this connection that "the limestone soils [of Malaya] do not bear dipterocarp forest" (Ashton 1964).

I prefer hypothesis b over a, but it will be very difficult to distinguish between them. Early in the evolution of the system, the two pathways should converge strongly. It is impossible to determine which Malaysian plant families could produce a major radiation of mast-fruiting species, though we might expect it to appear in one of those families with vegetation that is chemically very well protected and that constitutes a large part of the forest biomass for reasons other than being a mast-fruiting species (e.g., Anacardiaceae, Casuarinaceae, Ericaceae, Euphorbiaceae, Guttiferae, Lauraceae, Myrtaceae, Pinaceae, Podocarpaceae, Theaceae).

Finally, we are confronted with the question of why bad soil sites may have led to the evolution of mast fruiting in Malaysia, yet in the neotropics and mangrove swamps large poisonous seeds appear to be the primary reproductive strategy in forests with low tree species richness (see section IV). The answer may lie in the productivity of the vegetation, and hence animal biomass, near the site of low productivity, and the size of the low productivity site relative to the areas of higher productivity. For example, if a neotropical mangrove species were magically converted to a mast-fruiting species overnight, we could expect a mass migration of a diverse and large animal community into the mangrove forest at the time of seed production. It seems doubtful that a strip of mangroves a few hundred meters wide could come anywhere near satiating the large animal community found in the forest on latosols on slightly higher ground. The same argument may be applied on a larger scale to the British Guiana white sand and other edaphic vegetation types described by Davis and Richards (1933, 1934), and the Costa Rican *Prioria* and *Pentaclethra* forests to be discussed in section IV. The mystery remains, however, as to why the extensive Rio Negro drainage basin has not developed mast-fruiting species.

As will be remembered from previous discussion of the cues needed for supra-annual mast fruiting, it may be that South America does not offer a weather regime of the appropriate type.

Mast fruiting occurs in a few other types of tropical woody plants, and the bamboos are one of the most conspicuous. These forest grasses have large seeds which are highly edible (as appears to be the case for virtually all grasses). Perhaps the most dramatic case is that of Indian bamboos which are on a flowering cycle of 15 to 45 or more years. There is no suggestion that they are associated with regions of low primary productivity unless they are understory plants, and it is not obvious how the system became started. However, once started, it is fairly easy to visualize how the flowering cycle became so long, especially if humans have been involved as seed predators. If the bamboo is to satiate the local animal populations, it must have a large seed crop. To produce a large seed crop requires a long period of photosynthate storage. However, if the plant is to wait a long time, it must use a very intense cue for flowering (i.e., the weather event must be rare and conspicuous). A severe drought is about the only such cue available in a tropical area. However, when there is a severe drought, there are large numbers of starving animals, including humans. This result means that seed predation will be exceptionally heavy in the year of the flowering cue. Small wonder that the fruiting of the bamboo is regarded as a spiritual salvation to relieve the starvation caused by exceptionally bad droughts (Blatter 1929, 1930a,b; Santapau 1962). The same process is probably operative with *Strobilanthes* (Acanthaceae) with its 12-year gregarious flowering cycle in southern India (Matthew 1959, 1971; Robinson 1936; Steenis 1942); it is noteworthy that *Strobilanthes* is one of the two major understory plants in the low-diversity evergreen sholas forest of India mentioned in section IV (Lakshmanan 1968).

IV. FORESTS WITH LOW SPECIES RICHNESS

The previous discussion of forests rich in mast-fruiting Dipterocarpaceae leads immediately to thoughts about other tropical lowland evergreen trees that occur in stands where most of the large trees belong to less than about five species. The existence of such forests and the existence of dipterocarp forests falsify the dogma that diversity is mandatory for ecosystem stability in highly equitable climates.

These forests are thus very worthy of special scrutiny. With respect to seeds, most tropical lowland forests with low vegetative diversity may be placed in two categories. Either they are mast fruiting, or they rather continually produce large and toxic seeds, or seeds that are individually especially well protected in some other manner. Tannin-rich mangrove seeds, or seedlings in the case of *Rhizophora*, are a well-known example of the toxic seed category, but there are many others. Judging from my observations that rodents and insects on the Osa Peninsula (Costa Rica) do not attack the huge seeds of *Mora oleifera*, the *Mora* species (Leguminosae) that constitute large portions of the canopy in several Guiana and Trinidad swamp forest types (Beard 1945, 1946; Davis and Richards 1933, 1934) and on Suriname white sand soils (= *Dimorphandra*, Stark 1970) have very poisonous seeds. Fittkau (pers. comm.) reports the same for *Mora* on the Rio Negro. Brazilian *Mora* (= *Dimorphandra*) pods contain 8 percent rutin, a pharmacologically active flavone (Mors and Rizzini 1966), and large amounts of a dark red dye can be leached from *Mora oleifera* seeds. They appear to be very rich in alkaloids (E. A. Bell, pers. comm.). Beard (1946) reports that the seed contents can be eaten by humans if they are first ground and leached in water. In British Guiana and Trinidad, *Mora* produce dense stands of seedlings around their bases, a characteristic implying freedom from animal attack. The same high production of seedlings has been described for *Ocotea* and *Eperua* forest (Davis and Richards 1933, 1934; Takeuchi 1961, 1962a,b). *Eperua* flowers are extremely showy (Ducke 1940), and it seems unlikely that supra-annual mast flowering would have gone unobserved. As these and other authors have given no hint of mast fruiting by these species, it seems a reasonable inference that *Ocotea* and *Eperua* seeds will be found to be very toxic. *Pentaclethra macroloba* (Leguminosae) may make up as much as 40 percent of the canopy over some Costa Rican rainforest swampy soils (Hartshorn 1973), and is one of the most common subdominants on British Guiana white sand soil sites (Richards 1952) and in Trinidad *Mora* and other forests (Beard 1946). It has 5- to 10-gram seeds (fresh weight) that are ignored by almost all species of rainforest rodents and almost all insects (unpub. field notes). *Pentaclethra macrophylla* seeds in Africa are rich in an unidentified alkaloid (Pammel 1911), and *P. macroloba* seeds contain high concentrations of an unidentified alkaloid (Bell, pers. comm.). *Prioria copaifera* (Leguminosae), which forms nearly pure stands in coastal swamp forests of Costa Rica (Allen 1956), has large seeds in an indehiscent pod, with tannin-rich walls. *Raphia* and other palms form dense stands in fresh-water swamps in Africa and the neotropics (Richards 1952); while their seeds appear to contain no toxic compounds, at least in Costa Rica they fall into water and sink to the bottom where they are unavailable to most animals (unpub. field notes). While unexplored ecologically, tropical forested swamps are well known commonly to contain monospecific stands of trees, for example, *Pterocarpus officinalis* (Gonggryp and Burger 1948), *Parkinsonia aculeata* in swamps in Costa Rican deciduous forest (unpublished field notes), *Erythrina glauca*, and *Machaerium lunatum* (Lindeman 1953). It seems quite reasonable that in addition to being vegetatively well protected, these plants have seeds protected during part or all of the year by the water. It would not be surprising to find that some of the dipterocarps that form low-diversity forests in animal-rich areas have toxic seeds and reduced mast-fruiting behavior. It is of extreme interest in this context that *Shorea robusta,* which forms extensive deciduous forests in India, has exceptionally tannin-rich seeds (Burkill 1935) and apparently bears seed every year (in contrast to the Indian bamboos with their highly edible seeds and long mast cycles).

It is among tree species with large and toxic seeds that we might expect to find species that fruit annually and even more than once during the year. *Mora* forests bear fruits every year (Beard 1946), *Prioria copaifera* fruits irregularly throughout the year (Allen 1956), and Costa Rican *Pentaclethra macroloba* forest has several distinct fruit-bearing periods during the year (Hartshorn 1973).

If the seeds are very well protected, then any synchrony that does occur within the year should be set by constraints on pollination or the timing of seed germination. It is not surprising to find that mangroves in Costa Rica flower during much of the year, as they are insect-pollinated and thus conspecifics need not be shoulder to shoulder to insure cross-pollination; since new sites for colonization are available most of the year, fruit-bearing should be continuous. On the other hand, it is of interest that in the same region, the mangrove *Avicennia nitida* has a highly synchronized flowering period (Allen 1956) toward the end of the time when many other trees in the adjacent dry-land forest are in flower (Janzen 1967). *Mora oleifera,* growing in a narrow strip along the landward margin of Pacific Costa Rican mangrove swamps (Allen 1956), drops its huge seeds (up to one-half kilogram) just before the highest spring tides. The seeds are then floated and deposited just slightly ocean-ward from the parent trees (unpub. field notes).

A rainforest animal community may be severely reduced in ways other than low primary productivity at the site. Small islands commonly have forests with low tree-species richness within a given habitat, and they have severely reduced biomass and species richness of native herbivores (Janzen 1973b,c). Since a low habitat diversity on an island is not subject to potential invasion from adjacent habitats with high animal biomass, I expect plants to be able to grow in pure stands concomitant with relaxed selection for exceptional chemical defenses. This conclusion is suggested by the extreme ability of introduced mammals to decimate natural island floras (e.g., Howard 1967), and the observation that the forests of low diversity on tropical islands are not generally noted for commercial harvest of resins or other toxic secondary compounds. On tropical islands, we may also expect satiation of seed predators at the community level, as is the case in the *Euterpe globosa* palm forests of Puerto Rico (Janzen 1972). If there are no seed predators, then we expect a total lack of synchrony in fruit maturation and the production of a few huge and highly edible seeds, as is the case with the palm *Lodoicea maldivica* which grows in nearly pure stands in Seychelles Islands and has the largest seed in the world (McCurrach 1960). It is noteworthy in this context that the Philippines, Java, etc. are island communities, and the success of dipterocarps in their forests (Brown and Brown 1914; Serevo 1960; Slooten 1952; Whitford 1906) may be related to a reduction of animal communities among their vegetation for this reason, rather than due to poor quality soil.

With respect to the foliage, it is not only white sands that bear evergreen and monotonous vegetation that is particularly rich in secondary compounds. It appears that this is a general characteristic in habitats with very low primary productivity yet a climate favorable to animals. This is another way of saying, "in those habitats where the loss of a leaf causes a very great depression in plant fitness." The understory shrubs of evergreen lowland tropical rainforest are low in diversity when compared with the canopy and are a major source of spice-rich plants, have very reduced foliage-inhabiting insect biomass (Janzen 1973a,b), and appear on other grounds to have highly toxic foliage with a low rate of replacement (e.g., Bourliere 1973 stresses that rainforest vertebrates live almost entirely on canopy foliage or material that falls from the canopy). The evergreen ericaceous shrubs in the understory of eastern United States deciduous forest have highly toxic foliage (e.g., Forbes and Bechdel 1931; Pammel 1911) and long-lived leaves (Monk, pers. comm.). *Melaleuca* (Myrtaceae) forms large pure stands on acid, saline, or waterlogged soils in Malaya (Johnson 1967a), and its leaves are distilled for medicinal compounds. The pure stands of evergreen oaks (*Quercus* spp.), Lauraceae, and Ericaceae at high tropical elevations are probably protected from insect outbreaks by tannin-rich foliage just as are their temperate-zone relatives (Feeny 1968, 1969; Hathaway 1958) in forests of low diversity. The large pure stands of *Quercus oleoides* in the lowlands of Central America, usually on the worst soils of the region, are in the same category. In the evergreen-stunted Indian hill forest "sholas," the dominant tree families are Lauraceae, Myrtaceae, and Styracaceae with an understory of Rubiaceae and Acanthaceae (mast-fruiting *Strobilanthes* sp.); the dominant tree species belong to the Elaeocarpaceae, Euphorbiaceae, Flacourtiaceae, Magnoliaceae, Myrtaceae, Sapotaceae, Theaceae, and Thymelaeaceae. When denuded, sholas are occupied by nearly pure stands of myrtaceous or anacardiaceous shrubs (Lakshmanan 1968). As mentioned in the first section of this paper, *Mora* has especially durable wood (Allen 1956; Gonggryp and Burger 1948), and is a local commercial tannin source in Brazil (= *Dimorphandra*, Mors and Rizzini 1966). *Pentaclethra* has wood especially rich in tannins, and *P. macroloba* logs lie on the rainforest floor for many years before decomposing (Hartshorn 1973). *Prioria copaifera* wood is famous for exuding copious tannin-rich black gum when cut (Allen 1956; Mors and Rizzini 1966; Pammel 1911), and *P. copaifera* swamp forest in Panama was found to have the lowest caloric input of litter per hectare of six Panamanian rainforest types examined by Woods and Gallegos (1970). *Copaifera* (Leguminosae), another major resin producer, is a common small tree in Sarawak *Shorea albida* peat swamp forest (Anderson 1961; Burkill 1935). *Protium copal* (Burseraceae) is the second most common dicot canopy-member tree in the Costa Rican *Pentaclethra* forest mentioned above (Hartshorn, pers. comm.) and produces excessive amounts of resin (Allen 1956). Species-poor stands of desert perennials are notorious for toxicity. It is no accident that the temperate-zone evergreen conifer forests are rich in resins and other secondary compounds.

Mangroves, with their extraordinarily high tannin content (e.g., Allen 1956; Burkill 1935; Fosberg 1945; Nierenstein 1934; Pammel 1911; Standley 1920-1926), are perhaps an extreme case. As recently demonstrated in Viet Nam, a single defolia-tion of mangroves with chemicals is usually lethal to the tree, yet mangroves display some of the highest primary productivity of any forest (Golley, pers. comm.). It is tempting to postulate that the very conspicuous chemical defenses of mangroves are not only to avoid attack by marine animals, but to minimize the possibility of a mangrove being defoliated by herbivores. Frequent defoliation would be almost certainly unavoidable if the leaves in such a low-diversity evergreen foliage were highly edible. Their extreme sensitivity to defoliation could well be because the ever-present leaves are crucial in physiologically avoiding salt damage. Incidentally, if tidal flux did not result in frequent leaching and washing of litter out of the system, we would expect mangrove swamps to generate blackwater rivers. Indeed, a southern Mexican non-tidal mangrove swamp has been described as having a peat bog beneath it (Thom 1967).

The above comments should not be taken to imply that some species of evergreen trees on good soil will not be well defended or that deciduous species-rich forests will be devoid of species with chemically well-defended vegetative parts. Scattered through Costa Rican deciduous forest are evergreen tree species, which, if grown in a pure stand, would probably produce an ortstein, a podzol, and/or a deep litter layer. For example, *Hymenaea courbaril* (Leguminosae) produces copious amounts of resin (Burkill 1935; Langenheim 1973; Standley 1920-1926) and has extremely resinous and sclerophyllous leaves that remain on the tree about 11.5 months of the year. *Manilkara chicle* (Sapotaceae) is well known for its large quantities of latex (chicle of commerce) as are the evergreen figs (*Ficus* spp.). *Andira enermis* (Leguminosae) has bark and seeds that were used for a "pugative, vermifuge, febrifuge or anthelmintic but large doses are said to be dangerous, producing delirium or even death" (Standley 1920-1926). *Byrsonima crassifolia* (Malpighiaceae) has bark used for tanning and dyeing and "the plant is astringent" (Standley 1920-1926). *Curatella americana* (Dilleniaceae) has bark "rich in tannin and is used in Brazil for tanning skins" (Standley 1920-1926). The leaves of all of these evergreens are conspicuous in being almost entirely free of insect damage throughout the year. *Bursera simarouba* (Burseraceae) is another common tree in these forests; it is deciduous, but has a conspicuous green photosynthesizing layer directly under the stem epidermis and so is effectively evergreen. *B. simarouba* and congeners are prominent producers of medicinal gums, and "the Caribs employed it for painting their canoes to preserve them from the attacks of worms"

(Standley 1920-1926). *Jacquinia pungens* (Theophrastaceae), an understory shrub in the same forests, is an extreme case. It bears leaves only during the dry season, thus earning the misnomer "siempre verde," suffers virtually no defoliation, and has foliage rich in fish poisons (Janzen 1970).

Throughout this paper, I have talked as though the foliage of the common tree species in low-diversity tropical forests is nearly free of insect attack. In general, this appears to be the case, but there are a few recorded cases where apparently host-specific insects defoliate large areas of one species (Anderson 1961; Atkinson 1953; Gray 1972; Kalshoven 1953; Kapur 1958). The general explanation appears to be that some combination of weather events leads to exceptionally good reproduction of the herbivores. Severe defoliation follows before the insects eat all their host's leaves, have weather events depress their population, and/or parasites/predators build up on them. In this sense they appear to differ in no significant way from the analogous interaction in temperate-zone forests of low foliage diversity. Furthermore, it is likely that such defoliation leads to a delay or failure in production of mast seed crops at a later date, just as is the case in temperate-zone forests. In one example, it appears that a single defoliation by moth larvae (Lymantriidae) killed 12,000 hectares of *Shorea albida* forest with 400-480 trees per acre (Anderson 1961). Again, as in temperate-zone forests, I suspect that upon more detailed examination, we will find that herbivore outbreaks in dipterocarp forest occur at the time of production of new foliage or when the trees are weakened, rather than the trees being generally susceptible to insect attack.

DISCUSSION

The patterns dealt with here are not absolute categories. For example, a white sand soil need not generate a blackwater river, if the vegetation happens to be made up of species that use solely alkaloids for their defense against herbivores. A nutrient-rich soil may produce blackwater rivers if it becomes colonized by exceptionally tannin-rich species of plants. On the other hand, toxic foliage falling on exceptionally nutrient-rich soil may be competently degraded by the litter fauna, for example, see the discussion of the biogenesis of mull and mor soils by Handley (1954); the same argument applies to the blackwaters themselves, and we must remember that what is toxic on an empty stomach is just flavoring on a full stomach. Immigration of an aggressive mast-fruiting species of tree into a habitat may reduce the animal community even if the soil

quality is high.

A particularly conspicuous problem in the variability of interaction of well-protected plants and white sand soils is that much of the Brazilian white sand and other nutrient-poor soils (Goodland 1971) bear a low scrubby vegetation ("campo," "caatinga"), yet some of this does not appear to have an ortstein or even particularly dark deep humus layer (Arens 1963; Eiten 1963; Ferri 1960; Heyligers 1963; Sioli 1967; Spruce 1908; Stark 1970). These sands do, however, produce blackwater rivers. Some of this vegetation differs from the forest on the white sand soils of Sarawak, British Guiana, and Suriname in that it is occasionally burned. It is possible that a major part of the secondary compounds is lost through fire consumption of the litter rather than leaching by rainfall. This process also appears to be occurring in *Pentacme* (Dipterocarpaceae) forest on sandy soil in India (Ogawa *et al.* 1961). In fire-rich habitats we also expect the evergreen life form to be less prominent, as retention of leaves during the dry season raises the chances of losing a major part of one's nutrient reserves to the capricious herbivore fire. As the proportion of deciduous leaves rises in a tropical habitat, I expect that the absolute amount of secondary compounds in the litter and rainfall leachate will decline.

In bringing this paper to a close, it seems appropriate to comment briefly on some broader aspects of tropical ecology that it has brought to mind.

1) If the degree of evergreenness is an adaptive response to nutrient-poor soils, then the implications behind words such as "evergreen rainforest" and "sclerophyll scrub" should be subject to very close scrutiny.

2) If the rates of tropical vegetation replacement following fire or other perturbation really vary as much as they appear to over soil and elevational gradients, then attempts to describe and classify vegetation on low productivity sites seem especially futile; apparently "climax" vegetation may be the result of a fire two hundred years before, and in fact only be in the early stages of succession.

3) Low-diversity forests are commonplace in the tropics; apparently stable, they should lead us to search for the natural processes that lead to herbivores not building up on a food supply that appears unlimited to the appropriate specialist.

4) Litter depth should not be directly related to the productivity or the rate of litter input to a site, since it must be disappearing at the same rate that it is coming in. However, litter depth should be directly related to how long the litter has to lie on the

soil being leached by the rain, before its toxin content has fallen to the point where the litter fauna and flora can go to work on it.

5) From an animal's viewpoint, changes in productivity do not necessarily produce corresponding changes in available food; what counts far more is what the plant does with, for example, additional photosynthate, or how it re-allocates its resources when its overall productivity is lowered. A tropical rainforest might easily have three times the primary productivity of a temperate-zone forest, yet have only a tenth as much harvestable productivity from a herbivore's point of view. Seen the other way round, two tropical evergreen forests of the same standing crop may have drastically different productivities from the animals' viewpoint and, of course, very different primary productivities as well.

6) Toxicity of any compound is a relative thing. For example, on a nutrient-rich soil, a tree with very toxic leaves may have no effect on the litter fauna and soil-forming processes, while on a nutrient-poor site, a mildly toxic tree may have a huge effect.

In conclusion, it seems appropriate to point out that the type of analysis applied in this paper could be of use in understanding the dynamics of the development of temperate-zone peat bogs and their blackwater rivers (Conway 1949; Gimingham 1972; Gorham 1957; Heinselman 1963, 1965; Hutchinson 1957; Reader and Stewart 1972; Small 1972a), as well as temperate podzols and mors (Burges 1965; Dimbleby 1962; Handley 1954; Ponomareva 1969; Rode 1970; Skoropanov 1968; Waksman 1938). For example, many extra-tropical bog and mor organics have extremely low decomposition rates (e.g., Douglas and Tedrow 1959; Gorham 1957; Kendrick 1959; Perrin et al. 1964; Reader and Stewart 1972); if a bog is alternately drained and flooded, its woody plants will be replaced by fast-growing and often deciduous species that are not especially noted for toxic foliage (e.g., Gorham 1957; Lloyd and Scarth 1922). Presumably the changes in water level are repeatedly flushing out secondary compounds and oxygenating the substrate so that fungi and bacteria can decompose the phenolics and other toxic chemicals. A second example is provided by the observa-

tion that conifers and ericaceous shrubs on temperate-zone peats are very slow-growing plants, have foliage rich in toxic secondary compounds, and leaves with a long life span (Forbes and Bechdel 1931; Harley 1952; Knerer and Atwood 1973; Muenscher 1970; Pammel 1911; Tamm 1955; Williams 1970; etc.). They also possess "sclerophyllous" leaves which are generally interpreted as either adaptations to some as yet undiscovered type of physiological drought or to direct "symptoms" of nutrient deficiency (e.g., Arens 1963; Brünig 1969b; Loveless 1961; Small 1972a). By way of rebuttal, I would simply like to emphasize my strong agreement with Ponomareva (1969: 257) when she stated, "Coniferous tree species, each year shedding only part of their needles, perennial subshrubs, evergreen plants, etc. are all adaptive forms of live nature to conditions under which mineral nutrients are strongly leached out of the plant's environment," and add, "imagine what would happen to an evergreen bog during the winter if the foliage were highly edible to deer."

ACKNOWLEDGEMENTS

This study has been supported by NSF GB-25189, GB-7819, and GB-35032X, and my mother provided the travel expenses for the work at Bako National Park, Sarawak. I wish to express extreme gratitude to Prof. P. W. Richards for encouraging me to visit the white sand site at Bako. J. A. R. Anderson, P. A. Morrow, and T. C. Whitmore were of great help in the field work at Bako and in the development of these ideas. The manuscript has developed out of valuable discussions, aid and editorial comments by Drs. J. M. Anderson, P. Ashton, H. G. Baker, E. C. Bate-Smith, W. Benninghoff, M. Brinson, E. F. Brünig, P. Feeny, R. F. Fisher, E. J. Fittkau, L. L. Gilbert, A. Gomez-Pompa, G. Hartshorn, A. Hasler, R. F. Inger, D. Janos, C. Jordan, G. E. B. Kitaka, H. Klinge, K. E. Lee, H. Lieth, G. Marlier, R. McConnell, B. McNab, E. Medina, B. J. Meggers, M. Meselson, F. Nordlie, G. Orians, C. M. Pond, R. W. Richards, N. Scott, H. Sioli, S. Snedaker, N. M. Stark, J. H. Vandermeer, T. C. Whitmore, P. J. Zinke, and with classes in the Organization for Tropical Studies and at the University of Michigan. The help of Susan Abel is gratefully acknowledged in typing the manuscript and other logistic details.

This study is dedicated to those unfortunate tropical farmers whose governments cause them to attempt to survive on white sand soils either because they appear unexploited or for political reasons.

LITERATURE CITED

AITKEN, J. B. 1930. The Wallabas of British Guiana. Trop. Woods 23: 1–5.

ALEXANDER, M. 1964. Biochemical ecology of soil microorganisms. A. Rev. Microbiol. 18: 217–252.

ALLEN, P. H. 1956. The rain forests of Golfo Dulce. Univ. Florida Press, Gainesville. 417 pp.

ANDERSON, J. A. R. 1959. Observations on the ecology of the peat-swamp forests of Sarawak and Brunei. In Proc. Symp. Humid Tropics Vegetation, Tjiwani 1958, pp. 141–148. Publ. UNESCO Sci. Coop. Office South East Asia.

————. 1961. Destruction of *Shorea albida* by an unidentified insect. Emp. For. Rev. 40: 19–29.

————. 1963. The flora of the peat swamp forests of Sarawak and Brunei, including a catalogue of all recorded species of flowering plants, ferns and fern allies. Gdns' Bull. Singapore. 20: 131–228.

————. 1964a. Climatic changes in *Shorea albida* forests in Sarawak attributed to lightning. Commonw. Forest. Rev. (= Emp. For. Rev.) 43: 145–158.

————. 1964b. The structure and development of the peat swamps of Sarawak and Brunei. J. Trop. Geogr. 18: 7–16.

ANDERSON, J. M. 1973. The breakdown and decomposition of sweet chestnut (*Castanea sativa* Mill.) and beech (*Fagus sylvatica* L.) leaf litter in two deciduous woodland soils. II. Changes in the carbon, hydrogen, nitrogen and polyphenol content. Oecologia 12: 275–288.

ANONYMOUS. 1953. Bearded pig swim again. Malay. Nat. J. 8: 118–120.

ANONYMOUS. 1960. Lavantamento de reconnehmento dos solosdo estao de Sao Paulo. Brasil Boletim do Servicio Nacional de Pesquisas Agronomicas. 634 pp.

ANONYMOUS. 1972. Die Ionenfract des Rio Negro, Staat Amazonas, Brasilien, nach Untersuchungen von Dr. Harald Ungemach. Amazoniana 3: 175–185.

ARENS, K. 1963. As plantas lenhosas dos campos cerrados como flora adaptada as deficiencias minerais de solo. *In* Simposio sobre o Cerrado, pp. 287–303. Universidade do Sao Paulo, Sao Paulo.

ARNOLD, S. J. 1972. Species densities of predators and their prey. Am. Nat. 106: 220–236.

ARNOLD, G. W., AND J. L. HILL. 1972. Chemical factors affecting selection of food plants by ruminants. *In* J. B. Harborne (ed.). Phytochemical ecology. Academic Press, London, 272 pp.

ASHTON, P. S. 1964. Ecological studies in the mixed dipterocarp forests of Brunei State. Oxford Forestry Memoirs No. 25, 75 pp.

————. 1969. Speciation among tropical forest trees: some deductions in the light of recent evidence. Biol. J. Linn. Soc. 1: 155–196.

————. 1971. The plants and vegetation of Bako National Park. Malay. Nat. J. 24: 151–162.

ATKINSON, D. J. 1953. The natural control of forest insects in the tropics. Trans. 9th Inter. Congress Ent. 2: 220–223.

BAILEY, H. H. 1951. Peat formation in the tropics and subtropics. Proc. Soil Sci. Soc. Am. 15: 283–284.

BARNARD, R. G. 1956. Recruitment, survival and growth of timber tree seedlings in natural tropical rain forest. Malay. Forester 19: 156–161.

BATE-SMITH, E. C. 1962. The phenolic constituents of plants and their taxonomic significance. J. Linn. Soc. (Bot.) 58: 95–173.

————. 1973a. Haemanalysis of tannins: the concept of relative astringency. Phytochemistry 12: 907–912.

————. 1973b. Tannins of herbaceous Leguminosae. Phytochemistry 12: 1809–1812.

BAWA, K. S. 1974. Breeding systems of tree species of a lowland tropical community. Evolution 28: 85–92.

BEARD, J. S. P. 1945. A brief review of the vegetation of Trinidad and Tobago. *In* F. Verdoorn (ed.). Plants and plant science in Latin America, pp. 100–101. Chronica Botanica Co., Waltham, Mass.

————. 1946. The Mora forests of Trinidad, British West Indies. J. Ecol. 33: 173–192.

BENZING, D. H. 1973. Mineral nutrition and related phenomena in Bromeliaceae and Orchidaceae. Q. Rev. Biol. 48: 277–290.

BERRY, P. Y. 1964. The breeding seasons of seven species of Singapore Anura. J. Anim. Ecol. 33: 227–243.

————, AND G. VARUGHESE. 1968. Reproductive variation in the Torrent frog *Amolops larutensis* (Boulenger). J. Linn. Soc (Zool.) 47: 547–559.

BISHOP, J. E. 1973. Limnology of a small Malayan river, Sungae Gombok. Junk, Hague. 485 pp.

BLACK, A. P., AND R. F. CHRISTMAN. 1963. Characteristics of colored surface waters. J. Amer. Water Works Assoc. 55: 753–770.

BLATTER, E. 1929. The flowering of the bamboos. J. Bombay nat. Hist. Soc. 32: 899–921.

————. 1930a. The flowering of bamboos. J. Bombay nat. Hist. Soc. 33: 135–141.

————. 1930b. The flowering of bamboos. J. Bombay nat. Hist. Soc. 33: 447–467.

BOURLIERE, F. 1973. The comparative ecology of rain forest mammals in Africa and tropical America: some introductory remarks. *In* B. J. Meggers, E. S. Ayensu and W. D. Duckworth (Eds.). Tropical forest ecosystems in Africa and South America: a comparative review. Pp. 279–292. Smithsonian Institution Press, Washington, D.C.

BRIAN, P. W., H. G. HEMMING, AND J. C. McGOWAN. 1945. Origin of a toxicity to mycorrhiza in Wareham Heath soil. Nature, Lond. 155: 637–638.

BRINSON, L. G. 1973. A seasonal comparison of the subsystems of a tropical fresh-water delta. M.S. Thesis, Univ. of Florida.

BRINSON, M. M. 1973. The carbon budget and energy flow of the subsystems of a tropical lowland aquatic ecosystem. Ph.D. Thesis, Univ. of Florida.

BROWN, W., S. I. FALKEHAG, AND E. B. COWLING. 1967. Molecular size distribution of lignin in wood. Nature, Lond. 214: 410–411.

BROWN, W. H., AND D. M. BROWN. 1914. Philippine dipterocarp forests. Philipp. J. Sci. 9A: 314–568.

BROWNE, F. G. 1955. Forest trees of Sarawak and Brunei and their products. Government Printing Office, Kuching, Sarawak. 369 pp.

BRÜNIG, E. F. 1964. A study of damage attributed to lightning in two areas of *Shorea albida* forest in Sarawak. Commonw. Forest. Rev. (= Emp. For. Rev.) 43: 134–144.

——————. 1965. Guide and introduction to the vegetation of the kerangas forests and the padangs of the Bako National Park. UNESCO Symp. Ecol. Res. Humid Trop. Vegt. Kuching 1963. Tokyo.

——————. 1968. Some observations on the status of heath forests in Sarawak and Brunei. Proc. Symp. Recent Adv. Trop. Ecol. Pp. 451–457.

——————. 1969a. The classification of forest types in Sarawak. Malay. Forester 32: 143–179.

——————. 1969b. On the seasonality of droughts in the lowlands of Sarawak (Borneo). Erdkunde 23: 127–133.

——————. 1969c. Forestry on tropical podzols and related soils. Trop. Ecol. (= Bull. Int. Soc. Trop. Ecol.) 10: 45–58.

——————. 1973. Species richness and stand diversity in relation to site and succession of forests in Sarawak and Brunei (Borneo). Amazoniana 4: 293–320.

BURGES, N. A. 1965. Biological processes in the decomposition of organic matter. *In* E. G. Hallsworth and D. U. Crawford (Eds.). Experimental Pedology. Pp. 189–199. Butterworths.

——————. 1967. The decomposition of organic matter in the soil. *In* A. Burges and F. Raw (Eds.). Soil biology. Pp. 479–492. Academic Press.

——————, H. M. HURST, AND B. WALKDEN. 1964. The phenolic constituents of humic acid and their relation to the lignin of the plant cover. Geochim. cosmochim. Acta 28: 1547–1554.

BURGESS, P. F. 1965. Silica in Sabah timbers. Malay. Forester 28: 223–229.

——————. 1968. An ecological study of the hill forests of the Malay Peninsula. Malay. Forester 31: 314–325.

——————. 1969. Color changes in the forest: 1968-1969. Malay. Nat. J. 22: 171–173.

BURKILL, I. H. 1935. A dictionary of the economic products of the Malay Peninsula. The Crown Agents for the Colonies, London. 2402 pp.

CARTER, G. S. 1934. Results of the Cambridge expedition to British Guiana, 1933. The fresh waters of the rain-forest areas of British Guiana. J. Linn. Soc. (Zool.) 39: 147–193.

——————. 1935. Reports of the Cambridge expedition to British Guiana, 1933. Respiratory adaptations of the fishes of the forest waters, with descriptions of the accessory respiratory organs of *Electrophorus electricus* (Linn.) (= *Gymnotus electricus* Auctt.) and *Plecostomus plecostomus* (Linn.). J. Linn. Soc. 39: 219–233.

CARTWRIGHT, K. ST. G., AND W. P. K. FINDLAY. 1943. Timber decay. Biol. Rev. 18: 145–158.

CLAYTON, W. D. 1958. A tropical moor forest in Nigeria. Jl. W. Afr. Sci. Ass. 4: 1–3.

CLERFAYT, A. 1956. Composition des eaux de rivieres au Congo. Bull. Cent. belge Etud. Docum. Eaux 31: 26–31.

COLLINS, E. A., C. D. MONK, AND R. H. SPIELMAN. 1964. White-cedar stands in northern Florida. Q. Jl. Fla Acad. Sci. 27: 107–110.

CONWAY, V. M. 1949. The bogs of central Minnesota. Ecol. Monogr. 19: 173–206.

CORBET, A. S. 1935. Biological processes in tropical soils, with special reference to Malaya. Cambridge.

COULTER, J. K. 1950. Peat formation in Malaya. Malay. Agric. J. 40: 161–175.

CUMMING, J. E., AND B. MCKAGUE. 1973. Preliminary studies of effects of juvenile hormone analogues on adult emergence of black flies (Diptera: Simuliidae). Can. Ent. 105: 509–511.

DAVIS, T. A. W., AND P. W. RICHARDS. 1933. The vegetation of Moraballi Creek, British Guiana: an ecological study of a limited area of tropical rain forest. Part I. J. Ecol. 21: 350–384.

——————, AND ——————. 1934. The vegetation of Moraballi Creek, British Guiana: an ecological study of a limited area of tropical rain forest. Part II. J. Ecol. 22: 106–155.

DEL MORAL, R., AND R. G. CATES. 1971. Allelopathic potential of the dominant vegetation of western Washington. Ecology 52: 1030–1037.

——————, AND C. H. MULLER. 1969. Fog drip: a mechanism of toxic transport from *Eucalyptus globulus*. Bull. Torrey bot. Club 96: 467–475.

DIMBLEBY, G. W. 1962. The development of British heathlands and their soil. Oxford Forestry Memoirs, No. 23, 120 pp. Oxford, Clarendon Press.

DOUBOIS, T. 1959. Note sur la chimie des eaux du lac Tumba. Bull. Séanc. Acad. r. Sci. colon. (outre-Mer) 5: 1321–1334.

DOUGLAS, L. A., AND J. C. F. TEDROW. 1959. Organic matter decomposition rates in arctic soils. Soil Sci. 88: 305–312.

DUCKE, A. 1940. Notes on the wallaba trees. Trop. Woods 62: 21–28.

————, AND G. A. BLACK. 1953. Phytogeographical notes on the Brazilian Amazon. Anais da Academia Brasileira de Ciencias 25: 1–46.

DUNSON, W. A., AND R. R. MARTIN. 1973. Survival of brook trout in a bog-derived acidity gradient. Ecology 54: 1370–1376.

EITEN, G. 1963. Habitat flora of Fazenda Campininha, Sao Paulo, Brazil. I. Introduction, species of the "cerrado," species of open wet ground. *In* Simposio sobre o cerrado. Pp. 181–231. Universidade do Sao Paulo, Sao Paulo.

ERNEST, E. C. M. 1936. A test for the presence of natural preservative substances in wood. Forestry 10: 58–64.

EVANS, W. C. 1947. Oxidation of phenol and benzoic acid by some soil bacteria. Biochem. J. 41: 373–382.

FARMER, V. C., AND R. I. MORRISON. 1964. Lignin in sphagnum and phragmites and in peats derived from these plants. Geochim. cosmochim. Acta 23: 1537–1546.

FEENY, P. P. 1968. Effects of oak leaf tannins on larval growth of the winter moth *Operophtera brumata* J. Insect Physiol. 14: 805–817.

————. 1969. Inhibitory effect of oak leaf tannins on the hydrolysis of proteins by trypsin. Phytochemistry 8: 2119–2126.

FERRI, M. G. 1960. Contribution to the knowledge of the ecology of the "Rio Negro Caatinga" (Amazon). Bull. Res. Coun. Israel 8D: 195–208.

FITTKAU, E. J. 1967. On the ecology of Amazonian rain-forest streams. Atas do Simposio sobre a Biota Amazonica 3: 97–108.

————. 1971. Distribution and ecology of Amazonian chironomids (Diptera). Can. Ent. 103: 407–413.

FLORES, S. E., J. D. MEDINA, AND E. MEDINA. 1972. Compuestos humicos continidos en las "aguas negras" del Rio Atabapo. Acta cient. venez. 23 (Supl.): 84.

FOLDATS, E. 1962. La concentracion de oxigeno disuilta en las aguas negras. Acta biol. venez. 3: 149–159.

FORBES, E. B., AND S. I. BECHDEL. 1931. Mountain laurel and rhododendron as food for the white-tailed deer. Ecology 12: 323–333.

FOSBERG, F. R. 1945. Principal economic plants of tropical America. *In* F. Verdoorn (Ed.). Plants and plant science in Latin America. Pp. 18–35. Chronica Botanica Co., Waltham, Mass.

FOSTER, J. W. 1949. Chemical activities of fungi. Academic Press, New York. 647 pp.

FOSTER, R. B. 1973. Seasonality of fruit production and seed fall in a tropical forest ecosystem in Panama. Ph.D. Thesis, Duke University.

FOX, J. E. D. 1967. An enumeration of lowland dipterocarp forest in Sabah. Malay. Forester 30: 263–279.

————. 1968. Application of ecological research on silviculture in the lowland dipterocarp forests of Sabah. Proc. Symp. Rec. Adv. Trop. Ecol. 2: 710–773.

FOXWORTHY, F. W. 1927. Commercial timber trees of the Malay Peninsula. Malay. Forest Rec. 3. 195 pp.

FRANKIE, G. W., H. G. BAKER, AND P. A. OPLER. 1974. Comparative phenological studies of trees in tropical lowland wet and dry forest sites in Costa Rica. J. Ecol. (in press).

FRANKLAND, J. C. 1966. Succession of fungi on decaying petioles of *Pteridium aquilinum.* J. Ecol. 54: 41–63.

FREELAND, W. J., AND D. H. JANZEN. 1974. Strategies in herbivory by mammals: the role of plant secondary compounds. Am. Nat. 108 (in press).

FREEMAN, D. 1970. Report on the Iban. London School of Economics Monographs on Social Anthropology No. 41. Univ. of London, Athlone Press, New York. 317 pp.

GEISLER, R., H. A. KNÖPPEL, AND H. SIOLI. 1971. Ökologie der Süsswasserfische Amazoniens Stand und Zukunftsaufgaben der Forschung. Naturwissenschaften 58: 303–311.

GESSNER, F. 1964. The limnology of tropical rivers. Verh. int. Verein. theor. angew. Limnol. 15: 1090–1091.

GHASSEMI, M., AND R. F. CHRISTMAN. 1968. Properties of the yellow organic acids of natural waters. Limnol. Oceanogr. 13: 583–597.

GIMINGHAM, C. H. 1972. The ecology of heathlands. Chapman and Hall, London. 266 pp.

GLICK, Z., AND M. A. JOSLYN. 1970. Food intake depression and other metabolic effects of tannic acid on the rat. J. Nutr. 100: 509–515.

GOMEZ, E. D., D. J. FAULKNER, W. A. NEWMAN, AND C. IRELAND. 1973. Juvenile hormone mimics: effect on cirriped crustacean metamorphosis. Science, N.Y. 179: 813–814.

GONGGRYP, J. W., AND D. BURGER. 1948. Bosbouwkundige studiën over Suriname. H. Veen and Zonez, Wageningen, Netherlands. 262 pp.

GOODLAND, R. 1971. A physiognomic analysis of the 'Cerrado' vegetation of central Brasil. J. Ecol. 59: 411–419.

GORHAM, E. 1957. The development of peat lands. Q. Rev. Biol. 32: 145–146.

GRAY, B. 1972. Economic tropical forest entomology. A. Rev. Ent. 17: 313–354.

GRÜMMER, G., AND H. BEYER. 1960. The influence exerted by species of *Camelima* on flax by means of toxic substances. Brit. Ecol. Soc. Symp. 1: 153–157.

288

HANDLEY, W. R. C. 1954. A survey of previous work on mull and mor in relation to forest soils. Forest Commission Bull. 23. Her Majesty's Stationery Office, London. 115 pp.

HARDON, H. J. 1936. Podzol profiles in the tropics. Natuurk. Tijdschr. Ned.-Indië. 96: 25–41.

————. 1937. Padang soil, an example of podzol in the tropical lowlands. Proc. K. ned. Akad. Wet. 40: 530–538.

HARLEY, J. L. 1952. Associations between microorganisms and higher plants (Mycorrhiza). A. Rev. Microbiol. 6: 367–386.

HARRISON, J. L. 1955. Data on the reproduction of some Malayan mammals. Proc. zool. Soc. Lond. 125: 445–460.

————. 1962. The distribution of feeding habits among animals in a tropical rain forest. J. Anim. Ecol. 31: 53–64.

HARRISON, T. 1965. Some quantitative effects of vertebrates on the Borneo flora. Symposium on ecological research in humid tropics vegetation, Kuching, Sarawak, 1963. Pp. 164–169. UNESCO Science Cooperation Office for Southeast Asia.

HARSHBERGER, J. W. 1916. The vegetation of the New Jersey pine-barrens. C. Sower, Philadelphia. 329 pp.

HARTSHORN, G. S. 1973. The ecological life history and population dynamics of *Pentaclethra macroloba*, a tropical wet forest dominant and *Stryphnodendron excelsum*, an occasional associate. Ph.D. thesis, Univ. of Washington, Seattle.

HATHAWAY, D. E. 1958. Oak bark tannin. Biochem. J. 70: 34–42.

HEATH, G. W., AND H. G. C. KING. 1964. Litter breakdown in deciduous forest soils. Pp. 979–987. 8th Inter. Congr. Soil Sci., Bucharest, Romania.

HEINSELMAN, M. L. 1963. Forest sites, bog processes, and peatland types in the glacial Lake Agassiz Region, Minnesota. Ecol. Mongr. 33: 327–374.

————. 1965. String bogs and other patterned organic terrain near Seney, Upper Michigan. Ecology 46: 185–188.

HENDERSON, M. E. K. 1957. Metabolism of methoxylated aromatic compounds by soil fungi. J. Gen. Microbiol. 16: 686–695.

HERRERA, R. 1972. Suelos podzolicos tropicales en regiones de rios de aguas negras del territorio Amazonas. Characterizacion quimica y mineralogica de algunos perfiles. Acta cient. venezolana 23 (Supl.): 31.

HEYLIGERS, P. C. 1963. Vegetation and soil of a white-sand savanna in Suriname. N. V. Noord-Hollandsche Uitgevers Maatschappii. Amsterdam. 148 pp.

HLADIK, A., AND C. M. HLADIK. 1969. Rapports trophiques entre vegetation et primates dans la foret de Barro Colorado (Panama). Terre Vie 23: 25–117.

HOLTTUM, R. E. 1941. The uniform climate of Malaya as a barrier to plant migration. Proc. 6th Pacific Science Congress, Berkeley, California, 4: 669–671.

HON, L. Y. 1967a. Timber tests—damar minyak (*Agathis alba*). Malay. Forester 30: 140–144.

————. 1967b. Skin irritation caused by tembusu (*Fagraea fragrans*). Malay. Forester 30: 234.

HOWARD, W. E. 1967. Ecological changes in New Zealand due to introduced mammals. IUCN 10th Tech. Meeting, Lucern, Switzerland. IUCN Publ. No. 9 (New Series): 219–240.

HUININK, W. A. E. V. D. B. 1966. Structure, root systems and periodicity of savanna plants and vegetations in northern Surinam. Wentia 17: 1–162.

HURST, H. M., A. BURGES, AND P. LATTER. 1962. Some aspects of the biochemistry of humic acid decomposition by fungi. Phytochemistry 1: 227–231.

HUTCHINSON, G. E. 1957. A treatise on limnology, vol. 1. Wiley, New York. 1015 pp.

INGER, R. F. 1966. The systematics and zoogeography of the amphibia of Borneo. Fieldiana, Zool. 52: 1–402.

————, AND J. P. BACON. 1968. Annual reproduction and clutch size in rain forest frogs from Sarawak. Copeia 1968: 602–606.

————, AND P. K. CHIN. 1962. The fresh-water fishes of North Borneo. Fieldiana, Zool. 45: 1–268.

————, AND B. GREENBERG. 1963. The annual reproductive pattern of the frog *Rana erythraea* in Sarawak. Physiol. Zoöl. 36: 21–33.

————, AND ————. 1966. Ecological and competitive relations among three species of frogs (genus *Rana*). Ecology 47: 746–759.

————, AND ————. 1967. Annual reproductive patterns of lizards from a Bornean rain forest. Ecology 47: 1007–1021.

JACKSON, W. F. 1957. The durability of Malayan timbers. Malay. Forester 20: 38–48.

JANZEN, D. H. 1967. Synchronization of sexual reproduction of trees within the dry season in Central America. Evolution, Lancaster, Pa. 21: 620–637.

————. 1970. Herbivores and the number of tree species in tropical forests. Am. Nat. 104: 501–528.

————. 1971. Seed predation by animals. A. Rev. Ecol. Syst. 2: 465–492.

—————. 1972. Association of a rainforest palm and seed-eating beetles in Puerto Rico. Ecology 53: 258–261.

—————. 1973a. Sweep samples of tropical foliage insects: description of study sites, with data on species abundances and size distributions. Ecology 54: 659–686.

—————. 1973b. Sweep samples of tropical foliage insects: effects of seasons, vegetation types, elevation, time of day, and insularity. Ecology 54: 687–708.

—————. 1974. Epiphytic myrmecophytes: mutualism by ants feeding plants. Biotropica (in press).

—————, AND T. W. SCHOENER. 1968. Differences in insect abundance and diversity between wetter and drier sites during a tropical dry season. Ecology 49: 96–110.

JOACHIM, A. W. R. 1935. Studies on Ceylon soils II. General characteristics of Ceylon soils, some typical soil groups of the island, and a tentative scheme of classification. Trop. Agric. Mag. Ceylon agric. Soc. 84: 254–274.

JOHNSON, D. S. 1967a. On the chemistry of freshwaters in southern Malaya and Singapore. Arch. Hydrobiol. 63: 477–496.

—————. 1967b. Distributional patterns of Malayan freshwater fish. Ecology 48: 722–730.

—————. 1968. Malayan black waters. Proc. Symp. Rec. Adv. Trop. Ecol. (=Bull. int. Soc. trop. Ecol.) 1: 303–310.

JOHNSON W. E., AND A. D. HASLER. 1954. Rainbow trout production in dystrophic lakes. J. Wildl. Mgmt. 18: 113–134.

JONES, L. H. P., AND K. A. HANDRECK. 1967. Silica in soils, plants and animals. Adv. Agron. 19: 107–149.

KALSHOVEN, L. G. E. 1953. Important outbreaks of insect pests in the forests of Indonesia. Trans. 9th Inter. Cong. Ent. Pp. 229-234.

KAPUR, A. P. 1958. A report reviewing entomological problems in the humid tropical regions of south Asia. *In* Problems of humid tropical regions. Pp. 63–85. UNESCO.

KENDRICK, W. B. 1959. The time factor in the decomposition of coniferous leaf litter. Can. J. Bot. 37: 907–912.

—————, AND A. BURGES. 1962. Biological aspects of the decay of *Pinus sylvestris* leaf litter. Nova Hedwigia 4: 313–342.

KIRA, T., AND T. SHIDEI. 1967. Primary production and turnover of organic matter in different forest ecosystems of western Pacific. Jap. J. Ecol. 17: 70–87.

KLINGE, H. 1965. Podzol soils in the Amazon basin. J. Soil Sci. 16: 95–103.

—————. 1967. Podzol soils: a source of blackwater rivers in Amazonia. Atas do Simposio sobre a Biota Amazonica 3: 117–125.

—————. 1968. Report on tropical podzols. Unpubl. FAO Report. 88 pp.

—————. 1969. Climatic conditions in lowland tropical podzol areas. Tropical Ecology (=Bull. int. Soc. trop. Ecol.) 10: 222–239.

—————. 1973. Root mass estimation in lowland tropical rain forests of central Amazonia, Brazil. I. Fine root masses of a pale yellow latosol and a giant humus podzol. Tropical Ecology (=Bull. int. Soc. trop. Ecol.) 14: 29–38.

—————, AND W. OHLE. 1964. Chemical properties of rivers in the Amazonian area in relation to soil conditions. Verh. int. Verein. theor. angew. Limnol. 15: 1067–1076.

—————, AND L. A. RODRIGUES. 1968. Litter production in an area of Amazonian Terra Firme Forest. Part II. Mineral nutrient content of the litter. Amazoniana 1: 303–310.

KNERER, G., AND C. E. ATWOOD. 1973. Diprionid sawflies: polymorphism and speciation. Science, N.Y. 179: 1090–1099.

KWAN, W. Y., AND T. C. WHITMORE. 1970. On the influence of soil properties on species distribution in a Malayan lowland dipterocarp rain forest. Malay. Forester 33: 42–54.

LAKSHMANAN, N. K. 1968. The forest types of the Nilgiris and its ecological problems. Proc. Symp. Rec. Adv. Trop. Ecol. 2: 407–418.

LANGENHEIM, J. H. 1973. Leguminous resin-producing trees in Africa and South America. *In* B. J. Meggers, E. S. Ayensu and W. D. Duckworth (Eds.). Tropical forest ecosystems in Africa and South America: a comparative review. Pp. 89–104. Smithsonian Institution Press, Washington, D.C.

LEE, K. E., AND T. G. WOOD. 1968. Preliminary studies of the role of *Nasutitermes exitiosus* (Hill) in the cycling of organic matter in a yellow podzolic soil under dry sclerophyll forest in South Australia. Trans. 9th Int. Congr. Soil Sci. 2: 11–18.

—————, AND —————. 1971. Physical and chemical effects on soils of some Australian termites, and their pedologcal significance. Pedobiologia 11: 376–409.

LEVANIDOV, V. Y. 1949. The biology of aquatic organisms. The significance of allochthonic material as a food resource in a body of water using *Ascellus aquaticus* L. as an example. Acad. Sci. USSR, Proc. All-Union Hydrobiological Soc. 1: 100–117.

LEVIN, D. A. 1971. Plant phenolics: an ecological perspective. Am. Nat. 105: 157–181.

LINDEMAN. J. C. 1953. The vegetation of the coastal region of Suriname. Drukkerij en Uitgovers-Maatschappji v/h Kemink en Zoon N. V. Domplein 2—Utrecht, Netherlands. 135 pp.

————, AND S. P. MOOLENAAR. 1959. Preliminary survey of the vegetation types of northern Suriname. *In* I. A. de Hulster and J. Lanjouw (Eds.). The vegetation of Suriname. Van Eedenfonds, Amsterdam, Netherlands. 45 pp.

LLOYD, F. E., AND G. W. SCARTH. 1922. The bog-forests of Lake Memphremagog: their destruction and consequent successions in relation to water levels. Trans. R. Soc. Can. 16: 45–48.

LLOYD, M., R. F. INGER, AND F. W. KING. 1968. On the diversity of reptile and amphibian species in a Bornean rain forest. Am. Nat. 102: 497–515.

LONGHURST, W. M., H. K. OH, M. B. JONES, AND R. E. KEPNER. 1968. A basis for the palatability of deer forage plants. Trans. 33rd North American Wildlife and Natural Resource Conference, Houston, Texas. Pp. 181–192.

LOVELESS, A. R. 1961. A nutritional interpretation of sclerophylly based on differences in the chemical composition of sclerophyllous and mesophytic leaves. Ann. Bot. 25: 168–184.

LUTZ, L. 1928. Sur le role biologique du tanin dans la cellule vegetale. Bull. Soc. bot. Fr. 75: 9–18.

MANSINGH, A., T. S. SAHOTA, AND D. A. SHAW. 1970. Juvenile hormone activity in the wood and bark extracts of some forest trees. Can. Ent. 102: 49–53.

MARLIER, G. 1965. Etude sur les lacs de l'Amazonie centrale. Cadernos da Amazonia. 5. Instituto Nacional de Pesquisas da Amazonia, Manaus, Amazonia. 51 pp.

————. 1973. Limnology of the Congo and Amazon Rivers. *In* B. J. Meggers, E. S. Ayensu and W. D. Duckworth (Eds.). Tropical forest ecosystems in Africa and South America: a comparative review. Pp. 223–238. Smithsonian Institution Press, Washington, D.C.

MARSHALL, A. G. 1970. The life cycle of *Basilia hispida* Theodor 1967 (Diptera: Nycteribiidae) in Malaysia. Parasitology 61: 1–18.

MATTHEW, K. M. 1959. The vegetation of Kodaikanal grassy slopes. J. Bombay nat. Hist. Soc. 56: 387–422.

————. 1971. The flowering of the strobilanth (Acanthaceae). J. Bombay nat. Hist. Soc. 67: 502–506.

McCLURE, H. E. 1966. Flowering, fruiting and animals in the canopy of a tropical rain forest. Malay. Forester 29: 182–203.

McCONNELL, W. J. 1968. Limnological effects of organic extracts of litter in a southwestern impoundment. Limnol. Oceanogra. 13: 343–349.

McCURRACH, J. C. 1960. Palms of the world. Harper and Brothers, New York. 290 pp.

MEDWAY, L. 1969. The wild mammals of Malaya and offshore islands. Oxford University Press, Kuala Lumpur.

————. 1972a. Phenology of a tropical rain forest in Malaya. Biol. J. Linn. Soc. 4: 117–146.

————. 1972b. Reproductive cycles of the flat-headed bats *Tylonycteris pachypus* and *T. robustula* (Chiroptera: Vespertilioninae) in a humid equatorial environment. Zool. J. Linn. Soc. 51: 33–62.

MEIJER, W. 1969. Fruit trees in Sabah (North Borneo). Malay. Forester 32: 252–265.

————. 1970. Regeneration of tropical lowland forest in Sabah, Malaysia forty years after logging. Malay. Forester 33: 204–229.

————. 1973. Devastation and regeneration of lowland dipterocarp forests in Southeast Asia. Bioscience 23: 528–533.

MENON, P. K. B. 1956. Siliceous timbers of Malaya. Malay. Forest Rec. No. 19. 55 pp.

MILES, P. W. 1969. Interaction of plant phenols and salivary phenolases in the relationship between plants and Hemiptera. Entomologia exp. appl. 12: 736 744.

MIZUNO, T., AND S. MORI. 1970. Preliminary hydrobiological survey of some southeast Asian inland waters. Biol. J. Linn. Soc. 2:77–117.

MOHR, E. C. J. 1944. The soils of equatorial regions with special reference to the Netherlands East Indies. Edwards, Amsterdam. 766 pp.

————, AND F. A. VAN BAREN. 1954. Tropical soils. The Hague, Amsterdam. 498 pp.

MONK, C. D. 1966. An ecological study of hardwood swamps in north-central Florida. Ecology 47: 649–654.

————. 1968. Successional and environmental relationships of the forest vegetation of north central Florida. Am. Midl. Nat. 79: 441–457.

————. 1971. Leaf decomposition and loss of ^{45}Ca from deciduous and evergreen trees. Am. Midl. Nat.. 86: 379–384.

MORS, W. B., AND C. T. RIZZINI. 1966. Useful plants of Brazil. Holden-Day, San Francisco. 166 pp.

MUENSCHER, W. C. 1970. Poisonous plants of the United States. MacMillan, New York. 277 pp.

NIERENSTEIN, M. 1934. The natural organic tannins. Churchill, London. 319 pp.

NORRIS, D. O. 1969. Observations on the nodulation status of rainforest leguminous species in Amazonia and Guyana. Trop. Agric., Trin. 46: 145–151.

NYE, P. H. 1955. Some soil forming processes in the humid tropics. IV. The action of the soil fauna. J. Soil Sci. 6: 73–83.

————, AND D. J. GREENLAND. 1960. The soil under shifting cultivation. Tech. Comm. No. 51, Commonwealth Bureau of Soils, Harpendon. 153 pp.

NYKVIST, N. 1959. Leaching and decomposition of litter. Oikos 10: 190–224.

————. 1963. Leaching and decomposition of water soluble organic substances from different types of leaf and needle litter. Studia Forest Suecica, No. 3: 3–31.

OGAWA, H., K. YODA, AND T. KIRA. 1961. A preliminary survey on the vegetation of Thailand. Nature and life in southeast Asia 1: 21–158.

OHTAKI, T., R. D. MILKMAN, AND C. M. WILLIAMS. 1967. Ecdysone and ecdysone analogues: their assay on the fleshfly Sarcophaga peregrina. Proc. nat. Acad. Sci. 58: 981–984.

PAMMEL, L. H. 1911. A manual of poisonous plants. Torch Press, Cedar Rapids, Iowa. 977 pp.

PATRICK, R. 1964. A discussion of the results of the Catherwood Expedition to the Peruvian headwaters of the Amazon. Verh. int. Verein. theor. angew. Limnol. 15: 1084–1090.

PERRIN, R. M. S., E. H. WILLIS, AND C. A. H. HODGE. 1964. Dating of humus podzols by residual radiocarbon activity. Nature, Lond. 202: 165.

PIERCE, T. G. 1972. Acid intolerant and ubiquitous Lumbricidae in selected habitats in north Wales. J. Anim. Ecol. 41: 397–410.

POLAK, E. 1933. Über Torf und Moor in Niederlandisch Indien. Verh. K. Akad. Wet. 30: 1–85.

PONOMAREVA, V. V. 1969. Theory of podzolization. U.S. Dept. Commerce. 309 pp.

POORE, M. E. D. 1964. Integration in the plant community. Brit. Ecol. Soc. Jubilee Symp. Suppl. J. Ecol. 52: 213–226.

————. 1968. Studies in Malaysian rain forest. I. The forest on triassic sediments in Jengka Forest Reserve. J. Ecol. 56: 143–196.

QUISUMBLING, E. 1935. The occurrence of lactiferous vessels in the mature bark of Hevea brasiliensis. Univ. Calif. Publs. Bot. 13: 319–332.

————. 1951. Medicinal plants of the Philippines. Bureau of Print., Manila. 1234 pp.

READER, R. J., AND J. M. STEWART. 1972. The relationship between net primary production and accumulation for a peatland in southeastern Manitoba. Ecology 53: 1024–1037.

RETNAKARAN, A. 1970. Blocking of embryonic development in the spruce budworm, Cheristoneura fumiferana (Lepidoptera: Tortricidae), by some compounds with juvenile hormone activity. Can. Ent. 102: 1592–1596.

RIBEREAU-GAYON, P. 1972. Plant phenolics. Hafner Publishing Co., New York. 254 pp.

RICHARDS, P. W. 1936. Ecological observations on the rain forest of Mount Dulit, Sarawak. Part I. J. Ecol. 24: 1–37.

————. 1941. Lowland tropical podsols and their vegetation. Nature, Lond. 148: 129–131.

————. 1952. The tropical rain forest. Cambridge Univ. Press, Cambridge. 450 pp.

————. 1963. Soil conditions in some Bornean lowland plant communities. Symposium on Ecological Research in Humid Tropics Vegetation, Kuching, Sarawak. Pp. 198–205.

————. 1973. Africa, the "Odd man out." In B. J. Meggers, E. S. Ayensu and W. D. Duckworth (Eds.). Tropical forest ecosystems in Africa and South America: a comparative review. Pp. 21–26. Smithsonian Institution Press, Washington, D.C.

ROBERTS, T. R. 1973. Ecology of fishes in the Amazon and Congo basins. In B. J. Meggers, E. S. Ayensu and W. D. Duckworth (Eds.) Tropical forest ecosystems in Africa and South America: a comparative review. Pp. 239–254. Smithsonian Institution Press, Washington, D.C.

ROBINSON, M. E. 1936. The flowering of Strobilanthes in 1934. J. Bombay nat. Hist. Soc. 38: 117–122.

RODE, A. A. 1970. Podzol-forming process. U.S. Dept. Commerce. 387 pp.

ROTHSCHILD, G. 1971. Animals in Bako National Park. Malayan Nat. J. 24: 163–169.

ROUGERIE, G. 1958. Acidite des eaux en millieu forestier intertropical. Compte rendu hebd. Seanc. Acad. Sci., Paris 24: 447–449.

SANDWITH, N. Y. 1925. Humboldt and Bonplant's itinerary in Venezuela. Kew Bull. 1925: 295–310.

SANTAPAU, H. 1962. Gregarious flowering of Strobilanthes and bamboos. J. Bombay nat. Hist. Soc. 59: 688–696.

SCHNITZER, M., AND S. U. KHAN. 1972. Humic substances in the environment. Marcel Dekker, Inc., New York. 327 pp.

SCHULTES, R. E. 1969. De plantis toxicariis e mundo novo tropicale commentationes VI. Notas entotoxicologicas acerca de la flora Amazonica de Colombia. In Il Simposio y foro de biologia tropical Amazonica. Pp. 178–196.

SEAL, J. 1958. Rainfall and sunshine in Sarawak. Sarawak Mus. J. 8 (no. 11, new series): 500–544.

SENEAR, F. E. 1933. Dermatitis due to woods. J. Am. Med. Ass. 101: 1527–1532.

SEREVO, T. S. 1960. Silviculture of tropical rain forest with special reference to the Philippine dipterocarp forest. Proc. Fifth World Forestry Congress. Pp. 1986–1993.

SHANKS, R. E., AND J. S. OLSON. 1961. First-year breakdown of leaf litter in southern Appalachian forests. Science, N.Y. 134: 194–195.

SHAPIRO, J. 1957. Chemical and biological studies on the yellow organic acids of lake water. Limnol. Oceanogr. 2: 161–179.

SHELFORD, R. W. C. 1916. A naturalist in Borneo. Fisher, London. 331 pp.

SHERROD, L. L., AND K. H. DOMSCH. 1970. The role of phenols and B-glycosidase in the pathogeneity mechanism of *Gliocladium catenulatum* to roots of peas (*Pisum sativum* L.). Soil Biol. Biochem. 2: 197–201.

SIOLI, H. 1954. Gewasserchemie und Vorgäng in den Böden in Amazonasgebiet. Naturwissenschaften 19: 456–457.

————. 1955. Beiträge zur regionales Limnologie des Amazonasgebietes III. Ueber einige Gewässer des oberen Rio Negro Gebietes. Arch. Hydrobiol. 50: 1–32.

————. 1960. Estratificacao radicular numa caatinga baixa do alto Rio Negro. Bolm. Mus. para. 'Emilio Goeldi,' N.S., No. 10: 1–9.

————. 1964. General features of the limnology of Amazonia. Verh. int. Verein. theor. angew. Limnol. 15: 1053–1058.

————. 1967a. The Cururu Region in Brazilian Amazonia, a transition zone between hylaea and cerrado. J. Indian bot. Soc. 46: 452–462.

————. 1967b. Studies in Amazonian water. Atas do Simposio sobre a Biota Amazonica 3: 9–50.

————. 1968a. Principal biotopes of primary production in the waters of Amazonia. Proc. Symp. Rec. Adv. Trop. Ecol. 2: 591–600.

————. 1968b. Hydrochemistry and geology in the Brazilian Amazon region. Amazoniana 1: 267–277.

————, AND H. KLINGE. 1962. Solos, tipos de vegetacao e aguas na Amazonia. Bolm. Mus. para. 'Emilio Goeldi,' No. 1. 41 pp.

————, G. H. SCHWABE, AND H. KLINGE. 1969. Limnological outlooks on landscape ecology in Latin America. Trop. Ecol. (= Bull. int. Soc. trop. Ecol.) 10: 72–82.

SKOROPANOV, S. G. 1968. Reclamation and cultivation of peat-bog soils. U.S. Dept. Commerce. 234 pp.

SLOOTEN, D. F. VAN. 1952. Sertulum Dipterocarpacearum malayensium V. The Dipterocarpaceae of Eastern Malaysia (Celebes, the Moluccos and New Guinea). Reinwardtia 2: 1–68.

SLUD, P. 1964. The birds of Costa Rica. Bull. Am. Mus. nat. Hist. 128: 1–430.

SMALL, E. 1972a. Ecological significance of four critical elements in plants of raised sphagnum peat bogs. Ecology 53: 498–503.

————. 1972b. Photosynthetic rates in relation to nitrogen recycling as an adaptation to nutrient deficiency in peat bog plants. Can. J. Bot. 50: 2227–2233.

SMYTHE, N. 1970. Relationships between fruiting seasons and seed dispersal methods in a neotropical forest. Am. Nat. 104: 25–35.

SMYTHIES, B. E. 1960. The birds of Borneo. Oliver and Boyd, Edinburgh. 562 pp.

SNOW, D. W. 1962. A field study of the black and white manakin *Manacus manacus* in Trinidad. Zoologica, N.Y. 47: 65–104.

————. 1965. A possible selective factor in the evaluation of fruiting seasons in tropical forests. Oikos 15: 274–281.

SOMBROEK, W. G. 1966. Amazon soils. A reconnaissance of the soils of the Brazilian Amazon region. Centre for Agricultural Publications and Documentation, Wageningen. 292 pp.

SOUTHWOOD, T. R. E. 1972. The insect/plant relationship—an evolutionary perspective. *In* H. F. van Emden (Ed.). Insect/plant relationships. Symposia Roy. Ent. Soc. Lond. No. 6. Pp. 3–30.

SPIELMAN, A., AND C. M. WILLIAMS. 1966. Lethal effects of synthetic juvenile hormone on larvae of the yellow fever mosquito, *Aedes aegypti*. Science, N.Y. 154: 1043–1044.

SPRUCE, R. 1908. Notes of a botanist on the Amazon and Andes (2 vols.). A. R. Wallace (Ed.). London.

STANDLEY, P. C. 1920-1926. Trees and shrubs of Mexico. Contr. U.S. Natn. Herb. 23: 1–1721.

STARK, N. M. 1970. The nutrient content of plants and soils from Brazil and Surinam. Biotropica 2: 51–60.

————. 1971a. Nutrient cycling I. Nutrient distribution in some Amazonian soils. Tropical Ecology (= Bull. int. Soc. trop. Ecol.) 12: 24–50.

————. 1971b. Nutrient cycling II. Nutrient distribution in some Amazonian vegetation. Tropical Ecology (= Bull. int. Soc. trop. Ecol.) 12: 177–201.

STEENIS, C. G. G. J. VAN. 1942. Gregarious flowering of *Strobilanthes* (Acanthaceae) in Malaysia. Ann. Rev. Bot. Gdn., Calcutta (150th Anniversary Volume). Pp. 91–97.

STROSS, R. G., AND A. D. HASLER. 1960. Some lime induced changes in lake metabolism. Limnol. Oceanogr. 5: 265–272.

TAKEUCHI, M. 1961. The structure of the Amazonian vegetation. III. Campina forest in the Rio Negro region. J. Fac. Sci. Tokyo Univ. Ser. III, 8: 27–35.

————. 1962a. The structure of the Amazonian vegetation. IV. High campina forest in the upper Rio Negro. J. Fac. Sci. Tokyo Univ. Ser. III, 8: 279–288.

————. 1962b. The structure of the Amazonian vegetation. VI. Igapo. J. Fac. Sci. Tokyo Univ. Ser. III, 8: 297–304.

TAMIR, M., AND E. ALUMOT. 1970. Carob tannins-growth depression and levels of insoluble nitrogen in the digestive tracts of rats. J. Nutr. 100: 573–580.

TAMM, C. O. 1955. Some observations on the nutrient turn-over in a bog community dominated by *Eriophorum vaginatum* L. Oikos 5: 189–194.

THAPA, R. S. 1968. Observations on the cerambycid borer *Cyriopalus wallacei* Pasc. (Coleoptera: Cerambycidae) in dipterocarp forests of Sabah, E. Malaysia. Malay. Forester 31: 348–355.

THOM, B. G. 1967. Mangrove ecology and deltaic geomorphology: Tobasco, Mexico. J. Ecol. 55: 301–343.

UNGEMACH, H. 1969. Chemical rain water studies in the Amazon region. *In* Simposio y Foro de Biologia Tropical Amazonica. Pp. 354–358. Association pro Biologia Tropical.

WAKSMAN, S. A. 1938. Humus: origin, chemical composition and importance in nature. Baillier, Tindall and Cox, London. 526 pp.

WANG, T. S. C., S. CHENG, AND H. TUNG. 1967a. Dynamics of soil organic acids. Soil Sci. 104: 139–144.

————, T. YANG, AND T. CHUANG. 1967b. Soil phenolic acids as plant growth inhibitors. Soil Sci. 103: 239–246.

WARD, P. 1969. The annual cycle of the yellow-vented bulbul *Pychonotus goiavier* in a humid equatorial environment. J. Zool. Lond. 157: 25–45.

WEBB, L. J. 1968. Environmental relationships of the structural types of Australian rainforest vegetation. Ecology 49: 296–311.

WHALLEY, W. B. 1959. The toxicity of plant phenolics. *In* J. W. Fairbairn (Ed.). The pharmacology of plant phenolics. Pp. 27–37. Academic Press, London.

WHITEHEAD, D. C. 1964. Identification of *p*-hydroxybenzoic, vanillic, *p*-coumaric and ferulic acids in soils. Nature, Lond. 202: 417–418.

WHITFORD, H. N. 1906. The vegetation of Lamao Forest Reserve. Philipp. J. Sci. 1: 373–431.

WHITTAKER, R. H. 1970. The biochemical ecology of higher plants. *In* E. Sondheimer and J. B. Simeone (Eds.). Chemical ecology. Pp. 43–70. Academic Press, New York.

————, AND P. P. FEENY. 1971. Allelochemics: chemical interactions between species. Science, N.Y. 171: 757–770.

WILLIAMS, C. M. 1970. Hormonal interactions between plants and insects. *In* E. Sondheimer and J. B. Simeone (Eds.). Chemical ecology. Pp. 103–132. Academic Press, New York.

WILLIAMS, L. 1960. Little known wealth of tropical forests. Proc. Fifth World Forestry Congress. Pp. 2003–2007.

WILLIAMS, P. M. 1968. Organic and inorganic constituents of the Amazon River. Nature, Lond. 218: 937–938.

WILLIAMS, W. A., R. S. LOOMIS, AND P. DE T. ALVIM. 1972. Environments of evergreen rain forests on the lower Rio Negro, Brazil. Tropical Ecology (=Bull. int. Soc. trop. Ecol.) 13: 65–78.

WILLIS, J. C. 1966. A dictionary of the flowering plants and ferns. Cambridge Univ. Press. 1214 pp.

WOOD, G. H. S. 1956. The dipterocarp flowering season in North Borneo 1955. Malay. Forester 19: 193–201.

WOODS, F. W., AND C. M. GALLEGOS. 1970. Litter accumulation in selected forests of the Republic of Panama. Biotropica 2: 46–50.

WYCHERLEY, P. R. 1968. Climatological and phenological phenomena in Malaysia. Proc. Symp. Rec. Adv. Trop. Ecol. 1: 138–143.

YARDENI, D., AND M. EVANARI. 1952. The germination inhibiting, growth inhibiting and phytocidal effect of certain leaves and leaf extracts. Phyton, B. Aires 2: 11–16.

ZINKE, P. J., AND A. CASTRO. 1973. Fertility storage in forest soils of the Amazon basin, Brazil. Unpub. ms.

23

Recent Human Activities in the Brazilian Amazon Region and Their Ecological Effects

HARALD SIOLI

Like all other organisms, man, as long as he has lived on earth, has interacted with his environment and will continue to do so. In an ecological sense, "life" can be defined in the same way for man as for all other creatures: "Life" is not only the process of certain physicochemical reactions in the living matter, but it is also the constant interaction between the living organism, which has its internal laws (i.e., physiology, ethology, psychology, etc.), and the environment, which has internal laws according to which it is structured and operates. Between these two systems—organism and environment—there is a tension field that has to be overcome actively and passively by the organism for it to continue its life. That tension field is the stage on which the play of life is conducted by partners with equal rights, namely organism and environment, which influence, alter, and shape each other. Feedbacks are the rule so that any alteration produced in one partner reacts against the other. The result of that interplay is a functional unity on earth, for which we use the new, rather vague word: ecosystem.

The intensiveness of the interactions between organism and environment can be of very different degrees, ranging from almost imperceptible modifications inflicted on one or both partners to a ruthless struggle that may end with a total breakdown and death of one of them, a consequence that will

HARALD SIOLI, Max-Planck-Institut für Limnologie, Plön, West Germany.

then affect the survivor. Between these two extremes extends a whole scale of degrees of alteration. Their classification becomes a problem if we want to relate them to eventual advantages or disadvantages for the partner who causes them.

This interaction applies to man and his environment. The partner given to man by nature, is, sensu largo, the landscape, or, in modern scientific terms, the geosynergy or biogeocenosis, which he encountered when he made his appearance on the stage. From the beginning, man has dealt with his landscape-environment, has lived from it, has more or less altered it, and sometimes has even died with it. The natural landscapes he first encountered, as well as the altered landscapes produced by his interaction, are all more or less "favorable" or "hostile" for his existence. Thus, if we want to do more than describe the changes man has wrought in certain landscapes, and to evaluate them in regard to human life, we must try to construct a value-scale related to man's material as well as spiritual requirements.

A philosophical or aesthetic value would be too subjective, as well as too transitory temporally and geographically. Individual emotional values—ranging from the enthusiasm for nature to the demand that the "equilibrium of nature" not be disturbed —are equally arbitrary. A natural equilibrium, in the sense of an absolutely ideal "steady state," does not exist. Equilibrium in nature is neither stable nor labile, but indifferent. It resembles the behavior of a ball on the surface of an even table; again

295

and again the ball gets pushed from diverse directions, causing it to roll a bit before coming to rest at a new point on the table until the next impulse sends it on again. If too violent a stroke rolls it over the edge of the table, the whole play has come to an end.

Seen from the whole, it does not matter on what part of the table—the world stage—the ball rests. In the life of man, however, the different points are not of equal value, for they represent different portions of his environment, which, as we have seen, may be more or less "favorable" or "hostile." These values must be expressed in any scale we try to develop, and as a guiding line for our attempt, the "quality" of a landscape-as-living-space has been proposed (Sioli 1969b:309). Quality is to be understood more or less as a product (not a sum!) of the variety of the existing life (number of species of plants and animals), of the biomass (standing crop) and number of individual plants and animals, of the productivity (primary, secondary, etc., production) in organic matter and its energy-content, of the number of people who can live on the specific landscape, of their cultural level and content, and of the richness of their personal experiences. These factors cannot be expressed, at least at present, in quantitative terms, but perhaps they can provide a starting point and some tentative directives for evaluation of landscapes as habitats—as ecologists we would say: of biotopes—for mankind.

With these prefatory remarks, let us see what man has done with the landscape of Amazonia, what has been the result of his past activities in different regions, what are the present tendencies, and what are the outlooks and possibilities for the future.

Man's presence in the Amazonian lowlands dates back only a few millennia. Compared with other continents, man here is a newcomer, who probably entered the humid warm hylea in several waves, coming mainly from the northwestern direction and generally following the river courses. He evidently came as a hunter and fisherman and collector of wild plant foods, but most groups ultimately adopted agriculture to some degree. Aboriginal man existed in Amazonia in relatively small numbers, living only in tribal communities, and never building real cities. He collected, hunted, fished, and planted only to satisfy his own needs. Thus, he did not significantly interfere with the structure and dynamics of the natural ecosystems. The waters remained full of fishes, turtles, manatees, and caymans, and in the jungle there was plenty of game, as Carvajal, Rojas, Acuña (1941), and other early discoverers reported. Also, the small plantations, forming isolated spots in the storied forest cover and abandoned after a few years, did not damage the continuity of the great forest. Settlements were mostly located near the banks of the rivers, the preferred places being uninundated margins of the Amazon itself or its great shore lagoons (the várzea lakes), or along the courses of major affluents, such as, Rio Tapajós and Rio Xingú.

Some settlements either remained for considerable periods at the same place or were often reoccupied, to judge from the accumulation of by-products of their "metabolism": ashes, charcoal, bones, etc., along with broken pottery, transforming and overlaying the original yellow-brown forest soil, or sometimes the white sandy soil of "campina" vegetation, with "terra preta" (black earth), which exists in layers of up to 1 meter or more in thickness. In contrast to the original acid "brown loam," which is poor in nutrients for plant growth, or the even more sterile bleached sands, the terras pretas are more or less neutral and enriched in calcium, phosphate, etc. As a result of their renowned fertility, they are often utilized by the neo-Brazilian settlers for their plantations.

The terras pretas, however, are only small spots of local importance; relative to Amazonia as a whole, they represent nothing at all. The pre-Columbian Indians did not change the biogeocenosis, the landscape of their country, which did not offer favorable conditions for development of high cultures and greater population centers. No easily accessible metal ores are found within the hylea and, in most of Amazonia (that portion covered by the Pliocene-Pleistocene "Barreiras" sediments), even rocks and stones are lacking. Timber and palm-leaves were the raw material for the construction of shelters, which vanished soon after the departure of the builders. They left no ruins of abandoned cities or other monuments to testify to a life-form that had achieved relative independence from the original conditions of the surroundings. On the contrary, the life-form or cultural pattern of the aboriginal peoples of Amazonia seems to have been shaped thoroughly by the environment, even in its spiritual aspects, which filled both the celestial and underwater kingdoms with replicas of the jungle animals in the form of ghosts.

After the discovery of Amazonia and during the first centuries of Portuguese colonization, i.e., until the second half of last century, little changed in the general situation. The terra firme, which occupied most of the area, remained practically untouched and exploitation activities concentrated more on the waters and on the várzea of the Amazon, as the flood-plain of the great river is called. Let us start with the waters, which have suffered less from the effects of human interaction.

Since pre-European times, the rivers have been the main and until very recent years almost the only traffic routes. In spite of this, all watercourses still flow in their original beds and retain their natural shores. No industry has established itself in the Amazonian lowland, and consequently there is no problem of pollution with industrial wastes. Since no agricultural fields in Amazonia are treated according to modern methods with mineral fertilizers, run-off does not introduce such fertilizer salts into the waters, starting their eutrophication. Pollution by domestic (urban) sewage fortunately does not yet affect the waters to any important degree. In all rivers (with a few exceptions such as the waterfront of Manaus) one still can drink the water without danger of intestinal infections. The watermass of the Amazon system is so enormous, the final discharge into the ocean being an annual average of above 200,000 cubic meters per second, and the human population by comparison is still so sparse, that the self-purification capacity of the waters is more than sufficient to digest completely the end products of human metabolism in a short distance. It seems to me desirable that this situation, which is almost unique in the world, should be maintained, but it is endangered by the fact that many industries are looking for places where clean water is cheaply available and the expense of purification of the discharged waters can be avoided.

A few species of aquatic fauna have been heavily reduced in number by the neo-Brazilian settlers, who, unlike the natives, exploit the resources for exportation in addition to satisfaction of their own needs. Thus, the manatee (Trichechus inunguis) has practically vanished not only from the lower Amazon and its shore lakes but also from most other places where formerly it was plentiful. The great river turtle (Podocnemis expansa) has virtually disappeared from the same region, along with the smaller "tracajá" (Podocnemis dumeriliana). These animals were caught just when they emerged onto the sandy river beaches to lay their eggs, and then sold on the markets of Manaus and smaller cities; the eggs have also been collected by the millions. The great "pirarucú" fish (Arapaima gigas), which furnished when dried the main supply of durable protein food in the Amazonian interior, has become so decimated during the past 2-3 decades that its meat is now scarce and too expensive for general consumption. The fish is speared just after moving from the river into the várzea lakes, where it spawns, rears, and protects its young. The cayman or "jacaré" (Caiman niger), still abundant throughout Amazonian waters during the first half of this century, has now almost completely been transformed into belts, purses, billfolds, etc., and it is now rare luck to see one specimen even on a long trip through the rivers, lakes, and igapós. For years, the tanneries of Manaus alone have received 5 million or more jacaré skins annually. The disappearance of the jacarés from certain water bodies, however, has not, as one might expect, increased the number of fish; on the contrary, the fish seem to have decreased along with the jacarés. The ecological reason for this curious phenomenon has been analyzed by Fittkau (1970).

Passing now from the aquatic to the terrestrial biotopes, we must first speak about the ecological differences between terra firme and várzea. Terra firme designates terrain with sufficient elevation to escape inundation. It occupies by far the largest area and consists in central Amazonia of deposits laid down in the enormous Pliocene-Pleistocene inland lake that extended from the foot of the Andes to the Atlantic. These deposits, called "series of the Barreiras," are up to 300 meters thick. The soil was derived principally from the Guayana shield in the north and the central Brazilian upland in the south, both ancient regions composed of granitic and gneissic rocks and some sandstones. Both are bedrocks poor in inorganic nutrients for plant growth. The resulting weathering products were intensely leached by equatorial rains and then transported into the lake, where the water acted again on them as they settled on the lake bottom. After the disappearance of the lake, these soils and underlaying material were reexposed to the leaching effects of the heavy rains of the Amazonian climate. The result is extremely poor and acid soils, with almost no reserves of inorganic nutrients for plant growth.

In spite of these edaphic conditions, the terra firme is, with only few local exceptions, generally covered by high, closed forest, which gives the region an aspect of exuberant vitality falsely attributed to extremely fertile soils. The apparent fertility of that landscape, however, represents nutrients accumulated during centuries or millennia in the *living* matter, and their constant and uninterrupted circulation through generations of vegetation with a minimum of loss. The soil serves more as a mechanical substratum for the support of the tall trees and as a reservoir for water than as a supplier of nutrients. This statement is based on recent research on the waters and the soils of Amazonia. Only the western part of the Amazonian lowland, adjacent to the Andean range, has received younger sediments eroded from the Andes and consequently developed richer soils.

The ecological situation on the várzea is different from that of the central and lower Amazonian terra firme. When the glacial period ended, the rise in ocean level drowned the wide river valleys of lower Amazonia and the rivers started to fill with alluvium brought from their headwater zones, forming new flood plains related to the altered river level. The largest area was along the Amazon itself, because of its enormous discharge of turbid white water. The várzea shall be emphasized here since the sedimentation zones of the affluents are relatively unimportant. The Amazon's sediment load does *not* come from the ancient Guayana shield or central Brazil (as did the material composing the terra firme), but by far the greatest amount is derived directly or indirectly from the Andes with their complex lithology and young weathering crust, including abundant volcanic material. Therefore, the sediment particles carried by the Amazon (the same is true for such as the Rio Madeira) are less leached than the material of the terra firme and contain much larger reserves of plant nutrients. Every year during the flood period, the Amazon deposits a new layer of fresh, more or less neutral, and minerally rich sediments on the surface of the flooded stretches of its várzea, thus annually renewing the fertility of those areas. Along the lower Amazon, the várzea is covered by forest only near the banks.

The deepest depressions are filled by shallow várzea lakes and the shore-dam-forest is substituted by natural floodable grasslands or the várzea savannas. On the várzea of the upper Amazon, these vast campos are missing and the forest extends into the water of the várzea lakes.

Because of these differences, the effects of agricultural utilization on terra firme and the várzea are different, even when the same methods are employed, namely, cutting and burning the original forest and subsequently planting some crop for a shorter or a longer period. Agriculture was and still is the main occupation of the European colonizers and the subsequent neo-Brazilian settlers—interrupted only temporarily by the "rubber-boom," or locally by gold and diamond rushes. Only very recently has some industry been established in a few areas. Let us start, therefore, with the practices of agriculture in Amazonia and their effects on the ecology.

Naturally, ecological thinking has not been the basis for the expansion of agriculture or for the introduction of European agricultural principles into Amazonia. Some 25 years ago, however, the then Director of Instituto Agronômico do Norte at Belém, Felisberto C. de Camargo, began with his own insight into the ecology of the region and the results of studies by his collaborators to develop general ideas for a future, lasting, and productive utilization of the enormous area. Until after World War II, the "mentality of extractivism" dominated without restrictions the economic thinking in Amazonia. Although from the beginning of the Portuguese colonization and settlement, diverse attempts had been made to establish agricultural zones, until the 1880s aboriginal practices predominated. Wherever it seemed appropriate, small manioc plantations were made and then abandoned after one or two harvests, and a new "roça" or field was prepared by cutting and burning another small piece of jungle, used for the next two to three years, and then abandoned, etc.

This very extensive system of shifting cultivation was, however, well suited for the land and did no harm as long as the human population remained extremely low and the roças constituted only small and widely separated openings in the generally closed forest cover, which healed rapidly after their abandonment. The buffering capacity of the forest ecosystem was not exceeded, and after some 30 to 40

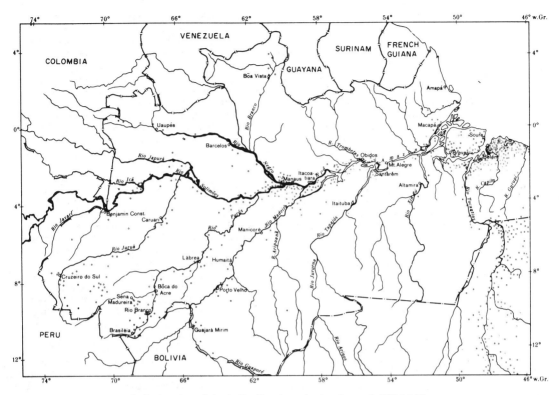

FIGURE 1. Distribution of population in Brazilian Amazonia. One dot equals 2500 inhabitants.
[After Sioli 1969 and Atlas Nacional do Brasil 1966.]

years only a botanist could distinguish former roças from the virgin forest by the species composition of the tree vegetation, the general aspect being the same. Even today, nothing can be said against maintenance, for the present, of the traditional system in regions where population density is still extremely low and the crop is only used to supply the local demands. Fortunately, the population density in most parts of Amazonia is in fact one of the lowest on earth: the State of Pará (excluding the capital of Belém) has only about 0.9 inhabitants per square kilometer. The State of Amazonas, omitting the capital Manaus (~200,000 inhabitants), has less than 0.3 inhabitants per square kilometer, the spaces between the river courses being practically empty (Figure 1).

Only once during the colonial period was agriculture attempted on a larger scale on the lower Amazonian terra firme. Between the end of the 17th and the middle of the 18th centuries, the Jesuit missionaries concentrated some 10,000 Indians around Vila Franca on the left bank of lower Rio Tapajós, where they taught them agricultural practices. It is doubtful whether this project would have had lasting success even if it had not been prematurely terminated by the famous edict of the Portuguese Foreign Minister, Marquês de Pombal, which expelled the Jesuits from Brazil in 1756. Today, the region contains many large and small areas in which the forest has been replaced by unproductive, meager savannas and there are many indications that they are of human origin. We now know how poor terra firme soils are in nutrient reserves necessary for plant growth, and in the Tapajós zone that deficiency is aggravated by a relatively dry climate, so that a series of four weeks without a drop of rain is a common annual phenomenon. Once the forest has been cleared on larger stretches, that "drought" period, together with increasingly sandy soil resulting from selective erosion, may be sufficient to disturb the water table to such a degree that young forest trees cannot develop and the terrain is consequently taken over by savanna vegetation similar to that of the Cerrado of Central Brazil.

Other new subsistence practices have been introduced only on the natural várzea grasslands along the lower Amazon. One of them was cattle raising, not only for the use of the local settlers, but also for consumption in newly founded communities like Santarém and Óbidos. Another was cocoa plantations of larger or smaller extent. These plantations were worked with slaves, and when slavery was abolished in Brazil in the 1880s these plantations were gradually abandoned. Only small "cacauais" can still be found, mainly in the region from just above Óbidos to below Santarém, and their production is economically insignificant. The majority of them rapidly reverted to forest, so that former human activity has left no definitive impact on the várzea landscape. Extensive cattle-raising, however, has persisted and is still of great economic importance in spite of the difficulties and the heavy losses inflicted on the herds by the annual floods.

During the 1880s, a new period began in the exploitation of Amazonia. First of all, those years marked the beginning of the rubber-boom, of the "golden rubber-time," in which the "mentality of extractivism" celebrated its greatest triumphs. Wild rubber was gathered in the farthest reaches of the interior of the vast country, and the price for raw rubber rose to one pound sterling for a pound (454 g) of rubber, which made other human occupations and activities uninteresting and forgotten. Thousands and thousands of "seringueiros" or rubber-tappers, were imported, mostly from the relatively crowded states of arid northeastern Brazil that suffered from periodic droughts, and sent into the interior of Amazonia to work on enormous estates sometimes extending over several thousand square kilometers of virgin forest, in which the *Hevea* trees grew wild. Nobody thought of establishing artificial

plantations of rubber trees, since the general idea was that the world would depend forever on the Amazonian rubber. Where the distance was too remote to warrant installation of a "center" for exploiting the natural rubber trees, expeditions attempted to get as much rubber as possible on one trip. That was not done by tapping the trees, so that they could recover quickly and remain productive for years, but by felling them to extract all the latex they contained. Thus, the forest was impoverished in native *Hevea* trees (as is occurring today in certain regions in regard to rose-wood trees and some good timber). The forest as a whole, however, was fortunately not destroyed.

When the first plantation rubber from Southeast Asia came onto the world market in 1912 and the rubber price dropped, the whole nightmare of expeditions and destruction of rubber trees far in the Amazonian interior came to an end. In that regard, the "golden age of rubber" was a transitional phase for Amazonia. The human tragedy was more enduring. Although the rubber tappers were able with time to adapt their life-style to the pattern of the forest-and-water ecosystem by adopting many former Indian practices, the two big cities, Belém and Manaus, which had grown and lived in luxury during that time, fell into decay.

At the beginning of the rubber boom Belém, the entrance to the Amazon, became a rapidly growing city, and the promise of economic wealth led the governors to view the apparent fertility of the forest country east of Belém as a potential permanent source of supply for that town. A colonization scheme was planned and administered by an official agency and a 300-kilometer long railroad was constructed from Belém to Bragança between 1883 and 1908 in order to facilitate access, to stimulate settlement by immigrants, and to provide rapid and reliable transportation of the products to the consumers. The areas of the "colônias" (i.e., of the settlements) were fixed by the government and in the course of time some ten thousand agricultural settlers—mostly Spaniards, Portuguese from the Azores, and Frenchmen—were brought into the so-called Zona Bragantina (Figure 2). In spite of all these efforts, however, the governmental colônias were not successful and the majority of the settlers wandered off again. Eugênia Gonçalves Egler (1961:533), has written: "The reasons for this fiasco are always seen, by the successive administrators, in

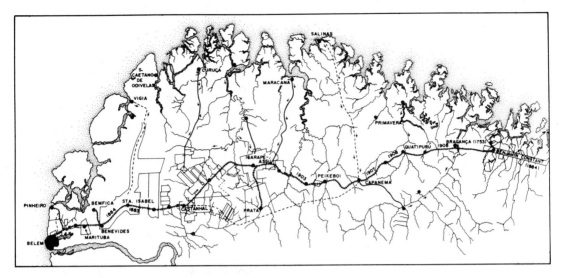

FIGURE 2. Agricultural colonization in the Zona Bragantina. [After Egler 1961.]

a lack of adequate administration of the 'colônias,' in insufficient official support and lack of propaganda in foreign countries for attracting greater numbers of immigrants. Local causes like low fertility of the soils, climatic difficulties or the appearance of pests in the plantations are never mentioned" (translation by H.S.).

Subsequently, even greater numbers of immigrants came on their own initiative, settling near the railroad, as did some 30,000 Cearenses after the great drought in the State of Ceará in 1915. They felled and burned the high forest in order to start plantations, and little by little 'they occupied the whole zone. But "the work of those people consisted in destroying gold for the production of silver," as a prudent Brazilian, Amaro Theodoro Damasceno, Jr., once expressed it, and Camargo (1948:128) who cites him, agrees: "Indeed, the felling of the forest was destroying gold, and producing manioc-meal, rice and other cereals, was producing silver. The physician Damasceno couldn't have expressed the situation better, but the worst is that the scandalous destruction of those forest riches still continues today" (translation by H.S.).

We must add that now, 22 years after Camargo's sad statement, annihilation of the high forest has

not only expanded over practically the whole area of the Zona Bragantina, i.e., about 30,000 square kilometers, but has also extended far beyond its southern border, which had been more or less along the Rio Guamá, where a strip about 20 kilometers wide is being destroyed along the new Belém-Brasília highway.

The method of utilization applied here has been and continues to be the same as elsewhere in the interior of Amazonia; namely, the already mentioned "shifting cultivation," involving cutting and burning an area of jungle, planting generally manioc (in the Zona Bragantina preferably corn, rice, sugarcane, cotton, and tobacco), and abandoning the roça after two or at most (and exceptionally) three harvests. Capoeira, a meagre secondary growth that takes over the abandoned roça, is cut and burned after 8-10 years, but yields only one harvest. After that, further utilization of the same area is generally not attempted in the interior of Amazonia. In the Zona Bragantina, however, it may be repeated about every 10 years.

Since colonization started in that zone, new areas of forest were always treated in the same manner. The ancient method of shifting cultivation had been tolerated by the forest when practiced only in small isolated spots, but here the rapid population growth (8 inhabitants/km^2 by 1950, excluding Belém; Soares 1956:187) brought the roças into increasing proximity and caused complete destruction of the landscape over vast continuous expanses. The jungle receded farther and farther from the traffic routes of the region, first the railway and later on the roads, giving place to a monotonous sequence of capoeiras. Luxuriant high forest was transformed into extensive stretches of stunted scrub and only a few skeletons, now becoming rarer and rarer, of isolated jungle trees still testify to the former exuberant growth in the region. The nutrient storage of the former forest community, as well as the water-holding capacity of the soil have been upset, and local changes of the climate in form of longer droughts have been produced as a result of large bare areas. The final result of this effort at "development" was reached in a relatively short time. The introduction of fiber plants, especially Malva (*Pavonia malacophylla*) and Uacima (*Urena lobata*), brought a short-term recovery in the general decline. But in general these new crops—as well as the use of the capoeiras for charcoal production—serve mainly to complete the process that produced a "ghost-landscape," as Eugenia Egler has called it, in less than 50 years.

Meanwhile, similar "agricultural colonies" were established in other parts of the lower Amazon, for example around Santarém, Monte Alegre, and Alenquer (Figure 3) and expanded toward the interior. Generally, they had no better success than the Zona Bragantina, but fortunately they were of smaller size so that damage to the soil and vegetation has been less extensive.

With regard to other agricultural experiments on the lower Amazonian terra firme, two private attempts to establish agricultural settlements with Japanese peasant immigrants should be mentioned. The first, on the Rio Uaicurapá (south of Parintins) (Figure 3) was dissolved after a few years, since the Japanese settlers despaired at their lack of success and revolted. The other center, Tomé-assú at Rio Acará-pequeno (150 km south of Belém; Figure 3), was initially dedicated to development of a great cocoa plantation which, however, also failed. The colony began to decline, a process that was intensified and accelerated by a severe malaria epidemic that killed many Japanese settlers and caused others either to move to southern Brazil or

to go back to Japan. Those who remained shifted to intensive production of vegetables for the Belém market.

The failures of these Japanese companies, however, were compensated and finally outweighed by the ultimate successes attained with new agricultural crops, namely pepper and jute. About 1930, a farsighted Japanese immigrant of the Tomé-assú group had taken black pepper seeds (*Piper nigrum*) with him to Amazonia. During World War II, when the Brazilian government transformed the former Japanese concession of Tomé-assú into a state colony and internment camp, the Japanese settlers started to expand the pepper plantations. By the end of the war, the product, which was a new one for Amazonia, was available to the Brazilian and soon also the world market. Pepper then became the basis of a completely new, intensive utilization of the Amazonian terra firme. Tomé-assú began to flourish, and pepper cultivation soon started to spread to the Zona Bragantina and the vicinity of Manaus.

Another Japanese immigrant had brought some seeds of jute (*Corchorus capsularis*) with him from India. After the decay of the colony at Rio Uaicurapá, he planted the jute near the small town of Parintins on the lower Amazon, but on the várzea rather than on the terra firme. The experiment was so successful that jute cultivation expanded rapidly along the várzea banks of the lower Solimões and the Amazon to below Santarém. Within less than 10 years, jute rose to second place among the local products of economic importance and changed the appearance of long stretches of the banks of the lower Amazon, replacing the original forest with rows of planted jute.

For centuries, settlers had established huts and small plantations of corn, beans, etc., on other parts of the várzea margin. But these alterations of the vegetation cover of the várzea, as well as the yearly burning of the natural grasslands in the dry season for quick production of fresh fodder for the cattle, did not and do not destroy the high productivity of the várzea as a whole. The great river itself is constantly altering the vast flood plain, eroding away one side, depositing fresh soils on the other side, and annually precipitating a new, fertile layer of silt and clay over all areas covered by the turbid white flood water. Thus, the fertility of that zone is periodically renewed.

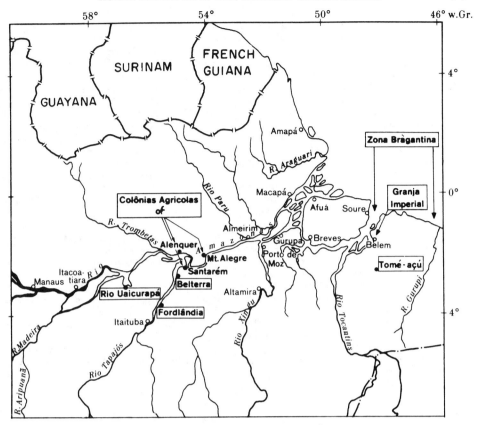

FIGURE 3. Agricultural colonies on the lower Amazon. [After Sioli 1969.]

In order to extend the area of fertile soil farther inland, Felisberto C. de Camargo developed a plan for increasing the amount of turbid, sediment-rich Amazon water that normally enters the várzea lakes either by overflowing the natural levees or through a few, narrow, natural water courses, called furos, which connect the river with the lakes. At the experimental station of Maycurú, a number of artificial channels have been dredged through the levee and now a strong current of Amazon water flows into the lakes, where it stagnates and deposits its sediment load, raising the level of the ground so that ultimately it can be used for plantations or pasture. The fissures that appear in the dry season give an idea of the absorption capacity of that fresh soil and may be taken as an indirect sign of its fertility.

What other effects this filling of parts of the várzea lakes by directed sedimentation may have in the future is still uncertain. The influx of a greater amount of water into the várzea lakes, which sometimes are of enormous extension (20 x 60 km and more), situated in a zone with extremely low gradient seems to cause a higher rise of the lake level on the downstream shores and to cause flooding of larger areas on that side. Whether the ultimate consequences of this great experiment will be more beneficial or more detrimental for the life of mankind on the várzea cannot yet be predicted. It remains true, however, that up to now human effects on the várzea are slight in comparison with the constant changes, destruction and rebuilding of the whole várzea terrain caused by the activity of the mighty Amazon itself. Only if man should carry out the fantastic project of the construction of a large dam across the lower Amazon, drowning an area of 300-400,000 square kilometers, would his interaction have a lasting and definite effect on the várzea, namely its disappearance. But I hope and I think we all hope and shall do our best to prevent this plan from ever becoming a reality.

There are many other possible human activities that do less harm to the Amazonian landscape (in the sense of the tentative value scale mentioned earlier) and to the productivity of the ecosystem. The basis for every intervention, however, must be an understanding of its ecology. The first ecologically based ideas for utilization of the Amazon region were developed and propagated by Felisberto Cardoso de Camargo some 20 years ago. In those years, the general poverty of most of the Amazonian terra firme soils was documented by many studies, mainly by Camargo and his scientific staff at Instituto Agronômico do Norte. The failures of agricultural efforts had been the first indicator, but soil analyses (Camargo 1948, 1958) and chemical analyses of natural waters (Sioli 1950, 1954, 1957b), both carried out at the Instituto, provided definite proof. The várzea, however, was built up by differ-

ent matter and is constantly being renewed, as we have seen.

Based on these findings, Camargo made a fundamental distinction between these two major terrestrial biotopes of Amazonia with regard to their prospective utilization. For each of them he worked out and presented a different practical plan for agricultural exploitation. Short-lived agriculture was to be restricted to the várzea, while the terra firme was reserved for long-term plantations and mainly for forestry, which over longer periods of time can produce a profit from the extremely scarce reserves of inorganic nutrients in the soil and out of the rainwater. The scheme designed by Camargo (1948, 1958) (Figure 4) for the region of Belém (Rio Guamá) is valid in principle for all of the Amazonian lowlands. The only difference is that the lower Amazon with its campo-covered várzeas permits extensive animal (cattle) husbandry by creation of artificial reserve pastures on nearby terra firme terrain for use during the flood season.

This scheme is to be considered a preliminary result of the application of landscape ecological thinking on an agricultural utilization of the region, a utilization that does not aim at a short-term exploitation according to the philosophy "après nous le déluge," but at a conservation of the natural habitability of the Amazonian landscapes and their productivity. The scheme has especially great

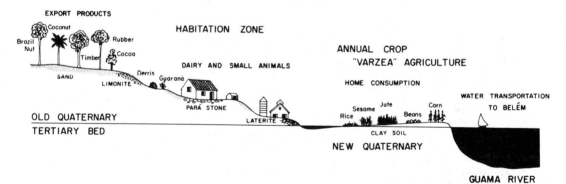

FIGURE 4. Scheme for differential permanent exploitation of terra firme and várzea environments recommended by Felisberto de Camargo. [After Camargo 1958.]

potential for cultivation of the várzeas in the estuary zone of the Amazon in regard to further development of rice culture (Lima 1956). Theoretically, there is the possibility, which Camargo has already envisaged, of surrounding the flat sediment islands by low dikes provided with floodgates. After removal of the várzea forest, vast areas would be available for wet rice; their productivity could satisfy half of the world's needs (Sioli 1966). The appearance of such a landscape would certainly be very similar to that of the Ganges Delta, which after some thousands of years still retains its fertility and productivity, and also a peculiar beauty.

The cultivation of the terra firme, however, is a different problem. Forestry, which would be the most advisable method of utilization for large extensions, is still in its infancy. The first large scale experiment was the famous rubber plantations initiated by the Ford Motor Company in 1926 at Fordlândia and Belterra on the lower Rio Tapajós (Figure 3). The first one, Fordlândia was founded for economic reasons: The price of rubber was controlled by the British and the Dutch, who at that time held a monopoly over plantation rubber, produced on their great estates in Southeast Asia and Indonesia. For the establishment of American-owned rubber plantations, Ford chose almost exactly the same spot from which 50 years before, in 1876, Sir Henry Wickham had secretly exported seeds of Hevea brasiliensis, brought them to Kew Gardens, and initiated the story of South Asiatic plantation rubber. Opposite the little village of Boim on Rio Tapajós, Ford secured from the Brazilian government a concession of 10,000 square kilometers, which was named "Fordlândia." In a short time, great areas of virgin forest were felled and burned, hygienic settlements were installed, and around 800,000 Hevea trees were planted. The project was envisaged as having a great future and planned for a final population of 100,000. Large amounts of capital were invested, including construction of a great saw mill (at that time the biggest in the whole of South America) making use of the forest trees. Àt the bank of the Rio Tapajós, a port was built suitable for ocean-going vessels, since the river is navigable during high waters for steamers of up to 10,000 tons.

In spite of these efforts, the ambitious enterprise failed for several reasons. First, the hilly terrain of Fordlândia, selected on the basis of historical rather than scientific criteria, was not particularly suitable for the collection of latex. Second, the usable timber was exhausted in a very short time, not only from the plantation itself, but also from the surrounding neighborhood from which trees could be brought by rafting, and within two years the sawmill had to be shut down. These two failures, however, did not cause abandonment of the Fordlândia project. The disaster began a few years later, around 1932, when the rubber trees were attacked by the "South American leaf disease." The fungus (Dothidella ulei) attacked the leaves, so that the trees were heavily damaged and finally died. This disease is unknown in the rubber plantations in the Orient. It occurs only in the American tropics in heavy epidemic form and only in plantations with a great number and high concentration of Hevea trees. It is insignificant among the wild rubber trees in the Amazonian jungle, which are a hundred or more meters distant from one another and thus isolated and protected against attack by the fungus spores by intervening trees of other species and genera.

In Fordlândia, however, the disease soon spread over the whole plantation, and could never be eradicated. But Ford did not give up; in 1934 he exchanged one quarter of the concession for another terrain, Belterra, situated 100 kilometers down river on a flat plateau about 150 meters above Tapajós level (165 m above sea level), where he started a new and even larger plantation, which finally reached 2,000,000 trees. The Dothidella disease appeared here also, however, and the plantation could only be saved by the trick of double grafting. Young trees were grown from seeds of native H. brasiliensis, adapted to the Amazonian soil conditions. After two years, they were cut shortly above the ground and scions of oriental clones of H. brasiliensis, selected for high rubber yield, were grafted on them. These Asiatic clones, however, were especially sensitive to Dothidella, so that after another two years the crowns had to be cut off and replaced by the crowns of another species of the genus Hevea, namely H. benthamiana or H. spruceana both resistant to Dothidella. Since the latex is extracted from the trunk, this did not inhibit production, and the yield of such trees, each composed of three individuals belonging to two different species, was comparable to that of pure H. brasiliensis individuals.

305

A rubber plantation requiring double grafting is not profitable in economic terms, however, and when the need for rubber declined after the end of World War II, Ford handed his plantations over to the Brazilian government, which consigned them to the Instituto Agronômico do Norte. Under the direction of Felisberto de Camargo, the first large-scale experimental plantations of Amazonian timber were initiated at Fordlândia, with somewhat promising results. In addition, large stretches of infected rubber trees were transformed into pasture and breeding of selected races and herds of zebu cattle and of water buffalos was introduced.

In Belterra, experiments on selection and crossing to produce *Hevea* varieties resistant to *Dothidella* continued, as well as the management of the rubber plantation. The latex yield of individual rubber trees was relatively good except for a much lower dry-rubber content; but the growth of the planted forest was noticeably slow. This may reflect the fact that latex contains a great amount of phosphate, which is drained from the trees by the periodic tappings, while the soil is extremely poor in this mineral. Camargo experimented with application of a complete (N-P-K) mineral fertilizer, and that year the rubber forest showed significant improvement. The experiment could not be repeated, however, because of its expense. The decline of the Ford Rubber Plantations on the Rio Tapajós could not be stopped. They remained a subsidized undertaking until they were separated from the Instituto Agronômico do Norte some ten years ago and since then have lost their significance as examples of utilization of the Amazonian terra firme by plantations.

It is interesting that up to now only two smaller scale efforts to exploit the terra firme have been successful in the sense of offering a new stability with economic value for mankind. Both are based on an intuitive or rational insight into the ecological realities and interrelations in the Amazonian landscape. One was a trial conceived and twice executed by Sakae Oti, a former member of the Japanese company that founded the Uaicurapá settlement. It was based on the ancient method of shifting cultivation, but it avoided the "shifting" and envisaged, instead of abandoned areas of capoeiras on impoverished soils, a new stable forest with a constant yield. To achieve this, a modest area (the second experiment comprised about 25

hectares) of virgin forest was cut and burned in the usual way for preparing a roça. The subsequent treatment was different, however, and consisted of simultaneous planting several crops: rice as densely as possible, manioc 1.5 meters apart, guaraná (*Paullinia sorbilis*) 6 meters apart, and Brazil nut trees (*Bertholetia excelsa*) 18 meters apart. The rice grew quickly, fertilized by the ash of the burned jungle, and produced a sufficient harvest within three months to pay the whole expense of preparation of the area. After $1\frac{1}{2}$ years, the manioc was harvested, paying the expense of maintaining the plantation clean of invading capoeira vegetation. The guaraná started to bear fruits in about 6 years, and finally, after about 15 years, the Brazil nut trees began to produce. The result was a new, planted, and productive forest replacing the original jungle, covering and protecting the soil, and yielding annually a good return for the human settler. Another favorable aspect of that small-scale enterprise was its economy. No large capital had to be invested for a long period, all stages paying for themselves by the consecutive harvests within short periods. This system can and should be varied by trying different crops and forest trees. Its main and lasting advantage, however, is maintenance of a stable forest ecosystem.

The second economically sound practice has been elaborated in recent years by a German engineer, Ernesto Rettelbusch, who was a son of an ancient peasant family. The goal was not forestry, but stable agriculture on terra firme terrain, and it was achieved—after years of failure with the usual ancient system—near Marituba, 20 kilometers east of Belém, in the Zona Bragantina (Figure 3). Since the knowledge and the methods developed here may serve as general guidelines for future intensive cultivation of other parts of the Amazonian terra firme, some details are relevant.

The basis for lasting agricultural production is, as is well known, a very simple arithmetical problem. A good harvest depends on several environmental factors, the most important ones being: land, solar energy, warmth, water, and inorganic nutrients for plant growth. Of the latter, what is removed from the soils with the harvests (and by other losses) must be replaced either from reserves remaining in the soils or by artificial introduction. On the Amazonian terra firme, the following are present in sufficient amount: land for the plantations, solar energy and warmth, and usually water. What are lacking are the inorganic nutrients, especially phosphate and potassium, nitrogen perhaps to smaller degree, and sometimes also trace elements (e.g., cobalt). As a result, a reasonable harvest requires supplying artificially those substances necessary for plant growth.

The most simple solution to this problem would theoretically be the application of mineral fertilizers, which naturally must correspond in their composition to the local necessities. Such a procedure, however, represents the transfer to Amazonia of a technique developed in and suited to conditions in the northern temperate climatic zone without taking into consideration the peculiarities of that region, particularly the physical properties of the soils and the quantities of rainfall. The lower Amazonian terra firme soils are not only chemically poor, they also contain almost no colloids for fixation of added fertilizer salts. The mineral fertilizers consequently would be leached out by the frequent heavy rains, and therewith lost to the plants. For the crops to profit, it would be necessary to repeat the fertilization every one or two weeks, a procedure that would be irrational and uneconomical. The agrotechnical problem, therefore, was to find another method of providing a lasting and economical supply of fertilizer.

That problem has been solved by Engineer Ernesto Rettelbusch in principle and in practice by a combination of intensive animal husbandry with intensive agriculture. His "Granja Imperial" (covering 240 hectares, of which at present only about one-third is in use) is a model for further enterprise. On the terrain are some 20,000 pepper plants, up to 20,000 chickens, 500 pigs, and around 200 cattle. Instead of mineral fertilizer, most of the fodder for the animals is imported, mainly from southern Brazil. The animals consume the fodder, utilizing the proteins, fats, and carbohydrates, while most of the minerals, including trace elements and organic colloids are released in the form of manure, which for centuries has been the basis for maintaining the fertility of European fields. When this animal manure is applied to the plantation, it lasts much longer as a supplier of nutrients than mineral fertilizers. Furthermore, the meat, milk, eggs, and other products of the animals pay most of the expenses of the farm, while the sale of pepper provides the profits.

Naturally, the interlocking of animal husbandry and agriculture is much more complicated in its details than can be described here. Also, readjustments to accommodate to the changing local and international market situation will always be necessary, so that no "permanent recipe" for lucrative agriculture in Amazonia can be offered. Such readjustments are, however, only variations on a general theme and do not change the basic principle for a sound utilization of the lower Amazonian terra firme, adapted to local ecological conditions. It takes into account the chemical and physical properties of the soil, as well as the climate. Since it aims at the management of relatively small areas, it represents a renunciation of the old colonial system of increasingly extensive exploitation. Consequently, it reduces the danger of erosion and enhances the possibility of preserving portions of the natural landscape, in form of national parks or something similar, for future generations when an increase in the human population will cause stronger pressure on the land. Finally, it transforms parts of the "wilderness," exuberant with life, into a "cultural landscape," which is equally stable because of constant regulation by man. It also meets our requirements for maximum utilization by offering to a greater number of human beings, not only food but, because of the variety of its internal structure, an enriched quality of life.

Granja Imperial is a model that is being increasingly adopted by other farmers in the formerly devastated Zona Bragantina, with results that have changed the aspect of the countryside. May it be a symbol that the future of humanity does not lie in a destructive battle against nature, but rather in working together with nature toward a new harmonic unity.

References

Bluntschli, Hans. 1921. Die Amazonasniederung als harmonischer Organismus. Geographische Zeitschrift 27(3/4): 49-67.

Camargo, F. C. de. 1948. Terra e colonisação no antigo e novo Quaternário na Zona da Estrada de Ferro de Bragança, Estado do Pará-Brasil. Bol. Mus. Para. Emílio Goeldi 10:123-147.

————. 1958. Report on the Amazon region. Problems of humid tropical regions, 11-24. Humid Tropics Research. Unesco. Paris.

Carvajal, Gaspar de, Alonso de Rojas, and Cristobal de Acuña. 1941. Descobrimentos do Rio das Amazonas. Translated and annotated by C. de Melo-Leitão. Brasiliana 203, Ser. 2, São Paulo.

Egler, E. G. 1961. A Zona Bragantina no Estado do Pará. Rev. Bras. Geogr. 23:527-555.

Fittkau, E.-J. 1970. Role of caimans in the nutrient regime of mouth-lakes of Amazon affluents (An hypothesis). Biotropica 2:138-142.

Hilbert, Peter Paul. 1968. Archäologische Untersuchungen am mittleren Amazonas. Marburger Studien zur Völkerkunde 1.

Lima, R. R. 1956. A agricultura nas várzeas do estuário do Amazonas. Bol. Tec. Inst. Agron. 33:1-164. Belém.

Meggers, Betty J. 1971. Amazonia: Man and culture in a counterfeit paradise. Aldine-Atherton. Chicago and New York.

Oti, Sakae. 1947. A historia da juta na Amazônia. A Vanguarda 9.7. Belém.

Sioli, Harald. 1950. Das Wasser im Amazonasgebiet. Forschungen und Fortschritte 26:274-280.

————. 1954. Beiträge zur regionalen Limnologie des Amazonasgebietes, II: Der Rio Arapiuns. Arch. Hydrobiol. 49:448-518.

————. 1956. Über Natur und Mensch im brasilianischen Amazonasgebiet. Erdkunde 10:89-109.

————. 1957a. Sedimentation im Amazonasgebiet. Geol. Rdsch. 45:508-633.

————. 1957b. Beiträge zur regionalen Limnologie des Amazonasgebietes, IV: Limnologische Untersuchungen in der Region der Eisenbahnlinie Belém-Bragança ("Zona Bragantina") im Staate Pará, Brasilien. Arch. Hydrobiol. 53:161-222.

————. 1966. Soils in the estuary of the Amazon. Pp. 89-96 in Scientific problems of the humid tropical zone deltas and their implications. Humid Tropics Research, Proc. Dacca Symp., Unesco, Paris.

————. 1969a. Zur Ökologie des Amazonas-Gebietes. In Biogeography and ecology in South America 1:137-170. Monogr. Biol. 18. W. Junk. The Hague.

————. 1969b. Entwicklung und Aussichten der Landwirtschaft im brasilianischen Amazonasgebiet. Die Erde 100(2-4):307-326.

————. 1969c. Die Biosphäre und der Mensch: Probleme der Umwelt in der heutigen Weltzivilisation. Universitas 24(10):1081-1088.

Sioli, H., and H. Klinge. 1966. Anthropogene Vegetation im brasilianischen Amazonasgebiet. Anthropogene Vegetation. Bericht über das Internationale Symposium in Stolzenau, 1961 der Internationalen Vereinigung für Vegetationskunde, 357-367. Den Haag.

Soares, Lúcio de Castro. 1956. Excursion guidebook No. 8: Amazônia. 18th Internat. Geograph. Congr. Brasil, Internat. Geograph. Union. Rio de Janeiro.

Valverde, Orlando, and Catharina Vergolino Dias. 1967. A rodovia Belém-Brasília: Estudo de geografia regional. Bibliotéca Geográfica Brasileira 22, Sér. A. Rio de Janeiro.

24

Reprinted from Agro-Ecosystems 3:291–302 (1977)

INTEGRATING FOREST AND SMALL-SCALE FARM SYSTEMS IN MIDDLE AMERICA

GENE C. WILKEN

Department of Economics, Colorado State University, Fort Collins, Colo. 80523 (U.S.A.)

(Received 5 January 1977)

ABSTRACT

Wilken, G.C., 1977. Integrating forest and small-scale farm systems in Middle America. *Agro-Ecosystems*, 3: 291—302.

Initial clearing and continued suppression of wild vegetation seem proof that farming and forest are incompatible. Yet many crop plants, like their wild relatives, fare better in species diverse, structurally complex communities. Some small-scale or traditional farmers in Middle America partially recreate forest conditions in their fields to improve edaphic and microclimate conditions. Farmers may either physically transfer forest products or simulate forest structures and nutrient cycles in cultivated fields.

The particular strategies adopted depend upon local climatic conditions and crop requirements. Generally, crops requiring high levels of sunlight must be grown in the open, in field rather than forest microclimates. In such cases nearby forests may supply soil conditioning materials such as leaf litter to improve soil structure and nutrient levels. Alternatively, long term forest-farm rotations allow regeneration of forest soils. Where slight reductions of solar radiation are permissible scattered leguminous trees may be tolerated in cultivated fields. If greater protection is desirable at early or all stages of crop plant growth, aspects of forest structure and microclimates may be emulated either individually (e.g., forest litter, undergrowth) or collectively in multi-storied commercial farms or dooryard gardens.

INTRODUCTION

A prevalent impression is that farming and forests are incompatible. Clearing practices such as slashing and burning, or permanent deforestation and continued suppression of wild vegetation reinforce the impression: destruction of native trees, shrubs, and grasses is a prerequisite to planting. And it is true that wild plants usually must be eliminated to create the open, sun-rich environments favored by most domesticates, reduce competition for soil nutrients and water, and facilitate cultivation.

However, it is also true that cultivated plant systems are neither self-sufficient nor particularly stable. Exposed fields and concentrations of a single or few species open possibilities for soil erosion, nutrient depletion, disease and pest infestation, and physical damage to plants. Many domesticates, like their wild relatives, fare better in species diverse, structurally complex communities. In fact, much of farming involves approximating or thwarting natural cycles and structures: cultivating to emulate naturally loose soils, weeding to terminate primary succession, irrigating to replace moisture evaporated from open field surfaces, and so on, with the farmer performing management and work roles in the system to maintain form and regulate energy and mass transfers.

In technologically advanced systems, energy-affluent farmers manipulate field forms and cycles with a variety of chemicals and powered equipment drawn from great distances through the supply linkages of spatially integrated economies. Production of energy and materials from local sources is of little concern. But in small scale or traditional farming systems, inputs characteristically originate in the immediate region and farm work is performed primarily by humans or animals that are fueled from local sources. Working within these energy and spatial constraints, small-scale farmers have learned over the years to recognize and utilize locally available resources.

Reliance upon local fuel and materials is used here to define small-scale farming*. The systems could be discussed topically in terms of management methods for specific resources (e.g., soil, water, space) or crops, or in terms of input sources. The latter option is used here with wild plant communities considered as sources for a variety of materials that can be transferred to farms intact or in modified form. Alternatively, native vegetation itself can be encouraged to grow or its structure simulated in cultivated fields.

Small-scale farmers work with a wide range of physical environments, native vegetation, and crops. There is an equally wide range of possibilities for using wild plant communities as farm resources. I shall limit this discussion to the utilization of forest resources in Middle America and select only a few examples of how wild and crop plant communities can be linked. Shifting cultivation will be ignored even though it is one of the most obvious examples of forest-linked farming systems. While this practice temporarily destroys the structure of native vegetation and alters energy, moisture, nutrient, and mass transfers, the whole process of clearing and crop production depends upon utilization of forest environments and soils. However, shifting cultivation has received perhaps more than its share of attention. Instead the focus here will be on permanent field systems where forest—field relationships are a result of deliberate management efforts by farmers. The goal is to support the argument that traditional farming practices include effective

---

* Although often used synonymously, such terms as "traditional", "peasant", and "small-farm" have specific meanings. To avoid confusion, particular situations or problems call for appropriate terms and operational definitions. The important characteristics of the systems discussed here are that they rely upon human or animal power and draw heavily from the immediate region for farm management resources. The terms "small-scale" and "traditional" will suffice to identify these nonindustrial, spatially-restricted farming systems.

techniques for resource management, environmental conservation, and food production.

SOIL STRUCTURE AND NUTRIENTS

Forest soils derive much of their celebrated structure from the steady incorporation of organic material. The process begins in the upper or litter layer where leaves and other debris fall and begin to decompose. Cultivated fields usually are less prolific producers of organic material and much of what is produced is removed to feed people and animals.* Although returns in the form of manure and household refuse are important, they do not fully compensate for nutrient losses or offset soil structure modifications caused by heavy soil use.

One solution is to transfer part of the forest floor to the fields. In parts of western Guatemala, for example, leaf litter (*broza, hojarasca*) is carried from nearby forests and spread each year over intensively cropped vegetable plots to improve tilth and moisture retention. Litter is raked up, placed in bags or nets (*bultos, redes*), and carried to fields by men, horses, or from more distant sources by trucks. Although quantities applied vary, farmers in Almolonga, Zunil, and Quezaltenango apply as much as 40 metric tons of litter/ha each year.**

Rough calculations made in mixed pine-oak stands near Quezaltenango (Xelajuj), Guatemala, indicate that a hectare of forest produces about 4,000 kg of litter annually. Thus, in this region, a hectare of cropped land requires the litter production from 10 ha of regularly harvested forest, or less if harvesting is sporadic. Although some farmers prefer leaves from certain trees, such as oak, most litter is a mixture as are the forests from which it comes and may include pine needles, oak and alder leaves, and debris from less common species.

After spreading, the leaf litter is worked into the soil with a broad hoe (*azadón*), the all-purpose tool of the region (Fig.1). In some cases litter is first placed beneath stabled animals and then, after a week or so, the rich mixture of pulverized leaves, manure, and urine is spread over the fields and turned under. The urine tends to offset field losses of nitrogen caused by bacterial action in the decomposing litter.

The production of two, three, or even four crops each year is not uncommon from hard-worked vegetable plots in Middle America. To maintain this level of output intensive use of chemical fertilizers is necessary. But fertili-

---

* In preconquest Middle America, and before the introduction of hoofed domesticated animals, it seems likely that more crop residues were returned to the soil than in present times when much goes to fodder.

** Data have been reduced to standard metric units for convenience and comparison. Vegetable farms in these regions are small, typically about 0.15—0.20 ha. Thus an average farmer might require only 6 or 8 tons of litter each year.

zers are "cold" in the minds of many farmers and, unlike "hot" animal or green manures, do not build "body" in the soil.* For this purpose organic materials are transferred from local forests to cultivated fields.

Use of litter is common, though much less systematic, throughout much of Middle America. Although leaves from houseyard or field-side trees, often mixed with household refuse, may be spread over fields, the purposeful harvesting of forests was observed only in the vegetable growing communities of the Department of Quezaltenango, Guatemala. In the *chinampa* district south of Mexico City leaves from abundant plot-side trees are combined in mulch piles with aquatic plants from adjacent canals and crop plant residue. Leaf litter occasionally is brought into the chinampa district. However, the source areas and extent of this practice were not determined.

Mulching with forest litter essentially involves transferring portions of forest floors to cultivated fields to enhance soil structure. Another practice involves transferring part of the forest itself to enhance soil nutrient levels. Alternating or intercropping such nitrogen-fixing crops as alfalfa or beans is common throughout Middle America. Less well-known is the custom of encouraging the growth of native leguminous trees in cultivated fields. But in Mexico from Puebla and Tehuacán south through Oaxaca, farms with light to moderately dense stands of *mesquite (Prosopis* spp), *guaje (Leucaena esculenta),* and *guamúchil (Pithecellobium* spp) are a familiar sight (Fig. 2). Stand density is extremely variable, from fields with but a few trees to veritable forests with crops planted beneath.

Farmers do not plant leguminous trees but simply tolerate and protect volunteer seedlings. Generally they are vague about the value of the practice

Fig.1. Working leaf litter (*broza*) into soil with broad hoes (*azadones*). Zunil, Quezaltenango, Guatemala.

---

* Identifying "hot" and "cold" effects of various soil amendments apparently is an extension of the widespread belief that different foods heat or cool the human body. According to classical Greek humorai pathology, where the tradition has its origins, a balance of "hot" and "cold" is desirable (e.g., Currier, 1966).

Fig. 2. *Mesquite* trees in a maize field. Tehuacán, Puebla, Mexico.

and, aside from ascribing pure custom, suggest only that the trees provide midfield shade and edible pods or seeds. But whereas these products do enjoy a brisk demand in local markets, the small harvests from scattered stands can hardly compensate for shade-lowered crop yields and obstructed plowing. Instead, it would appear that over the years farmers observed that maize fields with a few trees of certain types produced better harvests. The association became custom and resulted in mixed farms of crops and native leguminous trees. Removal of mesquite or guaje trees from fields converted to tractor cultivation and chemical fertilization supports this view. With the need for nitrogen satisfied by commercial fertilizer and the nuisance factor of mid-field obstructions increased by machine plowing, the custom lost its utility.

Fig. 3. Heavily pruned *sáuco* stumps in sandy potato fields. Ostuncalco, Quezaltenango, Guatemala.

Fig.4. Stands of pine on old wheat terraces. Totonicapán, Guatemala.

A slightly different practice is found near Ostuncalco, Guatemala, where rigorously pruned *sáuco* (*Sambucus mexicana*) stumps dot maize and potato fields (Fig.3). Leaves and small branches are removed annually, scattered around individual crop plants, then chopped and interred with broad hoes. Local farmers claim that crop quality and yields in the sandy volcanic soils of this region are dependent upon these annual applications of sáuco leaves.

What is perhaps a fuller development of forest—field integration seems to be operating around Totonicapán, Guatemala, where pine trees are allowed to colonize the narrow, hand-worked wheat terraces (*tablones*). Again, seedlings are not planted but neither are they chopped out with hoes during annual field preparation. Native conifers are in demand for fuel and lumber (Layton, 1973), and farmers consider the trees as a form of standing capital or second crop.

Fig.5. Terraced wheat fields and scattered pine groves. Totonicapán, Guatemala.

However, a survey of the region revealed a great range of stand densities, from plots with just a few seedlings or trees, to those where wheat cultivation is difficult even by hand methods and harvests are marginal, to fields where only trees grow on relic terraces (Fig. 4). It appears that a form of long-term rotation is practiced, perhaps not on all fields, but certainly on a fair proportion of the total agricultural lands, that periodically allows soils to regenerate under forest conditions. The result is a patchwork landscape of terraced wheat fields, mixed wheat-pine plots, and woodlots (Fig. 5).

If a long-term rotation is being practiced, it differs from other forest—fallow systems in that fields alternate from crops to trees with no transitional stages. As long as the terraces are cultivated, farmers weed out herbaceous pioneers leaving only tree seedlings. Eventually cultivation is abandoned on plots with existing dense stands of pine. When plots are later prepared again for planting (a stage not observed), they first would be cleared of all vegetation.

Lumbering is controlled under Guatemala law and requires authorization and payment of taxes before cutting. Taking single trees from private fields without permission, however, is not uncommon. Whether a farmer could clear an entire plot without fear of sanctions is not clear. It may be that the forest—field alternation suggested here is an old practice that has not yet adjusted to modern regulations.

FOREST STRUCTURE IN FIELDS

The interior of a forest is a benign region shielded from direct solar radiation, strong winds, and the impact of falling rain and hail (FAO/UN, 1962; Geiger, 1965). Some of these characteristics are inimical to farming, others advantageous. For example, the low energy levels on forest floors are inadequate for growth and development of most crop plants, and one of the main purposes of forest clearing is to create high-energy surface conditions conducive to crop growth. Yet many crop plants, especially in the seedling stage, benefit from the protection offered by asssociated vegetation. The problem thus becomes one of managing microclimates so that undesirable conditions are moderated or eliminated while desirable ones are preserved or created (Wilken, 1972). Forests provide both models and sources of material for recreating community structures in cultivated fields.

Perhaps the most common method of introducing aspects of forest microclimates into field areas is with windbreaks or shelterbelts. Windbreaks can significantly lower windspeeds downwind for a distance of 10 to 15 times barrier height and thereby reduce wind damage and erosion, and evaporation (Van Eimern et al., 1964). Although sometimes rows of cut brush or branches serve as windbreaks, more often narrow stands of living trees, in a sense strips of forest, extend along field borders. In some regions of Middle America windbreaks completely outline the checkerboard pattern of fields, forming a geometric open forest within farmlands.

Surface mulches do not look much like forests. Yet they function as litter layers by reducing soil radiation loads, inhibiting heat and moisture exchanges with the atmosphere, and absorbing the kinetic energy of rain or hail. Although an enormous variety of organic and inorganic mulch materials is employed in Middle America, it is not uncommon for farmers to bring grasses or litter from the forest itself to approximate forest floor conditions on field surfaces (Fig. 6).

Some plants require protection after they have grown above the immediate surface. Shelters of branches and leaves from native vegetation serve admirably for this purpose. For example, chile, tobacco, and coffee seedbeds usually are shielded by a variety of leaning branch or upright thatched arbors (Figs. 7 and 8). Leaf and branch arbors function much like natural brush or tree covers in reducing solar radiation loads and evaporative moisture losses, and shielding tender seedlings from driving wind, rain, and hail. In some cases, young plants are gradually acclimated to open field conditions by progressive removal of shelter materials or by natural withering and fall of arbor leaves.

Cultivation of plants that benefit from protection throughout their life requires another stage of forest structure mimicry. Protection from sun and wind is achieved with relatively dense stands of tall native trees, some of which themselves are productive. Thus the beverage plants of Middle America, coffee and cacao, and other crops such as bananas and vanilla produce a superior product or suffer less physical damage under an overstory of protective trees (Fig. 9). Shade trees also contribute leaf litter to plot surfaces. Viewed from above, a Middle American coffee *finca* or cacao field often appears more like a forest than a farm. It is only at the surface that the productive layer beneath the forest canopy can be distinguished.

Not all forest farm integrations fall neatly into single purpose divisions. For example it was mentioned earlier that leaf litter from plot-side trees is used to mulch the highly productive chinampas of Mexico. There and also in Tlaxcala where a similar system of intensive drained-field agriculture is

Fig.6. Light surface mulch of grass from nearby forest. San Agustin Atzompa, Puebla, Mexico.

Fig.7. Constructing arbor covered seedbeds. San Lucas el Grande, Puebla, Mexico.

practiced, thin willows spaced 2 or 3 m apart line the edges of many narrow, canal-bordered fields (Wilken, 1969). In addition to contributing leaves, they also act as windbreaks and their root systems help stabilize canal banks and drain excess moisture from the subsoil (Fig. 10). Furthermore, the trees support climbing crops such as *chayote* (*Sechium edule*) and supply twigs and branches for plant stakes.

The ultimate stage in forest emulation is reached in the multistoried dooryard gardens of Middle America, where farmers imitate the structure and species diversity of tropical forests by planting a variety of crops with different growth habits. Plots no more than 1/10 ha in size may contain two dozen different food plants, each of which has a different form and which together reproduce the well-known layered configuration of mixed forests.

Fig.8. Arbor covered *chile* seedbeds. Puebla, Mexico.

Fig.9. Multi-storied commercial farm: shade trees over bananas over coffee. Veracruz, Mexico.

Thus a garden may contain taller plants such as coconut or papaya, a lower layer of banana or citrus, a shrub layer of coffee or cacao, tall and low annuals such as maize and beans, and finally a spreading ground cover of vine plants such as squash. (Food production also takes place at still another level if one considers below-ground root and tuber crops.) Field arrangements may be orderly or chaotic: in extreme cases only the predominance of food plants and proximity to human settlement distinguishes a Middle America dooryard garden from wild vegetation (Fig. 11).

CONCLUSIONS

In opposition to the forest—field dichotomy, this paper identifies several systems where forests are valued as sources of farm inputs. The relationship,

Fig.10. Trees bordering drained fields. Tlaxcala, Mexico.

Fig. 11. Middle American mixed garden: all plants in photograph are economic. Tabasco, Mexico.

however, is only partially reciprocal. Although a variety of products is drawn from wild stands, little is returned except protection of forests and incidental tolerance of selected trees in cultivated fields. In the case of the Guatemalan broza system, for example, the steady export of organic matter may have long range detrimental effects on the supplying forests (Lutz and Chandler, 1946, p.188). Though other practices such as recreating partial forest structures in fields involve little or no drain on forests, neither do they provide incentives for maintaining natural stands. Nevertheless, the small-scale farmer can daily see the results of his exploitation or conservation of nearby forests.

In some ways, forest—field relationships in small-scale farming differ more in distance than in kind from those in technologically advanced, economically integrated systems. Industrialized farmers, for example, also draw heavily upon forests for such materials as lumber and chemicals. But they draw from distant sources and are alerted to overuse and depletion by economic signals transmitted through imperfect market systems. The small-scale farmer, on the other hand, also may be the forest manager. Potentially at least, proximity and dependence upon local wild resources should make him the more cautious custodian.

However, the techniques discussed here are practical in their present form only in small-scale hand- or animal-worked systems. The careful collection of forest debris or construction of shading arbors is possible only if human labor is priced below factory substitutes. Leguminous trees scattered in cultivated fields or the random distribution of crop plants in mixed gardens are not practical when technology demands uncluttered rows and single crops. Some attempts have been made to adapt traditional methods to modern systems, and more have been suggested (e.g., Greenland, 1975). Although results so far have been inconclusive, the possibilities warrant continued effort. Meanwhile, traditional methods serve the small-scale farmer well, and perhaps the real future lies in encouraging wider use of these effective techniques in the traditional farming world. Much of the world's cropland is, and will continue to be, managed by small-scale farmers (Wharton, 1969). Any improvement in productivity and resource conservation will directly contribute to the food supply in those areas where the need is greatest.

ACKNOWLEDGEMENTS

Research supported by grants from the Social Sciences Division of the National Science Foundation, and the Joint Committee on Latin American Studies of the Social Science Research Council and the American Council of Learned Societies. I am indebted to Dr. Dieter H. Wilken for plant identification, and to Dr. Glen D. Weaver for a careful review of the manuscript. An earlier draft of this paper was read at the 71st Annual Meeting of the Association of American Geographers, Milwaukee, Wisc., 20—23 April, 1975.

REFERENCES

Currier, R.L., 1966. The hot—cold syndrome and symbolic balance in Mexican and Spanish-American folk medicine. Ethnology, 5(3): 251—263.
FAO/UN (Food and Agriculture Organization of the United Nations), 1962. Forest Influences. FAO Forestry and Forest Products Studies No. 15., Rome.
Geiger, R., 1965. The Climate near the Ground. Harvard University Press, Cambridge, Mass., xiv + 611 pp.
Greenland, D.J., 1975. Bringing the green revolution to the shifting cultivator. Science, 190 (4217): 841—844.
Layton, R.L., 1973. The white pine lumber and furniture industry. Totonicapan, Guatemala. Paper presented at the Conference of Latin Americanist Geographers, Calgary, Alb. 28—30 June, mimeographed.
Lutz, H.J. and Chandler, R.F., 1946. Forest Soils. John Wiley and Sons, Inc., New York, N.Y., xi + 514 pp.
Van Eimern, J., Karschon, R., Razumova, L.A. and Robertson, G.W., 1964. Windbreaks and Shelter Belts. World Meteorological Organization, Technical Note No. 59, Geneva.
Wharton, Jr., C.R., 1969. Subsistence Agriculture and Economic Development. Aldine Publishing Company, Chicago, Ill., xiii + 481 pp.
Wilken, G.C., 1969. Drained-field agriculture: An intensive farming system in Tlaxcala, Mexico. Geogr. Rev., 59: 215—241.
Wilken, G.C., 1972. Microclimate management by traditional farmers. Geogr. Rev. 62: 544—560.

Editor's Comments
on Papers 25, 26 and 27

VALUES OF TROPICAL FORESTS OTHER THAN PRODUCTIVITY

Despite the evidence that tropical ecosystems have low productive potential for products immediately useful to man, there is no doubt that these ecosystems will be exploited. Developing countries frequently have large foreign debts and little capital, but they may have a natural resource such as a rain forest that, if exported as market products, could reduce the debt. Such countries can often be convinced to develop the resource, either alone or with the aide of an international agency and/or multinational corporation. Sale of lumber, pulp, and beef results in a large return of foreign exchange into the exporting countries. There is considerable pressure in Brazil and other areas of Latin America to utilize the forests as a marketable resource and to remove forests to accommodate cattle ranching. Tropical timber is the fastest growing export of the developing world. In 1976, $4.2 billion was earned by tropical wood exporting countries (Myers, 1979). Cattle ranching produced an exportable commodity that in Brazil brings $25 million/year in exports (Myers, 1979). As the beef trade grows, more and more tropical lowland forests will be converted to pasture.

The case for conversion is (1) the resources exist; (2) technical ability for conversion is available; (3) private wealth might be created; and (4) the amount and types of wealth so created represent the best uses of the land. That tropical rain forests are still present is attributable to their heretofore relatively low net value. However, modern technology has reduced the cost of forest conversion, and rising populations and rising individual expectations are increasing the demand for the marketable products of the forest as well as for the forest land.

The case for preservation argues that the current forest-conversion practices are wasteful in the short run and may jeopardize human survival in the long run. Particularly condemned is the practice of large-scale cutting and burning. This practice wastes valuable hardwoods, may lead to desertification, and may contribute to global weather changes (Stumm, 1977). Furthermore, the loss of unique tropical species is irreversible; lost is genetic information potentially useful in identifying new chemicals and pharmaceuticals and in invigorating agricultural plants through cross-breeding.

The arguments for conversion are given in well-understood monetary terms, while the case for preservation is expressed in less concrete terms. Thus, responsible government agencies must administer vast tracts of natural resources in the face of conflicting claims about the benefits and costs of their decisions. The major problem is that public decision makers lack conventionally acceptable objective criteria for weighing the trade-offs of particular decisions or overall strategies.

Research is needed that will test the applicability of new resource economic theory to tropical forests. Nonmarket values and short- and long-term and private and social benefits of tropical moist forests need to be evaluted in terms of conventionally accepted criteria—that is, in terms of dollars.

In Paper 25, Budowski discusses the values of tropical forests but in qualitative terms only. Lugo and Snedaker (Paper 26) have moved a step toward putting a market value on ecosystems by discussing how productivity of mangroves is essential to the productivity of local fisheries, which can be evaluated by traditional economics. Myers (Paper 27) as well as Westman (1977) and Ehrenfield (1976) discuss resource economics of threatened species, but they still fall short of actually placing a value on tropical forests and tropical species. Gosselink et al. (1974) have placed a monetary value on the productive function of a coastal ecosystem by evaluating the energy equivalent in calories of the total organic matter produced by the ecosystem. The energy-to-dollars conversion was made by dividing the caloric consumption of the United States by the gross national product of the United States. The approach was criticized by Shabman and Batie (1978) because it was not conventional, but as Odum (1979) pointed out in the rebuttal, what is needed is an expansion of conventional economic theory to place a value on nature's services that have traditionally been considered "free." These services are not really free but only appear free because they are not sold at the market place. They have a value to all people.

The interpretation of nonmarket values of tropical ecosystems is the major challenge facing tropical ecologists. Are there ways to express these values in terms of dollars so that public benefits can be weighed against private exploitative use of the forests? We need to ask questions such as, what is the replacement cost of tropical forests as a sink for pollution? What is the future market value of pharmaceuticals from yet undiscovered tropical species? Can the value of rare species be compared to the value of a rare painting? What is the value of a wilderness against which to evaluate a civilization?

REFERENCES

Ehrenfeld, D. W., 1976, The Conservation of Non-Resources, *Am. Sci.* **64**:648–656.

Gosselink, J. G., E. P. Odum, and R. M. Pope, 1974, *The Value of the Tidal Marsh*, Center for Wetland Resources, Louisiana State University Baton Rouge, LSU–SG–74–03, 32 p.

Myers, N., 1979, *The Sinking Ark*, Pergamon Press, New York 307 p.

Odum, E. P., 1979, Rebuttal of "Economic Value of Natural Coastal Wetlands: A Critique," *Coast. Zone Manage. J.* **5**:231–237.

Shabman, L. A., and S. S. Batie, 1978, Economic Value of Natural Wetlands: A Critique, *Coast. Zone Manage. J.* **4**:231–247.

Stumm, W., ed., 1977, *Global Chemical Cycles and Their Alterations by Man*, Abakon Verlag, Berlin, 346 p.

Westman, W. E., 1977, How Much Are Nature's Services Worth? *Science* **197**:960–964.

25

Reprinted from *Amazoniana* 4:529–538 (1976)

Why Save Tropical Rain Forests?

Some Arguments for Campaigning Conservationists.

by

Gerardo Budowski

There is a problem in communication.

Saving the tropical rain forest is presently high on the list of priorities of many international conservation organizations. In the United Nations Environment Programme (UNEP) it is a priority subject area. IUCN and WWF are also making big efforts in this regard, 1975 and forthcoming years will see considerable funds being devoted to this end.

Tropical rain forests (and tropical cloud forests) still cover large expanses throughout the world, but they are fast dwindling for a great variety of reasons. To be successful in the campaign, it is important that conservation officials be prepared for a series of questions, and often adverse reactions, from people with different backgrounds who sincerely do not think that it is really worthwhile spending so much energy on safeguarding the tropical rain and cloud forests, and indeed would rather see large tracts disappear - the sooner the better. The following short and incomplete analysis is intended to provide some answers which hopefully should be easy to convey to an unsophisticated audience. Needless to say, the answers should not be considered as testproof everywhere, much less comprehensive of the various prevailing conditions. Undoubtedly there are many more and often better arguments which can be advanced, depending on the specific prevailing conditions, the audience and other factors, and there will be further evolutions concerning arguments and emphasis with changing conditions. Therefore, great caution should always be exercised in the presentation of arguments, in providing satisfactory answers, and in pointing out the significance of tropical rain forests in their various aspects.

A great number of papers have been produced on the subject, some of them in various publications of IUCN and WWF. Examples are the Viewpoint in the August 1975 issue of the IUCN Bulletin, the background papers of the Yellowstone and Grand Teton National Parks Conference (published by IUCN), the published Ecological Guidelines for Development in Tropical Forest Areas of S.E. Asia (IUCN Occasional Paper No. 10) and The Use of Ecological Guidelines for Development in the American Humid Tropics (IUCN Publication New Series No. 31) resulting from the IUCN Bandung and Caracas meetings respectively on tropical rain forests, and the forthcoming similar publication on ecological guidelines for the whole of the humid tropics.

> For many people the tropical rain forests are considered as useless
> unless they are removed or modified and forestry is likely to
> achieve this goal.

It should never be forgotten that tropical rain forests, quite unlike animals like the panda, the vicuña, the tiger, the deer, or birds like the whooping crane, or the flamingoes do not enjoy a "love at first sight" sympathy from the public, and in fact there is a lot of suspicion in the minds of many people who sincerely believe that the tropical rain forest is basically "useless", it harbours "wild" and "dangerous" or "noxious" animals or at the least uncomfortable creatures such as leeches and mosquitos. For many city dwellers it is clearly linked with the notion of a "savage" and/or "uncivilized" world in opposition to the urbanized and "man-dominated" world. More than anything else, it has been for centuries the traditional area to be "opened up" to increase the area for crops and pasture for domestic cattle. Of course, harvesting timber and other products by extraction has also been practised since time immemorial. Some people including foresters, have advocated its replacement by other tree crops, because what is called exploitation amounts to "high-grading", that is, taking out the worthwhile species to leave the defective or non-commercial species. Millions of hectares of formerly timber-rich forests have been left in such conditions. Foresters intent on sustained yield production have devised other methods to replace high-grading, notably "refining" the heterogeneous mixed forest into a more easily managed forest with fewer species but of higher commercial value. This has been the basis of much past and on-going research: in fact, different silvicultural systems have been developed towards this very end by scientists and technicians. Their use is not yet widespread; successes, if objectively assessed, are extremely scarce, and failures (economic, technical, silvicultural) are much more common. The few cases which are usually considered as being successful by the forestry profession ("Malayan" system, "tropical shelterwood" system etc.) are far from being applied over vast areas.

A careful assessment of what can be promoted - and under what conditions - and what should be resisted, is badly needed.

The results so far appear to indicate the following trends: -

the more the tropical rain forest is heterogeneous, with a high proportion of non-commercial species, the less the silvicultural systems (of any type) are likely to succeed. Success stories refer to cases on waterlogged areas, river banks where special soil (and water) conditions prevail, bringing the number of different species in a unit of area to a relatively low number. Certain old secondary forests also fall into this category. This makes logging and other forms of exploitation an economically attractive operation, particularly if one single species makes up over fifty per cent of the timber. With so many seed trees, regeneration is also likely to be abundant. This is, for instance, the case with the Malayan dipterocarp forest of the tropical American Virola, Carapa or Irianthera trees;

the more the conditions of the forest are linked with drier (or deciduous) conditions, the more some very careful silvicultural interventions have a likelihood of succeeding. Conversely, the more rainfall increases, the fewer successes have been achieved. This latter premise is important because it disqualifies from any intervention large areas where annual precipitation is above 3000 or 4000 millimeters a year, and provides a good argument for opposing excessive extrapolation from success stories derived from different - drier - conditions.

Land-use - the key factor.

Many of the policies involving replacement and modification of the tropical rain forest are defended by people who sincerely believe that tropical rain forests - at least large extensions of them - are an "obstacle to development". This cannot be sufficiently stressed.

These arguments should under no circumstances be ignored by defenders of tropical rain forests. Any campaign to save tropical rain forests must, therefore, be linked with the best possible land-use for tropical areas where rain forests are - or were - found. It implies comparing rain forest in its original state with other uses, facing political, social, and economic realities, short and long-term goals, the possibility of retaining future options and an assessment of changing values, particularly those that relate to the somewhat elusive concept of "quality of life". Conservationists however possess certain vantage points. They know that above political, social and economic imperatives, which are man-made, there are ecological factors, natural laws that cannot be changed, which must be considered; they must obviously precede the man-made factors which at most can be grafted on them. After all, soil, water, plants, animals, form intricate relationships which obey certain rules. Interventions in nature lead to consequences, many of which we can foresee. How can these ecological interrelationships best be used for campaigns?

Strategies which lead to the conservation of tropical rain forests.

A first approach to our objective may therefore advantageously centre around the various land-use alternatives. This can often be very effectively achieved through a counter-question. Conservationists may, for instance, request that a careful assessment be made of alternatives which imply the removal of the forest in comparison to its maintenance. This has the merit that it discreetly challenges an assumption that has been passed on through generations by tradition, perhaps even by inertia, and may help to remove a deep-rooted "mental block". People who want to replace the forest by something else considered more "productive" should give as clear as possible an indication as to what they would like to have instead, and how they hope to achieve it. How much of it is based on fact, and how much on gamble? How does the end result compare with what was there before? Cannot the same objectives which are aimed at by these people be achieved by other programmes or initiatives which do not imply the removel or modification of the tropical rain forest? In a way, this puts the burden of demonstrating better land-use on those who oppose the tropical rain forest. This need not be presented in any polemical fashion. Promoters of rain forest removal may not have had a good chance to present their views. Sometimes they themselves will discover weaknesses, and they will be more willing to listen to the conservationists' alternatives. Let us examine a few, presented as positive suggestions, which may lead to advantageous positions for decision-making, such as the following: -

a) Instead of increasing agricultural output at the expense of tropical rain forest, has every effort been made to increase agricultural output on existing farm land? Often this can be achieved more effectively and more economically. For instance, good soils close to rivers are often used for poorly managed pastures when they could be converted into food crops. Have these solutions been sufficiently explored? How do they compare with more clearing of forests?

b) Can forest plantations on nearby degraded land (for instance very poor savannas) for production of timber, fuel or other forest products, achieve the desired objectives and actually relieve the pressure on the tropical rain forests? This could convert foresters, who take a dim view of the tropical rain forests, into strong allies.

c) Have sufficiently accurate long-term assessments been made as to the various physical, social, biological and economic consequences that rain forest removal or modification imply when seen in a local, national, regional and world context? Many examples throughout the tropical world show that the economics of national parks compete favourably with other alternatives. Universities, scientific programmes, even prestige,can be drawn in when weighing the alternatives.

Other questions can be asked in a similar fashion implying that past policies, traditional attitudes, a priori assumptions, e.g. the building of roads, are always a good thing for the people in the region, should be seriously questioned. However, these arguments alone will not change most mentalities. The root of the problem lies in avoiding the very reasons which presently lead to forest destruction.

A better knowledge of the reasons for tropical forest removal is essential.

A careful and objective analysis will often show that if the forest is replaced by crops or grazing, this may produce immediate profits - often small, for a short time only, and only for some people, while for others, often very large numbers, it may imply a considerable loss - and for a considerable period. Most of the time it has led to residual wasteland such as degraded, man-made savannas, and this can be witnessed in most of the tropical rain forest countries. These prospectives for short-term profits, even if they ruin the capital - virtually killing "the hen that lays the eggs" - are presently allowed because of prevailing land tenure practices allowing indiscriminate use and abuse, in fact often promoting it. In some countries destructive practices are actually stimulated by tax laws, the heavier tax being levelled on "virgin" or unused forest with lesser impositions on lands that are being "used" i.e. have been cleared. Rarely is a long-term comprehensive assessment made prior to the clearing action. And, at another level, who is willing to blame the poor peasant with a large and ever-growing family who needs more land for shifting agriculture to feed his family on a day-to-day basis? Actually, short and quick profits, regardless of the consequences, are at present the main reason for the widespread destruction of the Amazonian forest, while clearing by poor peasants to extend their areas for shifting cultivation - with population growth dominating this expansion - are most prevalent in the rest of the tropical rain forest regions of the world.

It must of course be admitted that all removal of tropical rain forest should not be condemned a priori. There are, sure enough, certain areas of good soils (fresh volcanic, alluvial) on relatively level land where rainfall is not excessive, but these are very scarce and most of them are already occupied by agriculture and grazing, some of which could possibly be improved considerably. And, even on these soils, it is extremely important to maintain some adequate samples of natural forest, precisely because their indicative value can be of great benefit for choosing the best land-use.

It may be worthwhile knowing, for instance, that in the whole Brazilian Amazon, probably less than two per cent, perhaps even less than one per cent of soils are considered as appropriate for permanent or sustained yield agriculture. However, there is still some discussion on this and it is not wise to advance definitive percentages. Many alluvial soils are situated close to the rivers and most of them are subject to more or less frequent flooding. This considerably limits their permanent utilisation. Others require very careful management practices to maintain their capacity to produce crops for an extended period.

Moreover, there is no doubt that in Brazil, at least, opening of roads allows mineral exploitation and may have other advantages, be they military, political, or "psychological" - the "conquest" of hitherto "unproductive jungle" and, as the Brazilians say, "the affirming of our sovereignty". These factors should not be ignored especially if a conservation campaign that has an origin or strong participation from outside Brazil is staged.

The problem is admittedly complicated. Road builders, for instance, are not exposed to ecologists and vice versa. Under our present system, they are not supposed to assess the consequences; neither are the medical doctors who save babies and children from such killers as measles, diarrhea, parasites and other infectious diseases, presently expected to deal with the consequences of population explosion - even if they should. Military and political considerations such as "we must populate our borders" and "affirm our sovereignty" are leading to decisions which, much too often, completely ignore ecological backlashes.

Agronomists and cattle specialists will of course continue to experiment with new techniques for increasing food production through trials with crops, new grasses, new breeds of cattle, new management practices, etc. Foresters will doubtless also try to devise better silvicultural methods. This should, of course, not be systematically opposed, but extreme caution should always be exercised when the results achieved on small experimental stations are extrapolated over large regions, a practice all too frequent which is politically very rewarding but can be disastrous when seen from an ecological angle.

These problems have been analysed carefully in various meetings recently, including three recent IUCN publications (1.2.3.) some are relatively short and are written in simple language, particularly designed to appeal to decision-makers and their advisers. These publications are presently being widely circulated in tropical Latin America and tropical South East Asia and are being very well received. More popular articles on tropical rain forests and their conservation have also been issued in the "New Scientist" and "Development Forum" (4.5.).

Conservationists who want to defend tropical rain forests may therefore be well advised to promote interdisciplinary discussions among people or organizations connected in one way or another with "development" projects affecting the rain forests - including of course other conservationists since conservation must be considered as a way of development.

A better use of ecological guidelines and a better assessment of ecological impacts prior to decision-making is always desirable.

The maintenance of the status quo as opposed to destructive alternatives.

In order to defend the maintenance of the tropical rain forest against other alternatives, it is necessary to have at hand the best possible arguments. The following is a list, but it is by no means exhaustive.

Forests as regulators of water regimes for maintenance of fertility and structure of the soil.

The presence of trees and the intricate life web upon and under the soil's surface proves to be by far the best possible mechanism for transforming rainfall into steady flows of water of the highest possible quality for human consumption, industries, as well as for the aquatic life within the forests and downstream, indeed for the coastal areas close to the river mouth.

Removal of the forest would greatly alter the water regime: stronger floods would result after heavy downpours since all the rainfall would flow down with only a small amount penetrating into the soil; landslides would be much more common, and erosion would be widespread. Relative droughts (leading to very low river levels) would appear after periods of lack of rain, since wells not sufficiently provided with underground water, would dry up. All these harmful effects would not take place under the protective presence of forests, or at least the negative impacts would be considerably reduced, especially on slopes.

Fish and other aquatic animals, both riverine and from nearby lagoons fed by rivers as well as coastal areas, would undoubtedly suffer from these extremes of water regimes. So would the vegetation exposed to much sharper and more intense fluctuations (both through greater floods and droughts). And so would many animals dependent on relatively stable vegetation communities and more stable water conditions.

Microclimatic changes are also likely to occur. The temperatures on exposed soils are greatly increased during the daytime and somewhat decreased during the night. Fluctuations in relative humidity are much sharper and wind and convectional phenomena and turbulence close to the soil are of course greatly increased.

Moreover, as a result of forest removal, a secondary vegetation of a completely different nature will come up. If cleared again and/or burned - and burning is possible after only a few weeks of drought, something which does not happen under the original rain forest which remains too moist - a drought resistant grass takes over. Such grass communities which replace former tropical rain forests are nowadays a most common phenomenon throughout the tropical rain forest region. They are very aggressive, fire resistant, but make very poor grazing land. It left alone and not burnt or grazed, they will eventually revert to forest after many years depending on the stage of degradation. But a complete return to the original rain forest is practically impossible.

As a result of such degradation processes, most tropical soils devoid of their original forest vegetation and being replaced by grasses will rapidly lose organic matter and transform the favourable "crumb" structure they maintain in the forest and which is particularly suited to root penetration, into a hard upper crust, particularly if exposed for longer periods. The soil thus becomes progressively less appropriate for plant growth, and what was formerly a very productive system, rich in productivity species, scientific, educational, and aesthetic features, is transformed into a biological desert.

Forests and rainfall: the case of the cloud forest.

There is still considerable discussion whether the presence or absence of tropical rain forest on level land will significantly affect rainfall. Much inconclusive debate has taken place and more research is needed. Most experts seem to agree that over large surfaces there will be some effect, but more hard evidence is needed badly. It is not wise at this stage to use this as an argument, neither is it sufficiently clear how the removal of rain forests might affect the earth's oxygen balance.

However, concerning rainfall, there is one very important exception: the various tropical cloud forests at elevations varying between 500 and 3000 meters, depending on latitude, land-mass and exposure. Here the "mist-trapping" property of the branches heavily covered with mosses, ferns, bromeliads and other epiphytic vegetation is known to considerably increase the condensation from clouds which pass through or otherwise produce a friction with the mechanical barrier offered. Differences of up to three times the precipitation have been recorded in cloud forests, in comparison with nearby open grasslands. Removal of the cloud forest therefore will considerably reduce the flow of water into the soil and the rivers in many mountain regions. This will obviously affect the lowland plains where precipitation may be much less and water is often a precious commodity, producing droughts whenever there is a prolonged dry spell, and floods and erosion when heavy showers occur in the mountains, since the water will run down without being slowly absorbed and led into the soil.

Forests and agriculture.

The presence of protective forests on the slopes is a safeguard for agriculture on the plains. The relationship is not sufficiently understood or taken advantage of by conservationists. In India, for example, the campaign to save the tiger has found a warm reception assuring the cooperation of ecologically-conscious agronomists. Saving the tiger means saving its forest habitat through the creation of a network of extensive reserves, often in hilly regions, and this in turn saves valuable watersheds covered by protective forests, assuring a steady flow of water that is used for irrigation and other useful purposes. While this example applies mostly to drier forests, it is equally true for wetter areas. Much of the rice agriculture in South East Asia is dependent on the forests of the wetter slopes. Many of the present floods and droughts could have been avoided, at least in intensity, if the protective cover of the slopes would have been maintained. Present deforestation trends could even worsen this.

Forests and reservoirs (or dams).

From the foregoing descriptions it is easy to visualise that the life span of a reservoir also depends largely on the presence of the forest cover. This is mainly because of the erosion problem, filling up a reservoir. The more erosion of the slopes, the shorter the life of the dam. The role of forests as "preventers" of erosion is only too obvious, but this very fact is not sufficiently exploited by conservationists. And yet, this is one of the simplest relationships that can be presented eloquently by numbers and economic realities.

Maintenance of the genetic capital.

This is one of the most common arguments advanced, particularly by the scientific community, but itself alone rarely produces a strong receptive feeling. The point is of course that important cultivated crops of domestic animals as well as a large variety of drugs, dyes, tannins, fibres, etc. come from the tropical rain forest. Many more remain to be discovered. To this must be added the ornamental value of many plants associated with the tropical rain forest; many of which are in use for house and garden plants or could be bred for such purposes. Moreover, there are good reasons to believe that many more applications will be found when better identification, chemical analysis and systematic testing take place. However, we are losing species at an increasing rate even before they can be discovered. This is obviously tragic but little has been achieved so far in using this fact to move decision-makers.

Maintaining the web of life.

A series of reactions are found to take place in tropical and cloud forests which effect life and conditions in often remote areas, not only water quantity and quality, but animals as well. Destroying one link of these reactions may have repercussions which we can sometimes foresee - migrant birds for example would no longer reach these areas - but most of which we do not know. Insects, rodents or other pests in nearby areas used for crops, grazing or human settlements may no longer be controlled - at least in part - when the forest is removed. Birds and other natural predators of these pests may need the forest for survival. Forests near human settlements play an important role for air purification, recreation, and education and inspiration.

The tropical rain forest as a source for scientific work.

Many basic processes in life can be better understood by studying the tropical rain forest, and in fact the amount of money spent towards such research already goes into huge sums and is constantly increasing. The scientific community has been extremely worried that the basic tools for research are being lost. A good part of UNESCO's Man and the Biosphere Programme is based, for instance, on the proper establishment and protection of biosphere reserves, with many of them located within the tropical rain forest, as basic tools for a multi-million dollar international programme. The whole exercise is aimed at producing better human conditions through the application of scientific knowledge to harness biological processes.

Educational implications

The tropical rain forest can be a marvellous live classroom as it has been increasingly discovered throughout the world. Nature trails and other devices can of course help in such education and training programmes but the main thing is the existence of the forest itself. Most schools situated in the tropics have long lost these valuable instruments and some tropical countries have to send their students elsewhere (eg. El Salvador).

Tourism, a possible alternative.

Scientific and mass tourism can be a valid land-use. It certainly proved to be extremely valuable in Puerto Rico where the Luquillo rain forest - most of it in fact of secondary nature - attracted about one million visitors in 1974. They all wanted to have a glimpse of the tropical rain forest.

Large scale scientific ventures were also undertaken in the area leading to fundamental knowledge of wide applications to understand biological processes and better adapt them to human needs.

The whole field of interpretation of tropical rain forest is in its infancy. A leading tropical forester, Dr. Frank Wadsworth, has recently written :-

" the natural tropical humid forests with their giant trees, spectacular animal life and background of mountains and rivers, are an undeveloped scenic resource of great potential economic value as an export to the entire world, possibly comparable to the animal life of Central-Africa. These forests, the world's most complex ecosystems, are prospective outdoor classrooms which could interest students from throughout the world and attract internationally financed scientific research projects on basic ecological problems of significance to all mankind".

Sure enough, one may not always see large animals, but the knowledge of interrelations between insects, lizards, frogs and vegetation can be very exciting. The various attack and defense mechanisms, the various associations between small creatures of one or different species and their feeding, nesting, or reproductive habits can be fascinating. How many people for instance, know that many birds feed or forage in groups with different species of birds actually enhancing with their presence at the same time in an area, the chance to find food for all? Or how systematically some insects and birds make the rounds to feed on nectar of certain plants?

A little knowledge of the fauna through simple publications will make all the difference to the visitor who walks through the tropical rain forest.

References

ALLEN, Robert (1975): The year of the rain forest. New Scientist, April 24: 178-180.
ALLEN, Robert (1975): Woodman, spare those trees! Development Forum, 3 (3): 6-7.
DASMANN, Raymond F. & MILTON, John P. & FREEMAN, Peter H. (1973): Ecological Principles for Economic Development. Published for IUCN and the Conservation Foundation by John Wiley &

Sons Ltd., New York and Chichester. 252 pp.

Ecological Guidelines for Development in Tropical Forest Areas of South East Asia, compiled by Duncan POORE. IUCN Occasional Paper No.10. Morges, 1974.

The Use of Ecological Guidelines for Development in the American Humid Tropics. Proceedings of International Meeting held in Caracas, Venezuela, 20-22 February 1974. IUCN, Morges, 1975. 249 pp. (Spanish version soon to be published. At this time a summary of the guidelines has been published in Spanish: Normas Ecológicas para el Desarrollo del Trópico Húmedo Americano, compiled by Duncan POORE. IUCN Occasional Paper No. 11. Morges, 1975. 39 pp).

Author's address: Accepted for publication in July 1975

Dr. Gerardo Budowski,
Head,
Forestry Sciences Department,
Centro Tropical de Investigación y Enseñanza
(CATIE),
Turrialba,
Costa Rica.

Reprinted with permission from pp. 39 and 51–56 of *Ann. Rev. Ecol. and Syst.*
5:39–64 (1974)

THE ECOLOGY OF
MANGROVES

Ariel E. Lugo

Department of Natural Resources, Commonwealth of Puerto Rico, P.O. Box 5887, Puerta
de Tierra, Puerto Rico 00906

Samuel C. Snedaker

Resource Management Systems Program, School of Forest Resources and Conservation,
University of Florida, Gainesville, Florida 32611

[*Editor's Note*: Material has been omitted at this point.]

Productivity

The photosynthetic process and associated metabolic activities result in a continual
recombination of mineral elements into organic matter. The degradation of organic
matter is accompanied by the release of those elements for possible recombination
at a later time. These two processes (recombination and degradation) in the cycling
of matter can be evaluated by monitoring the rates of fixation and release of carbon,
the basic building block, in terms of production and respiration. Measures of community metabolism are proving to be extremely valuable for the functional comparisons of ecosystems.

The productivity data available to us are summarized in Table 2. Each of these
studies was conducted using the same methodology (carbon dioxide exchange), and
together they offer an opportunity to compare mangrove forests representing different environmental conditions. The reader is referred to the original papers (10, 29,
48, 58, 79) for descriptions of the research techniques and the study locations. In
his work, Burns (unpublished) used the same equipment and techniques as Carter
et al (10). His research was performed on the southeast coast of Dade County,
Florida (see Figure 2). That site is characterized by very oligotrophic inputs of

seawater and shallow or nonexistent organic deposits. The productivity values for these mangrove sites are arranged in Table 2 in order of decreasing gross primary productivity.

Several agents can be identified as important regulators of mangrove productivity. However, these can be lumped into two factors (tidal and water chemistry), which Carter et al have subdivided into seven categories as follows (10):

1. Tidal factors.
   a. Transport of oxygen to the root system.
   b. Physical exchange of the soil water solution with the overlying water mass, removing toxic sulfides and reducing the total salt content of the soil water.
   c. Tidal flushing interacts with the surface water particulate load to determine the rate of sediment deposition or erosion within a given stand.
   d. Vertical motion of the ground water table may transport nutrients regenerated by detrital food chains into the root zone of the mangroves.
2. Water chemistry factors.
   a. Total salt content governs the osmotic pressure gradient between the soil solution and the plant vascular system, thus affecting the transpiration rate of the leaves.
   b. A high macro-nutrient content of the soil solution has been suggested (Kuenzler, 1969 [sic]) as enabling the maintenance of high productivity in mangrove ecosystems despite the low transpiration rates caused by high salt concentrations in sea water.
   c. Lugo et al (1973)[7] indicate that allochthonous macro-nutrients contained in wet season surface runoff may dominate the macro-nutrient budgets of mangrove ecosystems.

Carter et al (10) also suggested that the gradient of chloride concentration across the soil interface (expressed as the ratio of the chloride gradient between the soil and the overlying water mass to the chloride concentration in the soil water solution) could be considered an index that integrates the effects of both tidal and soil-water chemistry factors.

Applying this reasoning they found that, with an increase in the chloride ratio, the ratio of 24 hr respiration to gross primary productivity (a measure of energy used for maintenance) had a slow exponential decrease. Within the range of salinities studied (8–30‰), the gross primary productivity of mangroves increased as fresh water became available. Respiration rates along the same gradient, however, also increased. The increase in respiration is a reflection of the amount of physiological work associated with the problems of higher salinity environments. The work of Scholander et al (76, 77; see also 45, 90) shows that the rates of water loss are related to salinity adaptations in the plants. Those plants that grow in high salinity environments tend to transpire less than those growing in less saline conditions. Scholander et al (77) discussed some of the physiological costs of these adaptations and indicated that metabolic energy was involved in the process of transpiration. Since the supply of metabolic energy available to drive the translocation process is finite, a

[7]Lugo, Sell & Snedaker (49), available to Carter et al (10) as an unpublished manuscript in 1973. Kuenzler, 1968 (misdated by Carter et al as 1969) appears in this paper as (43).

Table 2 Summary of primary productivity and respiration data for mangrove ecosystems in several locations in Florida and Puerto Rico

Location	Date	Number Diurnals	Gross Primary Productivity	Net Primary Productivity	Total 24 hr Respiration	Reference
			gC/m^2 day			
Fahkahatchee Bay, Fla. Small tidal stream Red, Black & White Mangroves	Dec 72	10	13.9	4.8	9.1	(10)
Lower Fahka Union River Basin, Fla. Red, Black & White Mangroves	Dec 72	7	11.8	7.5	4.3	(10)
Upper Fahka Union River, Fla. Red, Black & Buttonwood Mangroves	Dec 72	10	10.3	6.6	3.7	(10)
Rookery Bay, Fla. Black Mangrove Forest	Aug 71 Jan/Feb 72	17	9.0	2.8	6.2	(48)
La Parguera, Puerto Rico. Red Mangrove	Jan 58 May 59, 60	sporadic hourly measures	8.2	0	9.1	(29)
Rookery Bay, Fla. Red Mangrove	Aug 71 May 59, 60	15	6.3	4.4	1.9	(48)
Kay Largo, Fla. Red Mangrove	Jun 68 Jan 70	6	˙5.3	0	6.0	(58)
Hammock Forest, Dade Co., Fla. Red Mangrove	Oct 73	3	1.9	1.3	0.6	Burns[a]
Scrub Forest, Dade Co., Fla. Red Mangrove	Oct 73	4	1.4	0	2.0	Burns[a]

[a] L. A. Burns, unpublished.

limit is set on the amount of water that a plant can effectively take up and transport against the osmotic gradient with its environment; this limit is also reflected in lower transpiration and higher respiration rates. For this reason, at very high salinities one would expect a decrease in the net productivity of mangroves. At lower salinities, competition with plant species adapted to less saline conditions would become increasingly pronounced. The trend toward high gross productivities (for mangroves) with increasing salinity, discounting nutrient availability, should not be expected to be a linear function at salinities beyond those measured by Carter et al (10). For example, respiration may overtake gross production and the species would consequently be eliminated from the community. Taking these factors into consideration, as well as their findings of greater nutrient availability in areas with lower salinities away from the sea, Carter et al (10) proposed a U-shaped relationship of mangrove metabolic dynamics along tidal and water chemistry gradients. Preliminary data shown as Figure 3 tend to support their contention. At the two extremes of the U the energetic costs of survival are high and most of the production is utilized in self-maintenance processes (respiration). The two extremes represent areas of either high nutrients and low amplitude tides, or low nutrients and high amplitude tides. Between the two extremes, or in the middle of the curve, nutrients and tidal amplitude are in some proper combination and net productivity is maximized.

It appears that environments flushed adequately and frequently by seawater and exposed to high nutrient concentrations are more favorable for mangrove ecosystem

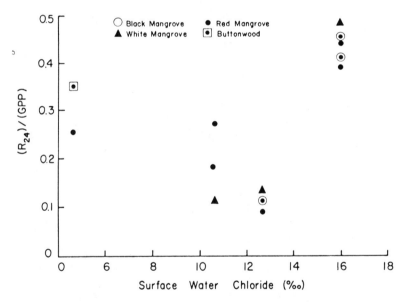

Figure 3 Maintenance metabolism ratio (24 hr respiration/gross primary productivity) of canopy leaves of four species of mangroves as a function of surface water chloride concentration [from Carter et al (10, p. III–28)].

337

development; forests in these areas exhibit higher rates of net primary productivity. From Table 2 one could classify the lower Fahka Union mangrove forests and the red mangrove forest at Rookery Bay as representing the most favorable mangrove environments, and the scrub forests on the east coast of Florida as representing the least favorable. The scrub mangrove systems apparently use a larger portion of their energy storages for respiration and low-loss recycling mechanisms, and thus a smaller proportion of their budget is available for growth.

The findings described above have implications with respect to the position of mangrove species along environmental gradients. In our own studies we have found that the rates of photosynthesis and respiration of mangrove species vary with their position in the classic zonation (48, 79).[8] A unit leaf surface area of red mangrove leaves exhibited a higher net primary productivity than those of black mangrove in the next zone or white mangroves interspersed with the black mangroves. The white mangrove exhibited comparable rates with the other species only when growing in conditions conducive to its dominance (10). When a species was out of its normal zone its primary productivity was low in comparison with that of the species characteristic of the zone. Among a wider range of mangrove environments studied by Carter et al (10) 1. the gross primary productivity of red mangrove decreased with increasing salinity; 2. the gross primary productivity of black and white mangroves increased with increasing salinity; 3. in areas of lower salinities and under equal light conditions the gross primary productivity of red mangrove was four times that of the black mangrove; 4. in areas of intermediate salinity the white mangrove had rates of gross primary productivity twice that of red mangrove; and 5. in areas of higher salinities, the white mangrove exhibited a gross primary productivity higher than that of black mangrove, which in turn was higher than the red mangrove. However, when respiration of the species was taken into account, the net productivity of these mangroves accorded well with their dominance in the zones where they were found. The authors feel that the principles emerging from these types of studies may eventually give new meaning to the classic species behavioral postulates of Lotka (47) and Shelford (78).

Quantitative studies of the secondary productivity of the consumers associated with mangrove ecosystems have not been performed, nor is it precisely known to what extent each animal species has a dependence (obligatory or facultative) on mangroves as a food resource. W. Odum (67), however, in his studies of the food habits of south Florida fishes, found that most of the sport and commercial finfish of the Gulf of Mexico were linked to food chains originating with mangrove detritus. His studies have been verified by our associates in the Fahkahatchee and Fahka Union Bays in south Florida (79) and by Austin & Austin (4) in Puerto Rico. These organisms also congregate in mangrove estuarine areas during the early stages of their life cycles where they derive protection from predators as well as food for

[8]The Rookery Bay mangrove forest where the metabolic interpretation for zonation was explored had aggregate exchange surface indices (m² of surface/m² of ground) of 4.3 for leaves, 0.67 for prop roots, and 0.066 for stems (79), and a complexity index (39) of 61. Data are not available for comparisons with other sites.

survival. An example of a species with considerable commercial value is the pink shrimp *Penaeus duorarum*, whose food habits and dependence on estuaries have been reported by Sastrakusumah (73) and Kutkuhn (44). In addition, birds utilize mangrove areas as rookeries and feeding grounds (11). Numerous other organisms spend part or the totality of their life cycles on the surface of red mangrove roots (5, 41, 57, 71). Similar functional characteristics were proposed for temperate marsh ecosystems by E. Odum & de la Cruz (63) and for a temperature estuary by Adams & Angelovic (1). That similar regional ecosystems exhibit similar functional mechanisms strengthens the statement of Macnae (54) that the tropical mangrove system is equivalent in geomorphology to the northern marsh shores.

[*Editor's Note*: In the original, material follows this excerpt.]

An Expanded Approach to the Problem of Disappearing Species

Species, as part of the common heritage, require common action to protect them.

Norman Myers

The problem of disappearing species has hitherto been tackled mainly from the standpoints of biology and ecology (*1*), with less attention to the economic and institutional factors that bring species under threat. Yet the principal destructive process at work now is modification or loss of species' habitats, which arises for the most part from economic development of natural environments. In this article I examine the problem from the economic and institutional perspectives, in order to identify opportunities for an expanded approach to conservation of species.

Animal species and subspecies at present known to science and recognized as threatened total slightly over 1000, with an extinction rate put at one per year (*2*), compared with a rate of roughly one every 10 years from 1600 to 1950 and a natural rate of perhaps one every 1000 years during the "great dying" of the dinosaurs. In addition, 20,000 flowering plants are thought to be at risk. The world's stock of all species is now estimated at 10 million, of which 8.5 million have still to be identified (*3*). The scale of the potential threat seems clear: if present land-use trends continue, and unless better conservation measures are implemented, society stands to lose a substantial part of its heritage in species and genetic resources within a few decades.

This article is based on the premise that such loss would be detrimental to human welfare. Our investigations of the genetic attributes of species are still in their early stages, but they have already contributed to modern agriculture, medicine and pharmaceuticals, and many industrial processes. Indeed the spectrum of species can be reckoned a repository of some of society's most valuable raw materials. Moreover, loss of species will affect generations into the indefinite future, whose options to utilize species in ways yet undetermined should be kept open. Yet certain sectors of the contemporary community have difficulty keeping themselves, let alone other species, in being, and I wish to emphasize this aspect of the situation. Whose needs are served by conservation of species, and at what cost to others?

Certain other trade-offs should be considered. Many temperate-zone environments have already been fundamentally modified. The disruption of North America's forests led to the extinction of the passenger pigeon, a species that could not adapt when its particular form of forest habitat disappeared, but it also led to the proliferation of the white-tailed deer, a successional species that thrives

The author is a consultant in conservation ecology, currently engaged on a project on Threatened Species and Genetic Resources for the Natural Resources Defense Council, Washington, D.C. His address is Post Office Box 48197, Nairobi, Kenya.

off disturbed vegetation. Which event has had a greater impact on man's well-being? There is benefit in maintaining genetic diversity not only among species but within species. I believe that we should keep as many options open as possible until, through research, we can reduce the areas of uncertainty.

A principal conservation need is to set aside sufficient representative examples of biotic provinces to extend protection to entire communities of species. Such measures have been urged by a number of international bodies. In particular, Unesco's Biosphere Reserves concept envisages a comprehensive network of protected areas and urges that sample biotopes be preserved for the benefit of present society and future generations (*4*). It does not consider socioeconomic factors or ways to motivate nations to implement the proposal.

A restricted approach of that sort may prove inadequate, especially in tropical regions of the developing world, which are thought to contain a large majority of the world's species (*5–7*). In these areas, the upsurge in human numbers and aspirations exerts increasing pressure on wildland habitats. Many nations in question are not in a position to designate extensive tracts of land as off limits to development (their present efforts to safeguard the bulk of the earth's species in effect subsidize the rest of the global community). A strategy to conserve the world's wealth of species must be made economically acceptable and politically practicable if it is to withstand competitive pressures from other forms of land use.

Tropical Rain Forests

Tropical rain forests are considered to contain more species than any other biome (*3, 5, 7*). Present patterns of forest exploitation suggest that few parts of this biome will escape gross disruption by the end of the century (*6–8*). A principal reason for this is the global demand for forest products, to which developing countries respond readily since it assists their trade earnings and immediate economic development. The principal consumers of processed forest products are developed nations, which thereby bear a degree of responsibility for what happens to tropical forests. By the end of the 1970's, shortages of forest products could precipitate a supply crisis to match the present shortages of fuel and food (*9*).

In Southeast Asia, the value of forest exports has now reached $2 billion per year (*10*). Most of the hardwood lumber goes to Japan, although the final product consumers are North American and European countries. To sustain this export trade (as well as some local consumption), 3 million hectares of forest are cleared each year. In West Malaysia, the value of output of veneer and plywood increased between 1960 and 1971 by 783 times (*11*). Within another 10 years, all accessible virgin forests of West Malaysia may have been logged (the same is true for the Philippines) (*11*). In Indonesia, timber exports make up 26 percent of the gross domestic product (*12*). Forest concessions in Indonesia totaling 570,000 km² have been assigned to over 500 timber firms, many of them foreign corporations with a total investment in 1973 worth $856 million. The Japanese are predominant, while a good number are American: Weyerhaeuser is among the largest, with an investment of around $35 million.

Of the Amazon Basin's 4 million km² of forest, as much as 100,000 km² are cleared each year, to open up virgin lands for livestock and crops as well as to exploit timber and other forest products (*13*). Many Amazon exploiters are multinational corporations, mostly American with some Dutch and Norwegian (*14*). Volkswagen do Brasil is about to go into cattle raising in the Amazon, with an investment of nearly $30 million for 10,000 km² of artificially created pastures. Japanese corporations are now offering investments for logging opportunities up to $500 million, and are to put $600 million into a pulp project (*14*).

In Central America, the area of planted pasture and number of beef cattle have doubled since 1960, almost entirely at the expense of natural forests, until two-thirds of the forests have now been cleared (*15*). As an instance of foreign-inspired exploitation, several fast-food chains in the United States find it cheaper to obtain hamburger beef from parts of Costa Rica that were recently virgin forests but are now cattle-raising grasslands than to purchase their meat from conventional sources in the United States. During the 1960's, Costa Rica's beef production increased by 92 percent while local consumption declined by 26 percent: almost all the extra output was exported to the United States (*16*).

These instances illustrate how the progressive depletion of tropical forests is not only attributable to unsound forestry policies on the part of developing nations. Developed nations also contribute, both directly and indirectly, to the destructive processes.

A principal way to safeguard species and genetic reservoirs is through parks and reserves. According to a recent review (*13*), however, parks and reserves in Southeast Asia comprise only 1.8 percent of the forest zone, in South America 0.67 percent, and in Africa 2.67 percent. By contrast, preliminary estimates suggest that as much as 20 percent of forests would have to be preserved, in selected localities covering distinct ecosystems, in order to ensure protection for endemic species. Moreover, since many species in this highly differentiated biome are characterized by localized distribution, a few large parks and reserves would not suffice to protect the range of biotic diversity. An extensive and strategically located network would be needed (*17*). Meantime protected areas are often unable to withstand economic pressures to put their lands to more useful-seeming purposes. A good number of parks and reserves have been violated by logging operations, hydroelectric power projects, highways, and settlement schemes. Throughout the rain forest biome, the conflict between conservation and development seems likely to grow critical for protected areas in years ahead, unless conservation policies and practices can better integrate parks with their socioeconomic and institutional environments.

Pragmatic Purposes of Species Conservation

Conservation of threatened species serves pragmatic purposes of immediate value. Genetic reservoirs make a significant contribution to modern agriculture (*18*), to medicine and pharmaceuticals (*19*), and to industrial processes in all parts of the world—especially in the advanced world, with its greater capacity to exploit genetic resources for a wide variety of purposes.

To consider one sector of agriculture, pest control is assisted through certain plants that produce chemicals to repel insects or inhibit their feeding. Pest control can likewise be advanced through selective breeding of adapted species of insects—a method that could prove more effective and economic in the long run, and result in less environmental disruption, than broad-scale application of persistent toxic chemicals. For instance, the little-documented ichneumonid wasps in tropical forests comprise at least one-quarter of a million species, certain of which could be used as predators and parasites of insect pests.

Plant species offer many starting materials for medicines and pharmaceuticals.

One group of significant drugs, alkaloids, are derived from nitrogenous substances found in certain plants (*20*). To date, a mere 2 percent of the planet's estimated 200,000 flowering-plant species have been screened for alkaloids, producing nonetheless about 1,000 different forms. The pyrrolizidine and acronycine alkaloids seem likely to prove active against several forms of tumorous cancer, while other recently discovered alkaloids are used to treat leukemia. The glycoside alkaloids are used for cardiac complaints, while still others show therapeutic promise against hypertension. The most abundant sources of alkaloid-producing plants are found in tropical forests.

Despite limited knowledge about genetic reservoirs, it seems a statistical certainty that tropical forests contain source materials for many pesticides, medicines, contraceptive and abortifacient agents, potential foods, beverages, and industrial products. Of particular value for human purposes are the specialized genetic characteristics of many localized species—yet these attributes are associated in many instances with restricted range, precisely the factor that makes them vulnerable to destruction.

Resource Economics of Threatened Species

Species can be considered an indivisible part of society's heritage now and forever. At least, that is how many people perceive species. The general public, notably in developed regions, expresses interest in the tiger, gorilla, vicuna, and many other species. It believes that the heritage of humanity, as well as that of Mauritius, has been impoverished by the loss of the dodo. Many people sense a degree of responsibility for species in other lands as part of mankind's patrimony.

At the same time, everybody's heritage is treated as nobody's business. However much the community may regard species as its estate, it has no effective way to express this interest through institutional devices such as ownership. An individual can own a cow, which enables and encourages him to take care of it and induces others not to use or misuse it. But species are not subject to readily identifiable property rights, and the same applies, of course, to species' habitats.

For centuries, the marketplace (*21*) and law (*22*) have tended to formulate and consolidate the rights of private property. (There are, of course, numerous exceptions to this sweeping statement, but this article deals primarily with the impact of Western marketplace mechanisms on natural environments of tropical regions, and the broad assertion concerning institutional rights of private property generally holds good for the countries in question.) The needs of common property—including not only species but the atmosphere and large water bodies—have suffered by default. This state of affairs has not mattered much until recently. Now that common-property resources are depleted and endangered through misuse and overuse, institutional mechanisms to safeguard them are in short supply (*23*).

Indeed the nature of species as common-property resources is not always recognized, even though it is central to the problem of conserving them. Two basic aspects of this situation are relevant. First, a species' intrinsic value is indivisible; if the tiger (like the atmosphere and the oceans) brings benefit to one person, it brings benefit to all. Second, since a species has no effective owner, it is subject to "open access" exploitation. Any individual can exploit the tiger for its skin, or the tiger's habitat for a variety of purposes. The consequence of this situation is that exploitation is almost always wasteful, and safeguard measures are almost always lacking (*24*).

So when species are endangered or extirpated, it is not such a clear-cut case of human short-sightedness and prodigality as is sometimes suggested. It is due more to the status of species and their habitats as common property. For example, the golden lion marmoset in southeast Brazil once inhabited 6500 km² of forest, but now the last 600 animals are confined to 550 km². This forest remnant will not, when exploited to elimination, produce overall economic benefits greater than those which could be generated were the golden lion marmoset found to lend itself, in the manner of the cotton-topped marmoset, to the development of anticancer vaccines (*25*). But forest clearing brings immediate profits to a limited number of individuals, hence the benefits to each are concentrated and appreciable; by contrast, protection for the golden lion marmoset brings benefits which, although they may extend over a far longer time, will be spread among many beneficiaries, hence will be diffused. Given the way society's institutions weight the choice, short-term benefit for private persons wins the day.

The rationale for the marmoset applies, on a larger scale, to the Trans-amazon Highway. This project may already have caused the extinction of a number of species of insects and plants with highly localized distributions in the rain forest (*26*). The benefits of saving these species would be dispersed and delayed, whereas the benefits of the highway are immediate, apparent, and quantifiable. Moreover, the highway constructors are almost certainly unaware of their impact on a range of species—thus, the irrational destruction of unique resources of potentially universal value.

Many of the factors enumerated above can be perceived at work through a single phenomenon—spillover effects, or effects external to the intended context of action (*27, 28*). Because of their common-property status, species are especially susceptible to the spillover impact of people's actions, which are directed at a hundred and one goals other than the destruction of species. The people who engender most spillover effects are those most engaged in economic activities, not only the exploiters but the consumers. With one-fifth of the world's population, the affluent nations account for four-fifths of raw materials traded through international markets—materials which frequently derive from tropical zones and whose exploitation or extraction causes modification of natural environments. Affluent sectors of the global community are thereby responsible for many spillover effects in other countries, and stimulate disruption of species' habitats in developing regions.

Proposals for Expanded Conservation of Species

A number of institutional initiatives could improve the situation. Three examples follow to indicate the scope for action.

1) Public organizations such as the U.S. Agency for International Development (*29*) and the United Nations agencies could be required to prepare environmental impact statements for their development projects, along the lines of those mandated for public works in the United States (*30*). Insofar as public organizations are not subject to the profit and loss considerations that govern private enterprises, it should be no unacceptable burden for these organizations to consider in more detail the environmental consequences of their activities. For example, the Jonglei Canal project in southern Sudan is designed to channel water away from several thousand square kilometers of Sudd swamp, an exceptionally rich biotic community.

The project could entail considerable—and hitherto largely unconsidered—consequences for species. Principal responsibility for its execution lies with international agencies. Sometimes a unique biotope with its genetic diversity can be protected at little extra project cost, provided the opportunity is identified in time. Sometimes, however, the ecological gain may be reckoned too limited or too diffuse or too long-term to offset an urgent need for, say, greater food output in the area in question (this is a major reason why environmental impact statements have not always been popular with developing countries or with international aid agencies). But at least the participants should be obliged to pay explicit attention to environmental costs.

2) Corporations could be required by law in their home countries to take account of environmental consequences of their activities overseas. They could be obliged to determine which wildlife communities would be affected either marginally or conclusively by their operations, and what reasonable steps they could take to avoid irreparable harm to endemic species. This measure would be in line with recent proposals (*31, 32*) that multinational corporations be obliged to disclose information on many of their activities, as part of an agreement to conform to standardized codes of conduct. Further, conservation measures on the part of multinational corporations could be encouraged through taxation systems in home countries. For example, were a forestry enterprise to leave part of its concession in an undisturbed state, it could be allowed a tax rebate; if it engaged in a form of exploitation that proved unusually damaging, it could be required to accept an extra tax burden. Its after-tax revenues would then reflect not only a corporation's private output but its net social impact (*33*).

3) Spillover effects could be adjusted by compensatory payments from the developed world to the developing world, through an international organization. The payments would be explicitly intended for use by developing countries to offset opportunities forgone to exploit natural environments containing exceptional diversity of species. (To avoid extortionist threats to exploit, compensation should perhaps be made available only to countries that put up significant funds of their own for the purpose.) The assistance should not be supplied as another form of foreign aid, but should reflect a recognition of joint responsibility for a deteriorating asset of the common heritage. This compensation idea accords with the "additionality prin-

ciple" adumbrated at the Stockholm Conference on the Human Environment (*34*), a principle that proposes that payments be made by the community at large to developing nations in order to offset adverse repercussions on their emergent economies from measures for environmental conservation (*35*).

This last proposal, like the idea of tax rebates for commercial enterprises that adopt conservationist practices, amounts to a mechanism to persuade communities whose land-use practices are destructive of species to desist through compensatory measures. "Bribes" of this sort are an accepted method of regulating sectoral economic activities in order to safeguard community interests (*21, 28, 32, 33, 36*). In the present case, they would have to be on a scale large enough to offset the economic activities that undermine species' survival. Sufficient funds could be raised only through some form of tax on the sectors of society that are affluent enough not only to pay but to register an interest in conservation of species and that bear some responsibility for the decline of species in developing regions.

Were this compensation proposal to be implemented, it would be subject to periodic appraisal. If the public in donor countries began to object to the financial strain, the scheme would have to be ended. Citizens of donor countries would thereby act with more explicit understanding of what prospects face species in developing regions, and would have a clearer recognition of their role in the situation. This would be in marked distinction from the present position, where the community in advanced nations has little opportunity to appraise its contributions—both negative and positive—in clear-cut terms. Moreover, the entire exercise would serve as a measure of the readiness of advanced-world citizens to pay for what they often suggest they want; it would permit them to put up or shut up—a response which is not generally available within present limited opportunities for safeguarding species. With refinements through experience, this institutional device could even serve as some sort of proxy pricing system to express people's minimal evaluation of the resources in question: it could develop into a framework that reflects costs and benefits as perceived by the participant parties.

These proposals are advanced as an initial review of opportunities for broadscale measures to meet the problem of disappearing species. They should be considered in the light of the options available. Difficult as community action

would be, the alternative is the present prospect, where support for threatened species amounts to a few fragmentary efforts—a limited response that represents an implicit decision by society to allow species to decline, even though the costs of protecting them need often not be exceptional in comparison with benefits to be derived.

Summary

Assistance for disappearing species is at present too localized and dispersed to make much impact on the problem with its growing dimensions. Species are threatened primarily because of their status as common property. Institutional deficiencies, notably those of free markets and property rights, promote depletion of species. Conversely, present institutional mechanisms offer little scope for society to express its preferences for goods without price or to establish responsibility for common-heritage resources. The situation postulates corrective measures on the part of collective authority at the international level. These measures would require a joint commitment by the developed and developing worlds, on a scale to reflect the increasingly interdependent needs and opportunities of the community at large. Whether the community perceives itself as a community or not, it functions as such in many of its ecological relationships and economic interactions. The community will sooner or later be obliged to respond to the problem of vanishing species: either sooner, through protective measures of sufficient scope, or later, when it finds that the disappearance of large numbers of species represents a loss through which it is indivisibly impoverished.

References and Notes

1. For example, K. Curry-Lindahl, *Let Them Live* (Morrow, New York, 1972); J. Fisher, N. Simon, J. Vincent, *Wildlife in Danger* (Viking, New York, 1969); J. Greenway, *Extinct and Vanishing Birds of the World* (Dover, New York, 1967); V. Ziswiler, *Extinct and Vanishing Animals* (Springer-Verlag, New York, 1967).
2. International Union for Conservation of Nature and Natural Resources (IUCN): *Red Data Book* (IUCN, Morges, Switzerland, 1974).
3. This figure is arrived at through extrapolation of species-abundance curves, leading to order-of-magnitude estimates. See P. H. Raven, *Syst. Zool.* 23, 416 (1974); _____, B. Berlin, D. E. Breedlove, *Science* 174, 1210 (1971).
4. Unesco, *Programme Man Biosphere (MAB) Rep. No. 12* (1974); *ibid.*, No. 22 (1974). Present networks of parks and reserves constitute no more than a starting point for conserving species. Although they cover almost 1 million km² (excluding Greenland and Antarctica), or 1.1 percent of the earth's land area, most parks and reserves protect the spectacular rather than the representative, with emphasis on unique scenery and major congregations of mammal and bird wildlife. Of the world's 198 terrestrial biotic provinces, more than one-quarter contain no protected areas, and a further 15 percent contain only

one. This limited degree of protection scarcely begins to match requirements for ensuring survival of gene pools of insects, reptiles, amphibians, fish, plants, and other genetic resources of little public appeal.

5. H. G. Baker, *Biotropica* **2**, 101 (1970); R. H. Lowe-McConnel, Ed., *Speciation in Tropical Environments* (Academic Press, New York, 1969); B. J. Meggers, E. S. Ayensu, W. D. Duckworth, Eds., *Tropical Forest Ecosystems in Africa and South America* (Smithsonian Institution Press, Washington, D.C., 1973); K. Stern and L. Roche, *Genetics of Forest Ecosystems* (Springer-Verlag, New York, 1974).

6. F. G. Farnworth and F. B. Golley, Eds., *Fragile Ecosystems* (Springer-Verlag, New York, 1974).

7. F. B. Golley and E. Medina, Eds., *Tropical Ecological Systems* (Springer-Verlag, New York, 1975).

8. A. Gomez-Pompa, C. Yazquez-Yanes, S. Guevara, *Science* **177**, 762 (1972); International Union for Conservation of Nature and Natural Resources, *IUCN Pub. New Ser. No. 31* (1975); *ibid., No. 32* (1975); *Proceedings of the 12th Technical Meeting of the International Union for Conservation of Nature and Natural Resources, Kinshasa, Sept. 8–20, 1975* (IUCN, Morges, Switzerland), in press; W. Meijer, *BioScience* **23**, 528 (1973); *Indonesian Forests and Land-Use Planning* (Univ. of Kentucky Press, Lexington, 1975); P. W. Richards, *The Tropical Rain Forest* (Cambridge Univ. Press, Cambridge, England, 1966); *Sci. Am.* **229** (No. 12), 58 (1973); Unesco, *Programme Man Biosphere (MAB) Rep. No. 3* (1972); *ibid., No. 16* (1974); *Nat. Resour. Res. No. 12* (1974); T. E. Whitmore, *Tropical Rain Forests of the Far East* (Oxford Univ. Press, London, 1975).

9. R. L. Bhargava, *Pulp and Paper Industries Development Programme, Phase I* (FO:DP/IN1/74/026, FAO, Rome, 1974); Food and Agriculture Organization, *Yearbook of Forest Products* (FAO, Rome, 1976); *U.S. For. Serv. For. Resour. Rep. No. 20* (1974).

10. Asia/Pacific Forestry Commission of the Food and Agriculture Organization, *Forestry Trends in the Asia-Far East Region* (FO:APFC/73/5, FAO, Rome, 1973).

11. E. Soepadmo and K. G. Singh, Eds. "Proceedings of the Symposium on Biological Resources and National Development, 5–7 May 1972, University of Malaya," special issue, *Malay. Nat. J.* (1972).

12. *Report on the Lili-Nas Workshop on Natural Resources, Jakarta, Indonesia, 11–16 September, 1974* (Lembaga Ilmu Benetahuan, Indonesia, and National Academy of Sciences, Washington, D.C., 1973), vol. 1.

13. M. J. Dourojeanni, in *Proceedings of the 12th Technical Meeting of the International Union for Conservation of Nature and Natural Resources, Kinshasa, Sept. 8–20, 1975* (IUCN, Morges, Switzerland), in press.

14. H. M. Gregersen and A. Contreras, *U.S. Investment in The Forest-Based Sector in Latin America* (Johns Hopkins Press, Baltimore, 1975).

15. T. B. Croat, *BioScience* **22**, 465 (1972); D. H. Janzen, *Nat. Hist.* **83**, 49 (1974).

16. A. Berg, *The Nutrition Factor* (Brookings Institution, Washington, D.C., 1973); J. C. Dickinson III, in *Latin American Development Issues*, A. D. Hill, Ed. (CLAG Publications, East Lansing, Mich., 1973); J. J. Parsons, *Rev. Biol. Trop.*, in press.

17. Biotic communities in isolated remnants of forest could steadily become depauperate, according to the theory of island biogeography [for a detailed discussion of the theory, see M. L. Cody and J. M. Diamond, Eds., *Ecology and Evolution of Communities* (Harvard Univ. Press, Cambridge, Mass., 1975)]. Although the migratory capacity of birds in general might seem to equip them for survival in a network of ecological islands, this is less the case for many tropical birds, whose comparatively sedentary habits and reluctance to cross even narrow belts of alien environments would leave them little able to move from one preserve to another. A protected area comprising 10 percent of a tropical forest biotope can ensure that no more than about half of the biotope's bird species will ultimately survive, and fewer still if the 10 percent of original habitat is split among a number of smaller preserves at some distance from each other.

18. See, for example, O. H. Frankel and J. C. Hawkes, Eds., *Plant Genetic Resources for Today and Tomorrow* (Cambridge Univ. Press, Cambridge, England, 1975); S. H. Wittwer, *BioScience* **24**, 216 (1974); *Science* **188**, 579 (1975).

19. S. V. R. Altschul, *Drugs and Foods from Little-Known Plants* (Harvard Univ. Press, Cambridge, Mass., 1973); T. Swain, Ed., *Plants in the Development of Modern Medicine* (Harvard Univ. Press, Cambridge, Mass., 1972).

20. T. Robinson, *Science* **184**, 430 (1974).

21. For a detailed treatment of the economics of private property vis-à-vis community resources, see A. N. Freeman, R. H. Haveman, A. V. Kneese, *The Economics of Environment Policy* (Dell, New York, 1973). For a preliminary attempt to put a dollar value on a few select species, see F. T. Bachmura, *Nat. Resour. J.* **11**, 674 (1971).

22. In the United States and other developed countries, community concern for community values has been extensively articulated through the public trust doctrine and similar legal mechanisms. Developing countries have not evolved legal procedures along these lines. Still less has the international community developed legal constraints to regulate use of environmental resources. See R. A. Falk, *This Endangered Planet* (Random House, New York, 1971); J. L. Sax, *Defending the Environment* (Vintage, New York, 1970).

23. A few institutional initiatives have been devised, such as the World Heritage Trust, the Convention on Wetlands of International Importance, and the Convention on International Trade in Endangered Species of Wild Fauna and Flora. With regard to species and wildlife in general, these measures are of such limited scale that they will make only a marginal impact on overall conservation needs. The Biosphere Reserves proposal (*4*) is broader in concept, but lacks consideration of socioeconomic constraints as described in this article.

24. For a classic statement of the case, see G. Hardin, *Science* **162**, 1242 (1968).

25. R. Laufs and H. Steinke, *Nature (London)* **253**, 71 (1975).

26. R. J. A. Goodland and H. S. Irwin, *Landscape Plann.* **1**, 123 (1974).

27. For a discussion of externalities and their relevance to conservation of natural resources, see, for example, R. U. Ayres and A. V. Kneese, *Am. Econ. Rev.* **59**, 282 (1969); E. J. Mishan, *Technology and Growth: The Price We Pay* (Praeger, New York, 1970).

28. For an analysis of externalities and conservation at the international level, see W. J. Baumol, *Environmental Protection, International Spillovers and Trade* (Almqvist & Wiksell, Stockholm, 1971).

29. Something of this sort has recently been proposed for the U.S. Agency for International Development; see *Fed. Regist.* **31**, 12896 (1976).

30. The World Bank evaluates some of its projects in terms of environmental impact, but the exercise is marginal as concerns species and their habitats. See World Bank, *Environment and Development* (International Bank for Reconstruction and Development, Washington, D.C., 1975).

31. *The Impact of Multinational Corporations on Development and on International Relations* (ST/ESA/6, Department of Economic and Social Affairs, United Nations, New York, 1974); *International Codes and Regional Arrangements Relating to Transnational Corporations* (E/C.10/9, Commission on Transnational Corporations, United Nations, New York, 1976).

32. T. N. Gladwin and J. G. Welles, in *Studies in International Environmental Economics*, I. Walter, Ed. [Wiley (Interscience), New York, 1976].

33. R. Marris, Ed., *The Corporate Society* (Macmillan, London, 1974); J. W. McKie, Ed., *Social Responsibility and the Business Predicament* (Brookings Institution, Washington, D.C., 1974); R. Nader, Ed., *The Consumer and Corporate Accountability* (Harcourt Brace Jovanovich, New York, 1973).

34. See also S. MacLeod, *IUCN Environ. Policy Law Pap. No. 6* (1974).

35. For a comprehensive presentation of the need for compensatory mechanisms at community level, see K. E. Boulding, *The Economics of Love and Fear: A Preface to Grants Economics* (Wadsworth, Belmont, Calif., 1974).

36. For a discussion of the principle of community action in a context of this type, see J. Buchanan, *The Demand and Supply of Public Goods* (Rand McNally, Skokie, Ill., 1968); E. T. Haefele, Ed., *The Governance of Common Property Resources* (Johns Hopkins Press, Baltimore, 1974).

37. I express thanks for helpful suggestions and criticisms of preliminary drafts of this article to H. F. Baker, J. A. Burton, F. R. Fosberg, W. Meijer, R. W. Risebrough, and F. H. Wadsworth. This article results from an investigation of the threatened species problem, funded by the World Wildlife Fund, U.S. National Appeal, Washington, D.C. The investigation has been expanded through a project of the Natural Resources Defence Council in Washington, D.C., supported by the Rockefeller Brothers Fund.

AUTHOR CITATION INDEX

SUBJECT INDEX

About the Editor

CARL F. JORDAN was introduced to tropical ecology in 1966 when he took a position with what was then the Atomic Energy Commission (AEC) to study the effects of irradiation on a tropical rain forest in Puerto Rico. After three years he transferred to another AEC laboratory, Argonne, near Chicago, Illinois, to deal with environmental problems of reactor siting. However, after living and working in the tropics, northern Illinois had little appeal. In 1974 Dr. Jordan reached an agreement with ecologists at a national laboratory in Venezuela (Instituto Venezolano de Investigaciones Cientificas) to initiate an ecological study in the Amazon Territory of Venezuela near the Brazilian frontier. The United States' part of the project was funded through the Insititute of Ecology at the University of Georgia, but Jordan and his family moved to Caracas, Venezuela where the offices and labortories for the project were located. As a result, he traveled extensively in tropical Latin America and became familiar with the wide range of ecosystem types in the American tropics. He returned to the University of Georgia in 1978, but ecological research in the tropics remained as his primary interest. Jordan's graduate students currently carry out doctoral research in Brazil, Venezuela, Puerto Rico, Costa Rica, and Mexico.